AF404721

High Resolution Laser Microprocessing

Implementation and techniques

Online at: https://doi.org/10.1088/978-0-7503-3239-2

IOP Series in Coherent Sources, Quantum Fundamentals, and Applications

About the Editor
F J Duarte is a laser physicist based in Western New York, USA. His career has covered three continents while contributing within the academic, industrial, and defense sectors. Duarte is editor/author of 15 laser optics books and sole author of three books: *Tunable Laser Optics, Quantum Optics for Engineers*, and *Fundamentals of Quantum Entanglement*. Duarte has made original contributions in the fields of coherent imaging, directed energy, high-power tunable lasers, laser metrology, liquid and solid-state organic gain media, narrow-linewidth tunable laser oscillators, organic semiconductor coherent emission, N-slit quantum interferometry, polarization rotation, quantum entanglement, and space-to-space secure interferometric communications. He is also the author of the generalized multiple-prism grating dispersion theory and pioneered the use of Dirac's quantum notation in N-slit interferometry and classical optics. His contributions have found applications in numerous fields, including astronomical instrumentation, dispersive optics, femtosecond laser microscopy, geodesics, gravitational lensing, heat transfer, laser isotope separation, laser medicine, laser pulse compression, laser spectroscopy, mathematical transforms, nonlinear optics, polarization optics, and tunable diode-laser design. Duarte was elected Fellow of the Australian Institute of Physics in 1987 and Fellow of the Optical Society of America in 1993. He has received various recognitions, including the *Paul F Foreman Engineering Excellence Award* and the *David Richardson Medal* from the Optical Society.

Coherent Sources, Quantum Fundamentals, and Applications
Since its discovery the laser has found innumerable applications from astronomy to zoology. Subsequently, we have also become familiar with additional sources of coherent radiation such as the free electron laser, optical parametric oscillators, and coherent interferometric emitters. The aim of this book Series in Coherent Sources, Quantum Fundamentals, and Applications is to explore and explain the physics and technology of widely applied sources of coherent radiation and to match them with utilitarian and cutting-edge scientific applications. Coherent sources of interest are those that offer advantages in particular emission characteristics areas such as broad tunability, high spectral coherence, high energy, or high power. An additional area of inclusion are the coherent sources capable of high performance in the miniaturized realm. Understanding of quantum fundamentals can lead to new and better coherent sources and unimagined scientific and technological applications. Application areas of interest include the industrial, commercial, and medical sectors. Also, particular attention is given to scientific applications with a bright future such as coherent spectroscopy, astronomy, biophotonics, space communications, space interferometry, quantum entanglement, and quantum interference.

A full list of titles published in this series can be found here: https://iopscience.iop.org/bookListInfo/series-in-coherent-sources-and-applications.

Turbulent Flows: an Introduction

Ian P Castro and Christina Vanderwel

University of Southampton, Southampton, UK

IOP Publishing, Bristol, UK

ISBN   978-0-7503-3619-2 (ebook)
ISBN   978-0-7503-3617-8 (print)
ISBN   978-0-7503-3620-8 (myPrint)
ISBN   978-0-7503-3618-5 (mobi)

DOI   10.1088/978-0-7503-3619-2

Version: 20212001

IOP ebooks

British Library Cataloguing-in-Publication Data: A catalogue record for this book is available from the British Library.

Published by IOP Publishing, wholly owned by The Institute of Physics, London

IOP Publishing, Temple Circus, Temple Way, Bristol, BS1 6HG, UK

US Office: IOP Publishing, Inc., 190 North Independence Mall West, Suite 601, Philadelphia, PA 19106, USA

To Lucy, my wife, for her unending patience and support.

Ian Castro

Contents

Preface

Turbulence is a subject of fundamental interest and importance within the field of fluid mechanics and has attracted wide and detailed attention for well over 150 years, not least from physicists and engineers in industry, academia, and government research laboratories of many kinds. Numerous books on the subject of turbulence have been variously aimed at late undergraduates, postgraduates, or full-time researchers. Some of these have become almost classical texts, containing much of the received wisdom that is of foundational importance to any students of the subject; such texts are often used as the basis for, or at least as adjuncts to, university courses in the subject. Other texts are essentially monographs, useful for deeper exploration of more specialised topics. However, although the number of research papers discussing turbulent flows in the contexts of aerodynamics, hydrodynamics, meteorology, oceanography, and even astronomy has grown exponentially over the last twenty years or so, hardly any modern books that can serve as general introductions to the topic have appeared since the very early years of this century. In this text, we have sought to provide such an introduction whilst, at the same time, giving students embarking on the subject a flavour of some of the more recent ideas. Like many earlier texts, we have unashamedly concentrated on what might be called canonical turbulent flows. This is not simply because they provide the necessary foundations for a basic understanding of turbulence, but also because it is largely those canonical flows which have attracted the most attention over the last few decades, particularly because of the increasing ability of computational approaches to yield accurate solutions to them under certain circumstances, thus allowing early hypotheses to be tested even more comprehensively than by physical experiments.

We have aimed our text at typical physics or engineering students who have already studied basic courses in fluid mechanics and are thus probably in their third or fourth year of undergraduate study, or perhaps starting postgraduate research work involving some kind of turbulent flow. In either case, such students might be engaging with their first formal course on turbulence. The flow of the book is similar to a typical order of presentation used in such courses (as, indeed, used by the authors in their own courses developed and taught at the University of Southampton). The level of mathematical understanding required is no more than the student would have acquired during their earlier studies. We purposely do not discuss experimental or computational methods. There are texts that cover both these topics, but the techniques involved generally move significantly faster than our level of understanding of turbulence so, in our view, the student is better served by keeping up to date with these techniques through the journal literature. However, this book does include a substantial section on the tools commonly used to interpret the data (which are usually time records of fluctuating quantities) produced by numerical or physical experiments; these tools are undoubtedly longer-lasting. We have also included short biographies of some of the 'Giants of Turbulence' at the end of each chapter, highlighting some of the important figures on whose shoulders so many subsequent researchers have stood in order to expand the understanding of turbulence over the last 150 years. Material for some of these sections has been partly culled from the splendid 2011 (CUP) book *A Voyage Through Turbulence* edited by Davidson *et al*, which provides much more extensive and fascinating reading for those particularly interested in the history of the subject and the development of the pivotal ideas.

Each chapter concludes with exercises which are designed to enhance the understanding of the chapter's subject matter. Many of these require the use of one or more of the data sets supplied in the associated 'GitHub' website freely available online at

https://cvanderwel.github.io/TurbulentFlows/

This online repository contains a copy of the exercises, all necessary data files, and partial worked solutions (mainly in MATLAB and Python). It will be maintained by CV and readers are encouraged to contact her with any queries or potential updates they would like to see included. The programming scripts in the repository are provided under the MIT License, which means that readers are free to use, copy, and modify the content to help them learn about turbulent flows. Our hope is that this will be useful not just to students but also to teachers who might like to expand the set of examples, use the codes provided as a basis to help with their own work, generate full worked solutions, etc.

Ian P Castro
Christina Vanderwel
August 2021

Acknowledgements

Our interest in and understanding of turbulent flows has naturally been influenced by numerous teachers and colleagues over many years. It would be impossible to acknowledge them all, but we would like to mention the few who were seminal in sparking our initial interest in the subject and setting us on our own paths of turbulence discovery: Peter Bradshaw and (the late) Les Bradbury (for IPC) and Stavros Tavoularis (for CV). We owe them an incalculable debt. We are also grateful to more recent colleagues, particularly Bharathram Ganapathisubramani and John Shrimpton, who, despite the restrictions of 'lockdowns' driven by the COVID-19 pandemic, during which the book was written, have provided helpful comments and encouragement.

Author biography

Ian P Castro

Ian P Castro is Emeritus Professor of Fluid Dynamics at the University of Southampton, UK. After graduating from the University of Cambridge (1968), he completed an MSc and then a PhD at Imperial College London while employed as the first Donald Campbell Memorial Fund Fellow. There followed six years as a Research Officer with the Central Electricity Generating Board (CEGB), after which he returned to academia, joining the University of Surrey in 1978. Various posts there led to appointment as Professor of Fluid Dynamics in 1990, and in 1993 he was appointed as Founder-Director of the National Power Environmental Flow Research Centre. He moved to Southampton, taking the new Chair in Fluid Dynamics, in January 2000. After serving as Deputy Head of School, from 2008 until formal retirement in late 2010 he was Head of the Aerodynamics and Flight Mechanics Research Group. He has served on various national and international scientific committees and although now Emeritus (from 2017), he continues some research and publishing activities which, over the years, have been concentrated on the experimental and numerical study of turbulent flows of environmental and industrial significance.

Christina Vanderwel

Christina Vanderwel is an Associate Professor in Fluid Dynamics at the University of Southampton, UK, and has worked there since 2014. She moved to the United Kingdom after completing an MSc (2010) and a PhD (2014) in Mechanical Engineering at the University of Ottawa, Canada, under the supervision of Prof Stavros Tavoularis. Since working at Southampton, she has been awarded a Marie-Curie Fellowship (2015) as well as a UKRI Future Leaders Fellowship (2020) to fund her research on turbulence and mixing. She has published on the topics of turbulent shear flows, boundary layers, the structure of turbulence, and turbulent diffusion.

Symbols

Roman

A, A_1, B, B_1	Nondimensional constants (used variously)
	or (for B) the Loitsyanskii integral ($\mathrm{m^7\ s^{-2}}$) (Equation (5.23))
a	Signal amplitude
	or nondimensional constant (used variously)
a_{ij}	The anisotropy tensor ($\mathrm{m^2\ s^{-2}}$) (Equation (5.37))
b_{ij}	Normalised anisotropy tensor (Equation (5.38))
C	Autocovariance function ($\mathrm{m^2\ s^{-2}}$),
	or Fourier integral of the fluctuating velocity (m),
	or C, C^*, additional log law constants
C_p	Heat capacity ($\mathrm{J\ kg^{-1}\ K^{-1}}$)
C_ϵ	The Kolmogorov constant
C_c	Corrsin–Obukhov constant (Equation (9.32))
C_d	Total drag coefficient
c	Scalar concentration ($\mathrm{kg\ m^3}$)
c_f	Local skin friction coefficient
D	Body drag
d	Jet nozzle, or cylinder, or sphere diameter (m),
	or zero plane displacement (m)
$E(\kappa)$	Energy spectrum function ($\mathrm{m^3\ s^{-2}}$)
E_{11}	One-dimensional spectral functions for the u'_1 velocity component ($\mathrm{m^3\ s^{-2}}$)
	and similarly for E_{22} and E_{33}
$E(f)$	Energy spectrum function ($\mathrm{m^2\ s^{-1}}$)
F	Force ($\mathrm{kg\ m\ s^{-2}}$)
f	Frequency ($\mathrm{s^{-1}}$),
	or nondimensional longitudinal correlation function,
	or friction factor,
	or f', normalised mean velocity similarity variable
f_i, f_o	Mean velocity similarity variables
$\bar{f}$	A normalised arbitrary flow variable
g	Gravitational acceleration ($\mathrm{m\ s^{-2}}$),
	or nondimensional transverse correlation function,
	or Richardson's constant (Equation (9.25))
H	Boundary layer shape factor
h	Roughness depth (m),
	or elevation of mass source (m),
	or channel half-height (m)
i, j, k	Tensor indices for the x_1, x_2 and x_3 directions
I_1, I_2	Boundary layer integral parameters (equations (8.36) and (8.37))
K	A (nondimensional) function of the longitudinal triple correlation
l, L	General length scales (m)
L_x, L_y	Longitudinal and transverse length scales (m)
M	Mach number,
	or Turbulence grid mesh size (m),
	or a puff of material (kg)
M_o	Jet momentum flux ($\mathrm{m^4\ s^{-2}}$)
m	Mass (kg)

$\dot{m}_i$	Mass flux (kg s^{-1})
N	Number of samples
P	Mean pressure (kg m^{-2} s^{-2}),
p	Instantaneous pressure (kg m^{-2} s^{-2})
p'	Fluctuating pressure (kg m^{-2} s^{-2})
p, P	or probability density and cumulative probability density functions
P_k	Turbulence energy production rate (m^2 s^{-3})
Pe	Peclet number (Equation (9.5))
Pe$_T$	Turbulent Peclet number
Pr	Prandtl number (Equation (9.3))
Pr$_T$	Turbulent Prandtl number (Equation (9.28))
Q	A scalar source mass flow rate (kg s^{-1}),
	or an arbitrary flow variable,
	or the second invariant of the velocity gradient tensor (s^{-2})
q	An arbitrary flow variable
q_i	Heat flux (m^3 K s^{-1})
R	Correlation function (m^2 s^{-2})
	or pipe radius (m)
R_T	Flow constant (entrainment parameter)
R_s	Scaling quantity for the Reynolds stress (m^2 s^{-2})
r	Spatial separation (m)
	or Radial coordinate (m)
Re	A general Reynolds number
S_{ij}	Rate of strain tensor (s^{-1}) (Equation (2.9))
Sc	Schmidt number (Equation (9.3))
Sc$_T$	Turbulent Schmidt number (Equation (9.28))
T	Energy transfer rate (m^3 s^{-3}),
	or Temperature (K),
	or timescale (s)
t	Time (s)
$\mathscr{T}$	Lagrangian integral turbulence timescale (s)
U, V, W	Mean velocities in the three coordinate directions (m s^{-1})
u, v, w	Instantaneous velocities in the three coordinate directions, usually x, y, z (m s^{-1})
u', v', w'	Fluctuating velocities in the three coordinate directions (m s^{-1})
U_B	Bulk velocity (m s^{-1})
U_c	Centreline velocity in a channel or pipe (m s^{-1})
U_m	Maximum velocity (m s^{-1})
U_s	A characteristic mean velocity scale (m s^{-1})
U_w	Wall velocity (m s^{-1})
U_∞	Freestream velocity (m s^{-1})
$\hat{u}$	A characteristic turbulence velocity scale (m s^{-1})
u_τ	Friction velocity (m s^{-1})
V	Volume (m^3) (see also the entry for U, V, W)
v	A Lagrangian velocity fluctuation (m s^{-1}) (see also the entry for u, v, w)
v_η	Kolmogorov velocity scale (m s^{-1}) (Equation (3.6))
$\overline{w}$	Boundary layer wake function (Equation (8.28)) (see also the entry for u, v, w)
$\overline{X^2}$	Ensemble variance of particle displacements (m^2) (Equation (9.16))
x, y, z	Cartesian coordinate directions
	or, (for x in chapter 4), any random process
y_o	Roughness length (m) (Equation (8.51))

Greek

β	Mean flow shear rate (s^{-1}) (Equation (5.28)), or flow spreading parameter, or Dissipation spectrum constant
Γ	Circulation (m^2 s^{-1})
γ	Molecular diffusivity (m^2 s^{-1}) (Equation (9.2)), or intermittency factor (Equation (6.51)), or normalised central moment of a probability distribution (Equation (4.10))
γ_T	Turbulent diffusivity (m^2 s^{-1}) (Equation (9.13))
Δ	Rotta–Clauser boundary layer integral thickness (m) (Equation (8.24))
ΔU^+	Roughness function (Equation (8.49))
δ	Flow width (m)
δ_{ij}	Kronecker delta -1 if $i = j$, 0 otherwise (Equation (A.3))
δ^*	Boundary layer displacement thickness (m) (Equation (8.10))
ϵ	Turbulence kinetic energy dissipation rate (m^2 s^{-3})
ϵ_U, ϵ_{σ^2}	Standard errors
ϵ_c	Dissipation rate of the scalar concentration (kg^2 m^6 s^{-1}) (Equation (9.31))
η	Kolmogorov length scale (m) (Equation (3.6)), or similarity coordinate (y/δ), or a normalised cross-stream location
η_B	Batchelor length scale (m) (Equation (9.30))
θ	Geometric angle, or Boundary layer momentum thickness (m) (Equation (8.11)), or as θ', temperature fluctuation (K)
κ	Thermal diffusivity (m^2 s^{-1}) (Equation (9.1)), or wavenumber (m^{-1}), or von Kármán constant
κ_1	Streamwise wavenumber (m^{-1})
λ	Wavelength (m) or the Taylor (dissipation) microscale (Equation (3.4))
μ	Fluid viscosity (kg m^{-1} s^{-1})
ν	Kinematic viscosity (m^2 s^{-1})
ν_T	Turbulent (eddy) viscosity (m^2 s^{-1}) (Equation (2.25))
Π	Kármán measure (diagnostic function) (Equation (7.32)) or the Coles' wake strength parameter (Equation (8.27))
ρ	Density (kg m^{-3})
σ	Standard deviation
σ_{ij}	The general stress tensor (kg m^{-2} s^{-2})
τ	Shear stress (kg m^{-1} s^{-2}) or time lag (s)
τ_η	Dissipation timescale (s) (Equation (3.10))
τ_w	Wall shear stress (kg m^{-1} s^{-2}) (Equation (8.8))
Φ	Mean velocity gradient profile function (and Φ_i, Φ_o)
ψ	Kolmogorov spectrum function
Ψ	Compensated Kolmogorov spectrum function or pressure-gradient parameter (Equation (8.47))
$\Phi_{ij}(\kappa, t)$	Three-dimensional wavenumber spectrum (m^5 s^{-2})
ω	Vorticity (s^{-1})
Ω_i	ith mean vorticity component (s^{-1})
Ω_{ij}	Rate of rotation tensor (s^{-1}) (equation (2.10))
ζ	A Lagrangian time (s)

IOP Publishing

Turbulent Flows: an Introduction

Ian P Castro and Christina Vanderwel

Chapter 1

Overall introduction

After some introductory remarks about turbulent flows, this first chapter outlines their general features in section 1.2, before discussing, in section 1.3, the more important effects that turbulence precipitates. But why does turbulence arise at all? This is answered in section 1.4, albeit in very general terms. In all sections of this chapter, the discussion is essentially qualitative, to give the newcomer a basic appreciation of the subject without immediately needing to consider the equations of motion. One cannot get far without the latter, of course, and the following chapter introduces these.

1.1 Initial remarks

Turbulence is a phenomenon which is common across the entire natural world and in many of humanity's engineering endeavours. The over-arching subject of fluid mechanics is, itself, one of the physical sciences that has an extremely wide range of application and its subset, turbulence, has often been rightly said to be the most natural state of fluid motion. The reasons for that will be explored in due course. Everyone must have watched, felt, or otherwise experienced the effects of fluid flow turbulence, although perhaps without quite realising it. Its all-pervasive existence is unquestionable and has been eloquently described by a long-standing practitioner:

'the turbulent motion of fluids has captured the fancy of observers of nature for most of recorded history. From howling winds to swollen floodwaters, the omnipresence of turbulence paralyses continents and challenges our quest for authority over the world around us. But it also delights us with its unending variety of artistic forms. Subconsciously we find ourselves observing exhaust jets on a frosty day; we are willingly hypnotised by licking flames in an open hearth. Babbling brooks and billowing clouds fascinate adult and child alike.

doi:10.1088/978-0-7503-3619-2ch1

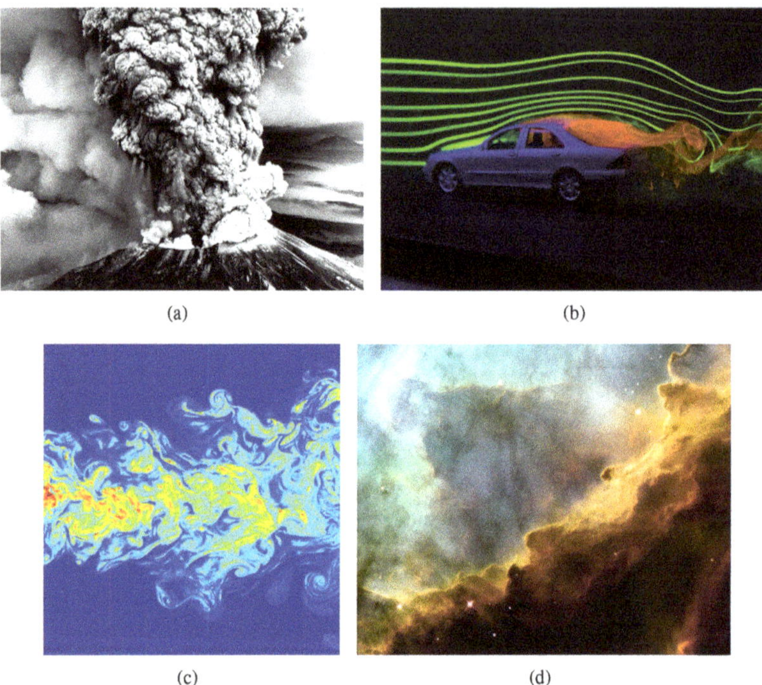

Figure 1.1. (a) The Mount St Helens eruption of 18th May 1980[2] (b) a vehicle wake visualised using dye in water[3], (c) a turbulent jet visualised using laser-induced fluorescence[4] and (d) 'a perfect storm of turbulent gases' in the Omega/Swan Nebula (M17)[5].

From falling leaves to the swirls of cream in steaming coffee, turbulence constantly competes for our attention.'[1]

One could add all manner of illustrative examples: the motion of erupting volcanoes; the smoky plume from an industrial chimney; the flow in the largest blood vessels; the wake behind moving cars, boats, and aeroplanes or, at galactic scale, a storm of gases in the Omega/Swan nebula. Some of these examples are illustrated in figure 1.1.

The practical effects of such turbulent motions can be extremely important. One often needs to know, for example: the way in which volcanic or chimney gases and particles are dispersed into the part of the atmosphere in which we live; the ease with

---

[1] William K George, with permission, from his 2013 notes for a student course on turbulence, found at: http://www.turbulence-online.com/Publications/Lecture_Notes/Turbulence_Lille/TB_16January2013.pdf. See also [4].

[2] Photo credit: R. Krimmel, USGS Cascades Volcano Observatory.

[3] Image credit: E. James, NASA Ames.

[4] Reproduced with permission from Westerweel, J., Fukushima, C., Pedersen, J., and Hunt, J. (2009). Momentum and scalar transport at the turbulent/non-turbulent interface of a jet. *J. Fluid Mech.*, 631, 199–230, copyright Cambridge University Press.

[5] Image credit: NASA, ESA, and J. Hester (ASU).

which blood clots are formed; whether or not we are likely to inhale noxious pollutants or viruses; the drag of moving vehicles (turbulence is a major contributor to the price of airline tickets); the mixing of sugar into our tea after stirring; the energy required to move fluids through pipelines.

Under the heading of 'Fluid Flow' in its entry for turbulence, the Oxford English Dictionary defines 'turbulent' as 'of, pertaining to, or designating flow of a fluid in which the velocity at any point fluctuates irregularly and there is a continual mixing rather than a steady flow pattern'. This is a not unreasonable statement and this book seeks to unpack its detail in ways which are appropriate to science and engineering students. It should be noted immediately that the definition implies, correctly, that turbulence is not a property of the fluid itself, but rather of the nature of the fluid's flow.

The governing equations for any fluid flow have been known and studied since the mid-nineteenth century, so it is perhaps surprising that so little about turbulence can be predicted with absolute certainty. If that *were* possible, life would, in some respects, be easier; for example, the weather forecasts on which our daily activities are so often ordered would be very much more accurate. Nonetheless, despite this general lack of precise predictive power, it is possible to explore and understand the more important features of turbulent flow. Over the years, this has proceeded along essentially three lines – theoretical analyses of various simplified forms of the equations of motion as applied to turbulent flow, consideration of the statistical features of turbulence, and undertaking experiments either in the laboratory (including the 'laboratory' provided by the atmosphere and the oceans) or using computers. In the following chapters, we refer to a mix of these, as appropriate, although it must be emphasised that neither experimental nor computational methods are discussed in this book in any detail. These are always developing, especially with the continuously improving power and miniaturisation of optical (and other) techniques along with the explosion in computational power which allows the equations of motions to be 'solved' over ever-increasingly large ranges of spatial and temporal scales. It is perhaps partly because of the latter that it has, in the past, been suggested that turbulence is not, as so often claimed, 'the last great unsolved problem in classical physics' but, rather (because we know the equations that describe it), simply a problem in computer design. Nonetheless, despite this explosion in computing power, solving the exact equations for any real flow of industrial or environmental importance is unlikely to be feasible for a very long time, for reasons we give later in the book. Such flows *are* widely computed, however, but this almost always involves modelling the equations in such a way as to make them more tractable. We do not consider turbulence modelling in any detail in this book; the subject already has a vast literature, which continues to expand.

Before proceeding in chapter 2 to consider the equations of motion, there are a number of central features and effects of all turbulent flows which are worth describing qualitatively. It is also worth outlining just why turbulence appears at all. These matters are summarised in the following sections and will be explored in more detail in the subsequent chapters.

1.2 General features of turbulence

1.2.1 Its chaotic nature

The first thing to say, noting the irregular fluctuations mentioned above in the Oxford English Dictionary definition, is that turbulent flow is an apparently random process which is both time- and space-dependent, with an infinite (or, at least, very large) number of degrees of freedom. We say 'apparently' because the equations are deterministic, so the entire flow could, in principle, be followed in time with any required accuracy, given precisely specified boundary and initial conditions, and sufficient computing power. However, the equations describe an inherently non-linear system and it is well-known that such systems require exquisite precision in the definition of these initial and boundary conditions if the system is to continue to appear deterministic. So exquisite, in fact, that it is not (nor perhaps ever will be) possible to follow precisely all the details of any large-scale turbulent flow of importance. Each realisation of the flow will very quickly differ in detail from any earlier realisation. In this sense, it would be better to describe turbulence as chaotic, rather than random, for since at least the 1970s many nonlinear dynamical systems have been known to yield chaotic solutions, even though they are fully deterministic.

However, the *statistics* of many turbulent flows (e.g. the mean velocity at an arbitrary set of points, or the energy in the fluctuations around those means) often *are* reproducible and thus predictable; they are not chaotic. We can illustrate this by considering figure 1.2, which shows two time records of the velocity, $U(t)$, measured within a turbulent jet at the same location but different times. They could have been obtained at a point somewhere in the middle of the jet shown in figure 1.1(c). Note that the velocity apparently fluctuates randomly and that these fluctuations occur on a range of timescales, with frequent relatively sudden excursions randomly inter-mixed with less rapid variations. The records look similar, but it is clear that they differ in detail – the two velocity values at the same specific time, as measured from the start of the records, are different. It would never be possible to reproduce these time traces exactly by repeating the measurement at other times. However, calculating, say, the mean velocity or the variance of the fluctuating velocity over

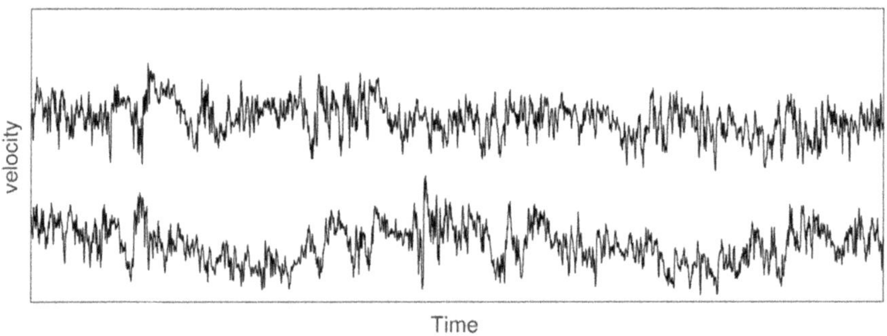

Figure 1.2. Two equal-period records of velocity within a turbulent flow, measured at the same spatial location but separated in time.

this same time period would normally yield the same results within a certain accuracy if the experiment were repeated and the time period were sufficiently long. (Just how long is a topic which we explore in chapter 4.) In the case of the two records shown, the mean velocities and variances over the time period shown differ by a little less than 1% and 2.3%, respectively. Because such statistical properties are normally reproducible, given the caveats outlined in chapter 4, we can make considerable progress in understanding turbulent flows by investigating them in terms of their statistical properties. Incidentally, the fact that a turbulent flow has inherent unsteadiness does not imply that the *mean* flow cannot be steady – it often is, as implied above. Flows that satisfy this condition are called *stationary* and by this we generally mean that the flow's statistical properties are independent of time and can be found by suitable averaging over sufficiently large times. Most jets, wakes, boundary layers and, in fact, nearly all the flows we discuss in this book can be assumed to be stationary. However, periodic flows such as those through engines, and transient flows such as the take-off or landing sequence of an aeroplane, or atmospheric flows affected by diurnal cycles, are examples of nonstationary turbulent flows; their analysis generally requires some different approaches to the averaging processes.

1.2.2 Its multiscale nature

The second major characteristic of turbulent flows is that they contain features at a wide range of spatial and temporal scales – smaller ones within larger ones, even smaller ones within the small ones, etc. For the temporal scale, this could be deduced by careful inspection of the velocity trace in figure 1.2. We will discover that the range of scales increases with the Reynolds number, Re, of a flow – a typical ratio of inertial to viscous forces. Flows in the atmosphere generally have an extremely wide range of scales because the associated Re values are very large and the spatial scales can typically range from millimetres to kilometres. A close inspection of figure 1.1(a), for example, shows numerous large eddies within the plume, each containing much smaller eddies, which themselves contain even smaller ones, etc. We will discuss this turbulence feature in chapter 3.

1.2.3 Its vortical and three-dimensional nature

Thirdly, turbulence is rotational; that is, it contains vorticity. Figure 1.1(c), for example, shows large-scale rotating motions within the jet, as can also be seen in the wake of the car in figure 1.1(b). In fact, the vortical motions occur at all scales, right down to the very smallest. Vorticity in a fluid is defined, in mathematical terms, by the curl of the velocity vector – i.e. $\omega = \nabla \times \mathbf{U}$. This quantity can be zero in laminar flows, but it has a special significance in turbulence and is non-zero everywhere and at all times. One of the main reasons that it is so significant is linked to the fourth general feature of turbulence – that it is always fully three-dimensional. Incidentally, this does not imply that the time-averaged mean flow has to be three-dimensional; often, it is not. The three-dimensional nature of turbulence means that at any point in space and time, the velocity vector (and thus the vorticity vector also) may be oriented in any one of the three spatial dimensions. Vorticity vectors can therefore be stretched and compressed in the directions along their axes by the velocity field they induce, which changes the value of the vorticity. This happens quite independently of viscous processes (except at the smallest scales); if non-zero vorticity is present in a laminar flow, it can only be diffused (or convected) but must generally remain unchanged. We could consider the turbulence seen in figure 1.1 as comprising, in each case, a tangled set of lumps of vorticity (eddies, in layman's terms), being stretched and twisted by the velocity field. This concept led Corrsin [2] to suggest a definition of turbulence as 'a spatially complex distribution of vorticity which advects itself in a chaotic manner in accordance with the vorticity equation. The vorticity field is random in both space and time and exhibits a wide and continuous distribution of length and time scales'. The vorticity equation referred to here is simply another way of expressing the fundamental equations of motion of a fluid and will be discussed in chapter 2, along with the related ideas introduced here. A later, alternative definition to Corrsin's, which includes the importance of boundary conditions and viscosity, is that of [1]: 'Turbulence is a three-dimensional time-dependent motion in which vortex stretching causes velocity fluctuations to spread to all wavelengths between a minimum determined by viscosity and a maximum determined by the boundary conditions. It is the usual state of fluid motion except at low Reynolds number.'

1.2.4 Its dissipative nature

It is important to recognise that turbulence dissipates energy. As smaller-scale motions drain energy from larger scales by the 'cascade' process discussed in chapter 3, they pass it on to motions at even smaller scales, until the smallest scales are so small that they become influenced by viscous effects and thus essentially lose energy to heat. There is a flux of energy down through all the scales, from the very largest to the smallest. This cascade process is fundamental to turbulent flows. But two questions immediately arise: where does the energy come from and where does it eventually go? In essence, the energy in the largest scales of motion is extracted from the mean flow and it is lost to heat (dissipated) at the smallest scales. If these two processes are not essentially in balance, then the flow cannot be steady. These various ideas will be explored in a little more detail in chapter 3. It is worth saying here, however, that there is a class of two-dimensional random flows which, although it has some turbulence-like features, is not true turbulence as generally understood. Crucially, in such flows, there can be no stretching of the vortices and it turns out that the energy cascade is up-scale – i.e. small eddies amalgamate into larger ones. Such flows can occur in the atmosphere or oceans, for example, where vertical gradients in density might be sufficiently large (from higher density at lower elevations to lower density above) that vertical motions are constrained so that the flow occurs in thin layers with little transfer between them. There is significant literature on two-dimensional 'turbulence' (see, for example, the works of Frisch [3] and Tabeling [6]), but the topic is not considered further in this book.

1.2.5 Its intermittent nature

Before considering the important effects of turbulence, which are in fact what makes the subject of such great significance, we mention a final point. Turbulent flows are often intermittent, meaning here that at a particular point in space the flow may sometimes be turbulent and at other times not. Consider a point somewhere near the outer edge of the turbulent jet shown in figure 1.1(c). As the large outer eddies roll by (from left to right), they will intermittently engulf that point, but at other times the fluid is essentially quiescent. Whether or not this happens is usually simply a function of the flow's boundary conditions. There is another, more important, kind of intermittency, which has to do with motion at the smallest scales; this will be discussed in chapter 3, but it is always present, whether or not the turbulence fills the domain, and it is characterised by the fact that the dissipative motions are not uniform in space, but occur randomly in a 'spotty' manner. Incidentally, these smallest scales might be very small indeed (depending on the Re) but they always remain many orders of magnitude larger than the molecular mean free path – except, perhaps, in the very unusual circumstances of rarified gas flows (say, in the extreme reaches of the atmosphere). This means that in all normal circumstances, it is completely valid to consider turbulence as a continuum phenomenon, described by the equations of fluid mechanics – the Navier–Stokes equations. These are essentially a result of regular Newtonian mechanics, ensuring that the fluid's motion satisfies conservation of mass, momentum, and energy.

1.3 The major effects of turbulence

There are two most important effects of turbulence; one is generally positive and the other more usually negative. Its most useful feature is that it is very efficient at mixing things. At the most prosaic level, this is why we stir our tea after adding sugar. If left to the action of molecular forces (i.e. diffusivity) alone, the tea would be cold long before the sugar was uniformly mixed within it. More importantly, industrial machines such as nuclear reactors, heat exchangers, and combustion chambers would be very much more inefficient without the greatly increased rate of mixing of mass, momentum, and heat which turbulence provides. Often, therefore, internal flows are deliberately made more turbulent to enhance mixing rates. A common way of illustrating the different mixing timescales in laminar and turbulent flows (e.g. Mathieu and Scott [5]) is to consider a cubical room of size L, with a small injection of heat introduced at some point within it. Assuming quiescent air, it would take a time of the order of L^2/κ, where κ is the thermal diffusivity, for the heat to be mixed throughout the room by molecular diffusion alone. On the other hand, if the room contains turbulent air with a characteristic velocity fluctuation of u', the time required for thorough heating would be of the order of L/u'. The ratio of these two times is Lu'/κ, which we could call a turbulent Peclet number Pe_T (analogous to the Re, with thermal diffusivity taking the place of molecular viscosity). With $L = 2\,\mathrm{m}$ and $u' = 0.1\,\mathrm{m\,s^{-1}}$, Pe_T would exceed 10^4, so it would take over 10000 times longer for the room to be heated without the introduction of some turbulence.

The negative effect of turbulence is simply that it leads to greatly increased frictional forces exerted by the fluid on solid surfaces. (As will be discussed in chapter 7, this is, in fact, linked to the increased rate of transfer of momentum within the region nearest the surface.) Transmitting gas, oil, water, etc., through pipes therefore usually requires far more energy than would be necessary if the pipe flow were laminar. Likewise, in typical external flows, turbulence usually has a deleterious effect. Most aircraft, under cruise conditions, typically use twice as much fuel than would be necessary if laminar flow could be maintained over all the external surfaces. This is essentially because the surface skin friction coefficient – a measure of the surface forces which oppose the motion of the aircraft – is 10–100 times as large as it would be in the latter circumstances.

But just why are most flows of importance fully turbulent? We now briefly discuss this in simple terms.

1.4 Why turbulence? Its source

As mentioned earlier, the Navier–Stokes equations, discussed more fully in the next chapter, are nonlinear. For simplicity, consider here a feature of the equation that describes the balance between applied forces in the x-direction and the resultant change of 'mass times acceleration' (or rate of change of momentum) in that direction – this is a straightforward application of Newton's second law of motion. There is a term that describes (the advective part of) the acceleration of the velocity component in that direction (U), which can be written simply as $U \partial U / \partial x$. This is clearly nonlinear, and it is this inertial term that leads to the possibility of instabilities that amplify and transfer energy to smaller and smaller scales of motion. Whether or not this happens in practice depends on whether the viscous term (one of the balancing forces) is large enough to damp the instabilities. Normally, the value of the Reynolds number (Re) is crucial. We may define the Reynolds number as $\mathrm{Re} = UL/\nu$, where U is a typical mean velocity, L is a typical length scale and ν is the kinematic viscosity; as noted earlier, Re is essentially a ratio of the inertial to the viscous forces. Any flow will be laminar at small enough values of Re, but for any particular flow there will be a critical Reynolds number (Re_c) above which the transition to turbulence begins, because the viscous forces are not large enough to damp out the effects (from the nonlinear terms) of the inevitable disturbances within the flow. These can have a number of sources; for example, slight thermal currents, imperceptible changes in boundary conditions or, perhaps, minute variations in the power driving the flow (e.g. the fan driving a wind tunnel or the pump pushing fluid through a pipe). In addition, the nature of the disturbances can effect Re_c. For a number of relatively simple flows, Re_c can be calculated exactly, but for most practical flows of importance this is not possible. Transition is an important and widely studied topic. It is a crucial matter in aeronautics, for example, since although the flow over the leading edge of a wing is usually laminar, it becomes turbulent further aft and exactly where this happens influences the aircraft's drag. There is therefore a very substantial literature devoted to transition in all kinds of circumstances, but we do not consider it further in this book. Nearly all flows of importance to physicists and engineers have a very large Re – many thousands or tens of thousands at least, or even (as noted earlier in the atmospheric context) many millions. Even much smaller-scale flows, like that over a car, for example, would have Re $\geqslant 10^7$ once the car exceeds a few miles per hour.

1.5 Subject giants: C-L-M-H Navier and G G Stokes

The Navier–Stokes equations were formulated over 150 years ago and yet they are the foundation for our study of turbulent flows and the basis of modern computational fluid dynamics. Prior to their formulation, fluid flows were analysed using Euler's equations, which can lead to much useful information. However, the Euler equations do not include any frictional terms and are therefore fundamentally incomplete. The full equations cannot be attributed to a single person but were actually derived separately and independently by the French mathematician Claude-Louis-Marie-Henri Navier (1785–1836) and the English physicist Sir George Gabriel Stokes (1819–1903).

Navier was born in Dijon, France, in 1785, and built his career in Paris working at the École Polytechnique as a university lecturer and scholar of engineering science. It was a time of great French mathematicians and, during his studies, Navier worked with and became a protégé of the famous mathematician Jean-Baptiste Fourier as well as superseding Augustin-Louis Cauchy at the École Polytechnique. During his research, Navier was the first person to derive the correct form of the equations of fluid motion, but the significance of his work was not realised until after his death. The reason was that although he had formulated the governing equations correctly, he had interpreted the additional term which needed to be added to Euler's equations as a correction for intermolecular forces, rather than the frictional shear stresses that were actually the missing mechanism. It was not until seven years after his death that his contemporary, Adhémar Jean Claude Barré de Saint-Venant (1797–1886), reworked Navier's derivation and published a paper in 1843 that properly identified the importance of the coefficient of viscosity, giving credit to Navier for the original idea.

Stokes was born in Skreen, Ireland, in 1819, and built his career as an academic at Pembroke College, Cambridge, studying physics and mathematics. He actually made substantial contributions to the field of optics and is credited with developing the principles of spectrum analysis, winning the Rumford Medal of Britain's Royal Society in 1852 for his observations of the fluorescent properties of quinine and coining the word 'fluorescence'. From 1849 until his death, he held the title of Lucasian Professor of Mathematics at Cambridge, a prestigious post whose former holders also include Sir Isaac Newton and Stephen Hawking. In 1885, he served a five-year term as president of the Royal Society and, in 1887, he also represented his university as one of two members for the Cambridge University constituency in the British House of Commons. Stokes was the only person ever to hold all three positions simultaneously; Newton held the same three, although not at the same time. Stokes was made a baronet in 1889. It was Stokes' research in fluid mechanics that led him to derive the governing equations of fluid flows, beginning correctly

Figure 1.3. (Left) Claude-Louis-Marie-Henri Navier (1785–1836). (Right) George Gabriel Stokes (1819–1903).

with the idea of internal stresses, and he published his derivation in 1845, two years after Saint-Venant, having been unaware at the time of his or Navier's work in France.

By 1850, the world was aware of the governing (partial differential) equations of fluid mechanics and solution of these equations then became the challenge. However, computers were hardly even dreamt of for almost another century and this is one reason why so much focus was, and continues to be, applied to the attempt to simplify the equations, developing physical analogies and finding approximate solutions. Of course, in the modern age, full solutions to the governing equations can now be obtained using direct numerical simulations (if the computational resources allow and the Reynolds number is low enough), but the engineering insights that can be gained by studying the simplified flows are still invaluable in our understanding and command of turbulence. See exercise 1.1.

Sample exercises

1.1 Make a list of industrial and environmental flows for which turbulence is a crucial component.

1.2 The file 'TurbulenceSample.txt'[6] contains a time history of the streamwise velocity measured in a wind tunnel using hot-wire anemometry. It was sampled at 60 kHz for a total time of 30 s. Plot the signal of the velocity U (m s^{-1}) versus time t (s). Take this opportunity to zoom in and out and consider the apparent 'randomness' of the signal. Calculate the mean and variance of the signal.

---

[6] https://github.com/cvanderwel/TurbulentFlows/blob/main/data/TurbulenceSample.txt

1.3 Estimate the time taken for molecular processes to mix a source of carbon monoxide fully throughout a household kitchen, considering it be a cubical room with sides $L = 3$ m and assuming a diffusivity rate on the order of 10^{-5} m^2 s^{-1}. How much quicker would it be if a fan were introduced to give an average fluctuating velocity (rms) of about $u' = 0.1$ m s^{-1}?

1.4 Use the literature to explore the critical Reynolds number for transition in a pipe. On what factors might this depend?

References

[1] Bradshaw P 1971 *An Introduction to Turbulence and Its Measurement* (Oxford: Pergamon)

[2] Corrsin S 1961 Turbulent flow *Am. Sci.* **49** 300–25

[3] Frisch U 2002 *Turbulence* (Cambridge: Cambridge University Press)

[4] George W K 1990 The nature of turbulence *Forum on Turbulent Flows* ed W M Bower, M J Morris and M Samimy (New York: American Society of Mechanical Engineers)

[5] Mathieu J and Scott J 2000 *An Introduction to Turbulent Flow* (Cambridge: Cambridge University Press)

[6] Tabeling P 2002 Two-dimensional turbulence: a physicist approach *Phys. Rep.* **362** 1–62

IOP Publishing

Turbulent Flows: an Introduction

Ian P Castro and Christina Vanderwel

Chapter 2

The governing equations

This chapter introduces the basic Navier–Stokes equations, expressed initially for a laminar flow in section 2.2 but which embody some particular limitations, specified in section 2.1. Definitions of two of the important properties of fluid flows – vorticity and the rate of strain – follow in section 2.3. Thus far, the material should be familiar to readers who have studied basic courses in fluid mechanics. The remaining sections first discuss the way in which the equations can be averaged for a turbulent flow, introducing the crucial concepts of 'Reynolds averaging' and the consequent closure problem in section 2.4. Subsequently, transport equations for turbulence quantities are presented and discussed in sections 2.5–2.7. Together, these various equations allow a significant understanding of turbulent flows to be developed, usually in the context of specific types of flow, i.e. flows determined by specific boundary conditions, and this forms the bulk of the book from chapter 5 onwards.

2.1 Limiting assumptions

We begin by identifying the limiting assumptions made throughout this book. Firstly, we consider only single-phase flows. In many situations, mixtures of different fluids, or fluids and solids, occur. These include, for example, the flow of oil, water, and/or air in industrial pipelines, or combusting flows of air and propellant, or volcanic eruptions of a mixture of gas, molten rock, steam, etc. Such flows are clearly more complicated than those involving only a single phase, but nonetheless can be usefully studied starting from appropriate extensions and additions to the single-phase equations of motion. Some of the easier scenarios involve the flow of a fluid containing small particles – small enough not to influence the nature of the flow – or the flow of a fluid containing an immiscible contaminant (e.g. carbon dioxide in a car's exhaust jet). We discuss fundamental mixing processes in chapter 9, but multiphase or multiple-fluid flows are not discussed. Clarification of the following assumptions requires a little background, which will be familiar to those who have studied basic fluid mechanics.

                2-1

We assume, secondly, that the fluid's density is constant throughout the flow. If entropy is constant along a streamline then it can easily be shown that

$$\frac{d\rho}{\rho} = -M^2\frac{du}{u}, \tag{2.1}$$

where ρ is the density, M is the Mach number and u is the velocity. Fractional changes in density are therefore very small compared with fractional changes in velocity if $M \ll 1$. (Changes in density because of the flow are exceedingly small, even in strong winds!) This means that the flow can be considered incompressible, so that everywhere in the fundamental equations where density and velocity are multiplied together inside a derivative, the density can be moved outside. Note that fractional changes in pressure are likewise small, but this does *not* mean that we can assume constant pressure – the flow dynamics are influenced by pressure *gradients*, which may not be small. For a liquid, whose density is normally orders of magnitude higher than that of a gas, flow is nearly always very close to incompressible. On the other hand, a gas flow at a significant Mach number is compressible. We do not consider such flows in this book.

Thirdly, we consider only Newtonian fluids. For a solid, Hooke's law, discovered in the seventeenth century and strictly a constitutive equation for the solid rather than a 'law', linearly links stress and strain via a constant of proportionality – the elasticity. At around the same time, Newton realised that in a fluid flow, there is an equally straightforward, linear relationship between stress and the *rate* of strain (rather than the strain itself). Simply stated, the shear stress between adjacent fluid layers is proportional to the rate of shear, i.e. the velocity gradient. (Note, incidentally, that it is shear stresses which set a fluid in motion. In fact, this could amount to a definition of a fluid; shear stresses cannot set a solid in motion.) So, for example, the shear stress, τ_x, in the x-direction acting on a small volume of fluid, may be written as

$$\tau_x = \mu\frac{dU}{dy} \tag{2.2}$$

where μ is the viscosity of the fluid. This is sometimes known as Newton's viscosity law but, again, it is really a constitutive equation for the fluid and defines the viscosity. Equation (2.2) is just one component of the local stresses on a small parcel of fluid. The other components of the full stress/rate of strain relationship will become clear from what follows later (equation (2.5)). Note that if the viscosity, rather than being constant, varies for any reason (e.g. because of large changes in temperature or because of the stress itself), then there are additional non-linearities in the system which naturally complicate things further. Throughout this book, we assume that viscosity is constant in time and space – the definition of a Newtonian fluid. Like density for incompressible flows, viscosity inside a differential (which appears when developing the equations for flow of any fluid) can thus be taken outside the differential. It is a happy fact of nature that by far the most common fluids (air and water) can be considered Newtonian to a high degree of accuracy for most purposes.

2.2 Basic equations for laminar flows

Given the restrictions discussed above, the equation describing the balance between applied forces and the resultant fluid motion (i.e. Newton's second law, $F = ma$, applied to our single-phase, incompressible, Newtonian fluid flow system) can be expressed as

$$\frac{\partial u_i}{\partial t} + u_j \frac{\partial u_i}{\partial x_j} = -\frac{1}{\rho_o} \frac{\partial p}{\partial x_i} + F_i + \nu \frac{\partial^2 u_i}{\partial x_j \partial x_j}, \tag{2.3}$$

written in standard tensor notation, where p denotes pressure and with the 'mass $\times$ acceleration' (i.e. the 'rate of change of momentum') on the left-hand side, which arises from the sum of the applied forces on the right-hand side. This expresses the balance for the ith component of the velocity; there are obviously three of these, choosing i to be 1, 2, or 3 for the velocity in the orthogonal x, y, and z directions, respectively. Here, we have moved the reference density ρ_o from the 'mass $\times$ acceleration' terms on the left-hand side over to the right-hand side, so that the viscosity μ becomes the kinematic viscosity, $\nu = \mu/\rho_o$. Note the presence of the all-important nonlinear term on the left-hand side. Readers who have undertaken basic fluid mechanics courses should be familiar with a typical derivation of this equation from first principles, perhaps by using a straightforward control volume analysis, but may be more familiar with the equations in vector form, or even written out fully without using either vector or tensor notation. For those not familiar with tensor notation, the basics are given in appendix A.

We have included a force term, F_i, in equation (2.3), for generality and to emphasise that, in addition to pressure and viscous forces, there may be additional forces arising from a number of situations which we do not consider in this book. Typical amongst these are the electromagnetic forces which may arise if the fluid is electrically conductive. One of the most common, however, particularly in the atmosphere, is the buoyancy force which arises when there are sufficiently large vertical variations in density. In this case, $F_i = (\Delta\rho/\rho_o)\mathbf{g}$ where $\rho = \rho_o + \Delta\rho$, $\mathbf{g} = (0, g, 0)$, and g is the gravitational acceleration, so that the term only appears in the equation for the vertical velocity. If these density variations are caused by temperature (rather than solutes in a liquid, say) then we can write $\Delta\rho = -\alpha\rho_o\Delta T$, where α is the fluid's coefficient of expansion and T is temperature. Very often, the density variations, although large enough to affect the dynamics of the flow, are sufficiently small that the density itself can be moved from inside the differentials on the (original) left-hand side of (2.3), as mentioned above. The fluid can thus still be considered as incompressible, and the density only becomes important because of its gradients – as in F_i. Such flows are often termed 'Boussinesq'. Notice too that p in equation (2.3) contains the hydrostatic pressure (i.e. $\rho_o z$, where z is the vertical coordinate) but, provided that the pressure does not appear in the boundary conditions, this makes no difference in uniform density cases. As mentioned above, we do not generally consider variable density flows in this book and thus set $F_i = 0$.

The equations of motion given above (2.3) can alternatively be written (with $F_i = 0$) as

$$\frac{\partial u_i}{\partial t} + u_j \frac{\partial u_i}{\partial x_j} = \frac{1}{\rho} \frac{\partial \sigma_{ij}}{\partial x_j}, \tag{2.4}$$

where σ_{ij} is the full stress tensor, which is given by

$$\sigma_{ij} = -p\delta_{ij} + 2\mu S_{ij} = -p\delta_{ij} + \mu \left(\frac{\partial u_i}{\partial x_j} + \frac{\partial u_j}{\partial x_i} \right). \tag{2.5}$$

S_{ij} is the rate-of-strain tensor (see section 2.3) and δ_{ij} is the Kronecker delta – unity for $i = j$ and zero otherwise.

With $F_i = 0$, there are four unknowns in the three equations inherent in equation (2.3). Arguably, the most fundamental requirement is that mass be conserved, and this supplies the necessary fourth equation. Its general form can be expressed as $\partial\rho/\partial t + \partial(\rho u_i)/\partial x_i = 0$. For an incompressible flow, this reduces to a simple kinematic condition on the velocity field:

$$\frac{\partial u_i}{\partial x_i} = 0. \tag{2.6}$$

Equations (2.4) and (2.6) provide a completely closed system (i.e. four equations, four unknowns – u_i for $i = 1$–3, and p, all generally functions of time as well as the three space coordinates). By now, the reader will appreciate that only for very simple boundary and initial conditions can the system be solved analytically and then only for laminar flows, for which the variables are not functions of time – like a simple parabolic flow in a pipe, for example. Nonetheless, every possible kind of turbulent flow, limited only by the assumptions discussed in 2.1, is fully contained within this set of equations. Although there is no general analytical theory that describes turbulence, there are many situations that can be helpfully explored, which yield some basic understanding of its nature. Some of these will be discussed in later chapters.

There is a further equation, important for many contaminant-laden flows, that expresses the balance of convection and diffusion of passive scalars within a flow. Denoting the concentration of the latter by c, we can express this as

$$\frac{\partial c}{\partial t} + u_i \frac{\partial c}{\partial x_i} = \gamma \frac{\partial^2 c}{\partial x_i \partial x_i}, \tag{2.7}$$

where γ is the diffusivity of the scalar within the fluid. A common case is that of temperature, when c is actually T, the temperature, and γ is the fluid's thermal diffusivity, κ. Equation (2.7) is, in that case, essentially an energy conservation equation, on the assumptions that (i) the fluid has a constant heat capacity per unit volume (ρC_p) and (ii) there is no internal heat generation (e.g. by chemical reactions). Unlike the Navier–Stokes equations above, equation (2.7) is linear in the dependent variable (c). In the case of temperature, if the temperature variations are sufficiently small that the resulting buoyancy force, F_i, remains negligible, then temperature acts as a passive scalar – i.e. the temperature variations do not affect the velocity field. This means that the temperature field can be determined once the velocity field is known. For larger temperature variations, the buoyancy force F_i must be included and the temperature variations – equivalently the density variations – affect the velocity field, so that equations (2.3) and (2.7) (with $c = T$) are coupled and must be solved together.

2.3 Vorticity and the rates of strain and rotation

For a constant density flow, the rates of strain given by $\partial u_i/\partial x_j$ can be decomposed into symmetric and antisymmetric parts, viz:

$$\frac{\partial u_i}{\partial x_j} = S_{ij} + \Omega_{ij} \tag{2.8}$$

where S_{ij} is the symmetric rate-of-strain tensor

$$S_{ij} \equiv \frac{1}{2}\left(\frac{\partial u_i}{\partial x_j} + \frac{\partial u_j}{\partial x_i}\right) \tag{2.9}$$

and Ω_{ij} is the antisymmetric rate-of-rotation tensor

$$\Omega_{ij} \equiv \frac{1}{2}\left(\frac{\partial u_i}{\partial x_j} - \frac{\partial u_j}{\partial x_i}\right). \tag{2.10}$$

The vorticity (which, recall, is defined as the curl of the velocity, $\boldsymbol{\omega} = \nabla \times \boldsymbol{u}$) and the rate of rotation are related, using suffix notation, by

$$\omega_i = \varepsilon_{ijk}\Omega_{ij} \quad \text{and} \quad \Omega_{ij} = -\frac{1}{2}\varepsilon_{ijk}\omega_k. \tag{2.11}$$

ε_{ijk} is the alternating symbol – unity if ijk are in cyclic order (i.e. 123, 231, or 312), -1 if ijk are in anti-cyclic order (i.e. 321, 231, or 132), but zero otherwise (Appendix A). Note that whilst Ω_{ij}, like S_{ij}, is a second-order tensor, the vorticity ω_i is a first-order tensor (i.e. a vector, or, more strictly, a pseudovector; see appendix A). Note also that S_{ij} and ω_i are not independent quantities – gradients in the strain field are related to gradients in the vorticity field (because it can be shown that $2\partial S_{ij}/\partial x_j = -[\nabla \times \omega]_i$). Despite this relationship, a uniform strain field can exist without vorticity and vice versa.

Because of the particular importance of vorticity in turbulence, which we explore more fully later, we give below the vorticity equation, which is straightforward to derive by taking the curl of the Navier–Stokes equations. The result is

$$\frac{\partial \omega_i}{\partial t} + u_j\frac{\partial \omega_i}{\partial x_j} = \omega_j\frac{\partial u_i}{\partial x_j} + \nu\frac{\partial^2 \omega_i}{\partial x_j \partial x_j}. \tag{2.12}$$

It is important to note that pressure does not appear in this equation. The left-hand side describes the convection of the ith vorticity component due to fluid motion, while the final term on the right-hand side denotes the molecular diffusion of that component of the vorticity. The first term on the right-hand side has no direct counterpart in equation (2.3) and is, in fact, the most interesting. It is commonly known as the vortex stretching term, which expresses how the vorticity is amplified (or damped) by velocity strains oriented along its axis. Actually, this interaction between vorticity and the rate of strain (velocity gradient) consists of both stretching *and* tilting of the vorticity components. If $i=1$, for example, $\omega_2 \partial u_1/\partial x_2$ and $\omega_3 \partial u_1/\partial x_3$ describe how ω_1 is influenced by the tilting of the ω_2 and ω_3 components due to the gradients $\partial u_1/\partial x_2$ and $\partial u_1/\partial x_3$, respectively. More obviously, perhaps, $\omega_1 \partial u_1/\partial x_1$ accounts for the effect of stretching due to the gradient $\partial u_1/\partial x_1$. In a purely two-dimensional flow (one where, for example, $u_3 = 0$ and $\partial/\partial x_3 = 0$), only one component of the vorticity is non-zero and equation (2.12) reduces to a simple balance between convection and diffusion of the vorticity – there can be no vortex stretching. In turbulence, by contrast, the stretching/tilting term plays a major role, as discussed more fully in chapter 3.

In equations (2.3) (or (2.4)), (2.7), and (2.12), the time-dependent term, which expresses the rate of change in time of the dependent variable, is identically zero in steady flows and, in fact, we will only consider such flows in everything that follows, except for one classic case (decaying turbulence, in section 5.2).

2.4 The averaged mean flow equations for turbulent flows

We come now to the crucial matter of how we think about these equations when the velocity (and pressure) behaves, typically, as shown in figure 1.2 – i.e. it appears random in time (and space). Recall first that in turbulent flows, all four variables in equation (2.3), i.e. u_i for $i = 1 - 3$ and p, vary continuously in space and time over a very wide range of temporal and spatial scales and thus for any realistic flow there was, until the advent of computers, no hope of solving the equations in order to follow all these motions exactly. Even now, accurately computing a full solution for a given set of initial and boundary conditions is only possible for quite low Reynolds numbers, as we will see in chapter 3.

The major, groundbreaking development, on which nearly all later considerations of turbulence have been built, was made by Osborne Reynolds in his 1895 paper published by the Royal Society. His arguments might have seemed strange at the time (even to referees), but are now second nature to everyone in the field. He first wrote down mathematically – and very simply indeed – his belief that all the dependent variables could be broken down into a time-averaged mean value and a value which fluctuates about that mean. This might now be seen as almost self-evident. So, defining u and U as the instantaneous and time-mean velocities, with u' as the velocity that fluctuates about the mean, and similarly for the other two velocity components and the pressure, we have:

$$u = U + u', v = V + v', w = W + w', p = P + p'$$

or, for the velocities in tensor notation,

$$u_i = U_i + u_i' \quad \text{for } i = 1\text{–}3. \tag{2.13}$$

This is generally called a Reynolds' decomposition. His second step was to substitute these expressions into equations (2.3) and (2.6) and time average the resulting equations. Note that, in general, mean quantities should be obtained by ensemble averaging (i.e. averaging over numerous repeated ensembles of the process; for an ergodic, stationary process, this average is equal to the time average – all this is discussed fully in chapter 4). By definition, the time averages of all the fluctuating variables are zero. It follows immediately from (2.6) that

$$\frac{\partial U_i}{\partial x_i} = 0 \tag{2.14}$$

and thus that

$$\frac{\partial u_i'}{\partial x_i} = 0. \tag{2.15}$$

Both the mean and fluctuating velocities therefore satisfy the incompressibility condition. The equation of motion, (2.3), becomes, after substituting equation (2.13), time-averaging, using equations (2.14) and (2.15), noting that the time average of the product of a fluctuating and average quantity is zero, and some manipulation:

$$\frac{\partial U_i}{\partial t} + U_j \frac{\partial U_i}{\partial x_j} = -\frac{1}{\rho}\frac{\partial P}{\partial x_i} + \nu \frac{\partial^2 U_i}{\partial x_j \partial x_j} - \frac{\partial \overline{u_i' u_j'}}{\partial x_j}. \tag{2.16}$$

Throughout this text, overbars (as in the $\overline{u_i' u_j'}$ above) indicate a time average. Equation (2.16) is often termed (for short) the Reynolds equation; the Sample Exercises section at the end of this chapter includes an exercise (2.1) that asks readers to derive it for themselves. Note that the general, large-scale, external force term, F_i, has been neglected (see discussion above) and the density subscript has been dropped for convenience. Like the basic Navier–Stokes equation, the Reynolds equation describes the transport of momentum resulting from the applied (internal) forces, but unlike equation (2.3) (with continuity) it no longer represents a closed system, because an additional term is present – the last one on the right-hand side. This term arises from the nonlinear convective term on the left of equation (2.3) and can be collected with the penultimate term and the pressure term, so that equation (2.16) can be written as

$$\frac{DU_i}{dt} \equiv \frac{\partial U_i}{\partial t} + U_j \frac{\partial U_i}{\partial x_j} = \frac{1}{\rho}\frac{\partial}{\partial x_j}\left[\mu \frac{\partial U_i}{\partial x_j} - P\delta_{ij} - \rho\overline{u_i' u_j'} \right]. \tag{2.17}$$

The right-hand side clearly contains the sum of the viscous stresses (the first term in the brackets), the isotropic stress (pressure acts equally in all directions) and what are called the Reynolds stresses, $\rho\overline{u_i' u_j'}$. We repeat again that the equation is a balance between the applied forces on the right-hand side and the resulting motion – the momentum flux – on the left. Note that the total turbulent kinetic energy, k, is $\frac{1}{2}\overline{u_i'u_i'}$, so we can define what is commonly called the anisotropy tensor as

$$a_{ij} \equiv \overline{u_i'u_j'} - \frac{2}{3}k\delta_{ij}. \tag{2.18}$$

We can therefore express the sum of the last two contributions in equation (2.17) as

$$-\frac{\partial\overline{u_i'u_j'}}{\partial x_j} - \frac{\partial P}{\partial x_i} = -\frac{\partial a_{ij}}{\partial x_j} - \frac{\partial}{\partial x_i}\left(P + \frac{2}{3}k\right), \tag{2.19}$$

which shows that the isotropic component $2k/3$ can be absorbed into the pressure. The importance of this is that it shows that only the anisotropic component is effective in transporting momentum. The Reynolds stress tensor $\rho\overline{u_i'u_j'}$ contains six of these additional stresses; each one represents an average momentum flux due to turbulent velocity fluctuations. Clearly, the turbulent motions influence the mean flow (and, as will be seen below, vice versa). They are all unknowns, so time-averaging the equations of motion has led to significant complications.

We can further explore the physical significance of the shear stresses by considering the simple shear flow shown in figure 2.1. The flow has just one non-zero mean velocity component, U_1, which is a function only of the direction normal to it, x_2. It follows that $\partial/\partial x_1 = \partial/\partial x_3 = U_2 = U_3 = 0$, and in steady flow, the equation of motion (2.17) reduces in this simple case to

$$\mu\frac{\mathrm{d}U_1}{\mathrm{d}x_2} = -\rho\overline{u_1'u_2'}. \tag{2.20}$$

The first term is the viscous shear stress due to the shear rate of strain, implying immediately that the second term, $-\rho\overline{u_1'u_2'}$, is also a shear stress, which can clearly be interpreted as a turbulent momentum flux generated by the vertical fluctuations (u_2') that transfer axial momentum ($\rho u_1'$) vertically. Note that if, as in this case, the mean velocity U_1 increases with x_2, then a parcel of fluid moving upwards, i.e. with a positive u_2', will carry a lower U_1 velocity than its new surroundings, so that u_1' will be negative and $-\rho\overline{u_1'u_2'}$ will be a positive quantity, which will in fact be the case under most circumstances in turbulent shear flows. So the Reynolds stress in equation (2.17) will normally add to the viscous stress. In this particular simple case, the other Reynolds shear stresses ($\rho\overline{u_1'u_3'}$, $\rho\overline{u_2'u_3'}$) are zero. On the other hand, the three Reynolds stresses given by $\overline{u_i'^2}$, are certainly not zero; they sum to give twice the total kinetic energy in the turbulence motions, as noted above.

These are called the normal stresses, because, like the pressure, they act normal to the faces of the small control volume one might use to derive the equation of motion (2.3). One can imagine, at least in a very loose sense, that they are analogous to the individual viscous stresses given by the three separate (normal) components of $\mu\partial U_i/\partial x_i$. Note, however, that $\overline{u_i'u_i'}$, which is twice the total kinetic energy of the turbulent motions ($k = \frac{1}{2}\overline{u_i'u_i'}$) cannot, under any circumstances, be proportional

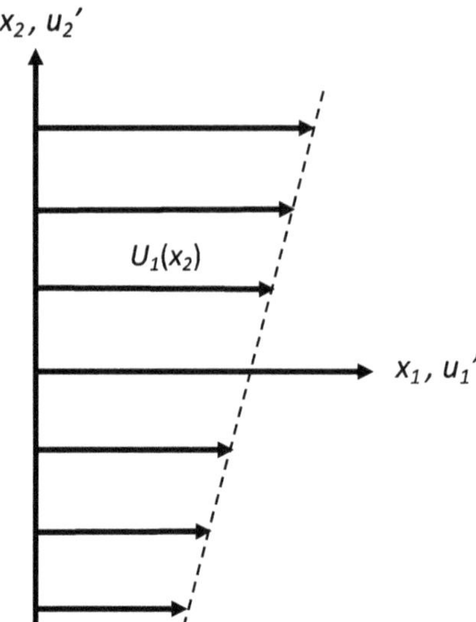

Figure 2.1. A simple shear flow.

to $\partial U_i/\partial x_i$, because then there would be no turbulence, as the latter is zero (by continuity, equation (2.14)). Indeed, the three individual normal stresses $(\overline{u_i'^2})$ can, themselves, never be zero, whatever the value of the normal rates of strain. In the simple shear flow example given here, the two remaining normal strains are actually zero ($\partial U_1/\partial x_1 = \partial U_3/\partial x_3 = 0$), whereas the individual normal stresses $\overline{u_1'^2}$ and $\overline{u_3'^2}$ are certainly not. We explore this flow in greater detail in section 5.2 of chapter 5.

Readers will probably be familiar with the idea that in a solid undergoing a general load, one can determine principal coordinate axes for which the shear stresses (proportional to shear strain) are all zero. A similar situation pertains for the Reynolds stresses. Imagine the $x - y$ plane, with instantaneous velocity components u_1, v_1 in those directions. Now rotate the axes anticlockwise by an angle θ. The velocity components, u_2 and v_2, along those new axes are simply given by

$$u_2 = u_1 \cos(\theta) + v_1 \sin(\theta) \quad \text{and} \quad v_2 = v_1 \cos(\theta) - u_1 \sin(\theta). \tag{2.21}$$

It can thus be shown that the shear stress in the new coordinate system is given by

$$\rho\overline{u_2'v_2'} = \rho\overline{u_1'v_1'} \cos 2\theta - \tfrac{1}{2}\rho(\overline{u_1'^2} - \overline{v_1'^2})\sin(2\theta). \tag{2.22}$$

Clearly, a shear stress in one coordinate system corresponds to a combination of normal and shear stresses in another, which emphasises that there is no fundamental difference between normal stress and shear stress. Given the values of the stresses in the original system, one could obviously choose θ in equation (2.21) to ensure that the $\rho\overline{u_2'v_2'}$ stress is zero, and this defines the principal axes (just as in solid mechanics).

Note, however, that the total turbulence energy ($k = \frac{1}{2}\overline{u_i'u_i'}$) *is* independent of the axis system, unsurprisingly, and is thus a scalar quantity. Note also that the individual stresses ($\overline{\rho u_i'^2}$), when divided by 2ρ, are the three mean-square components of the turbulence energy. These are often rather loosely presented as root-mean-square (rms) values, i.e. $u_{i,\mathrm{rms}}' = \sqrt{\overline{u_i'^2}}$, but the transport equations for stresses and energy that we explore in the next sections do not contain rms quantities, so at least in that sense the latter have less physical significance.

Just as for the mean flow (Navier–Stokes) equation, the turbulent flow version of the mean vorticity equation (2.12) can be obtained by substituting a Reynolds decomposition for vorticity, i.e. writing $\omega_i = \Omega_i + \omega_i'$. Note that this mean vorticity, Ω_i should not be confused with the rate-of-rotation tensor defined by (2.10). The decomposition and subsequent time-averaging eventually yields, after some algebra (exercise 2.3),

$$\frac{\partial \Omega_i}{\partial t} + U_j\frac{\partial \Omega_i}{\partial x_j} = \underbrace{\Omega_i\frac{\partial U_i}{\partial x_j}}_{I} + \underbrace{\overline{\omega_j'\frac{\partial u_i'}{\partial x_j}}}_{II} + \underbrace{\frac{\partial}{\partial x_j}\left(\nu\frac{\partial \Omega_i}{\partial x_j} - \overline{u_j'\omega_j'}\right)}_{III}. \tag{2.23}$$

Notice the two additional terms that have arisen from the Reynolds averaging procedure in this case. First, the final term in *III* contains diffusive (transport) of vorticity by the turbulent motions, $-\partial/\partial x_j(\overline{u_j'\omega_j'})$, augmenting the first term, which is the viscous diffusion of vorticity. In homogeneous turbulence (where, by definition, there are no spatial gradients of any averaged quantity) this first term is necessarily zero. Secondly, the mean vorticity reorientation and stretching process described by term *I* is accompanied by term *II*, which describes the same processes but is caused by turbulent fluctuations. If the *mean* flow is two-dimensional, term *I* disappears, but term *II* is always necessarily non-zero in turbulent flows and, as we will see, is crucial to the cascade process in which energy is transferred from large to small scales.

If we apply the Reynolds decomposition to the scalar equation (2.7) by writing $c = C + c'$ and $u_i = U_i + u_i'$, it is straightforward to show that scalar equation in a turbulent flow becomes

$$\frac{\partial C}{\partial t} + U_j\frac{\partial C}{\partial x_j} = \frac{\partial}{\partial x_j}\left(\gamma\frac{\partial C}{\partial x_j} - \overline{u_i'c'}\right) + Q \tag{2.24}$$

where we have added, for generality, a source density Q in units of scalar per volume per second. One usually wishes to predict the average scalar field $C(\mathbf{x}, t)$, and this rests on being able to predict the scalar turbulent flux $\overline{u_i'c}$. This is a similar closure problem to that faced in solving the Reynolds equations (2.16). Recall, however, that the scalar equation is linear and therefore easier to solve. If the scalar is temperature with sufficiently large variations to require inclusion of the force term in equation (2.16), it is necessary to solve both equations (i.e. (2.16), with F_i added to the right-hand side, and (2.24)) together because of the resulting dynamical linkages between them. Common approaches for simpler situations when, given a velocity field, equation (2.24) can be considered on its own, are discussed for typical model cases in chapter 9. It is straightforward to develop a transport equation for the scalar flux, $\overline{u_i'c'}$, analogous to that for the momentum flux $\overline{u_i'u_j'}$ which we discuss in the following sections. Like the other transport equations, this is sometimes used in turbulence modelling for flows in which one needs more information than that given by the scalar mean field itself (e.g. the statistics of c' in order to estimate peak concentrations), but we do not pursue that topic in this book.

We now repeat our earlier remark that the Reynolds stresses appearing in equation (2.16) are all unknowns, so that the equation set is no longer closed. The benefit of restricting attention to a statistical description of the flow, i.e. to time-averaged forms of the conservation equations for the instantaneous velocities and pressure, has therefore been weakened considerably by a fundamental problem – closure of the resulting equations. Given that, for high-Reynolds-number real turbulent flows, full temporally and spatially resolved solutions of the Navier–Stokes equations (2.3) with (2.6) for given initial and boundary conditions are impractical, a vast amount of effort has been expended in seeking solutions of the Reynolds equations, thus allowing at least the mean flow field to be determined. Except in simple cases (some of which are discussed in later chapters) this generally requires a computational solution but, in any event, it crucially requires some kind of 'closure model', that replaces the unknown Reynolds stresses by functions of the mean velocity field. This process is termed 'turbulence modelling' and there is a vast literature associated with it, often alongside a general discussion of computational fluid dynamics (CFD) because most of the latter activity requires some kind of turbulence modelling. The subject is not explored in this book; some fairly modern summaries can be found in the books by Pope [5] and Wilcox [8], for example.

One can learn a lot about turbulence by considering the behaviour of the Reynolds stresses themselves and, in particular, the turbulent kinetic energy, k $(=\frac{1}{2}\overline{u_i'u_i'})$. It is straightforward, although tedious, to derive equations describing the transport of these quantities. Many turbulence models use at least one of them, often two, or even more. Note here, however, that they all contain quantities such as $\overline{u_i'u_i'u_j'}$ and $\overline{p'u_i'}$; the former, for obvious reasons, are called triple velocity products. Exact transport equations for these can also be derived, but they, in turn, contain unknown fourth-order velocity products – and so on. The system can only become closed (so no modelling is required) by using the infinite set of such higher-order transport equations, which is clearly impossible. Turbulence modelling therefore rests on deciding how many of these equations to use and how to model the remaining (higher-order) terms in such a way that the physics of the flow in question is captured in some optimal way. Such modelling is often informed by a consideration of the transport equations and, although we do not discuss modelling per se, we do discuss what can be learned about the physics from the simpler of the various transport equations – particularly those for the second-moment quantities such as the Reynolds stresses and the turbulence kinetic energy.

Before proceeding to discuss these transport equations, however, we mention one of the simplest turbulence models, but only because we use it in chapter 6 to develop some simple solutions for classical shear flows. This model was first introduced by Boussinesq in 1887, or perhaps by St Venant in 1843, who hypothesised that the (anisotropic) Reynolds stresses defined by equation (2.18) could be taken as proportional to the mean rates of strain, so that

$$-\overline{u_i' u_j'} + \frac{2}{3}k\delta_{ij} = \nu_T\left(\frac{\partial U_i}{\partial x_j} + \frac{\partial U_j}{\partial x_i}\right), \tag{2.25}$$

where ν_T is a *turbulent, or eddy, viscosity*. This relation is analogous to the exact (Newtonian) relationship between the viscous stresses and the mean rates of strain, viz

$$\tau_{ij}/\rho = \nu\left(\frac{\partial U_i}{\partial x_j} + \frac{\partial U_j}{\partial x_i}\right), \tag{2.26}$$

discussed above and which, for the viscous shear stress in the $i = 1$ (x) direction, leads to equation (2.2) $- \tau_x/\rho = \nu\partial U/\partial y$ – in the common situation in which $\partial V/\partial x \ll \partial U/\partial y$. Since the isotropic part of the turbulence stress, $\frac{2}{3}k\delta_{ij}$, can be absorbed into the pressure as noted explicitly in equation (2.19), it is clear that using this model in the Reynolds equation (2.16) leads to the Navier–Stokes equation for laminar flow (2.3) but with u_i replaced by U_i, ν by $\nu_{\text{eff}} = \nu + \nu_T$, and p by $P + \frac{2}{3}k$; ν_{eff} is often called the effective viscosity and at a high enough Reynolds number it dominates the laminar viscosity (so that the latter can conveniently be ignored). Any closure model that relies on this hypothesis to solve the equations of motion clearly requires full knowledge of the $\nu_T(\mathbf{x}, t)$ field, which, we emphasise, is a property of the *flow*, not the fluid. Even today, most commercially used turbulence models, however complicated they are in terms of specifying ν_T, rely on the gradient hypothesis encapsulated by equation (2.25). For a typical discussion of its limitations, see [4, 5], but one can see immediately a couple of basic failings: (i) if the mean strain rate changes suddenly, it implies that the turbulence stress does as well, and (ii) it predicts isotropy between the various normal stress components in any flow dominated by a single rate of strain. These predictions are nonphysical. Nonetheless, there are situations in which the eddy viscosity model performs quite well, as we will discover.

2.5 The Reynolds stress equations

Obtaining the transport equations for the Reynolds stresses and the other second-order quantities consists of calculating the time averages of appropriately weighted versions of the basic equation of motion (2.3). Specifically, first multiply (2.3) for the ith velocity component by the jth component, u_j, then use the Reynolds decomposition (given by equation (2.13)) and expand all the terms, then interchange i with j throughout to produce a second equation, and finally time average the result using the continuity equation. Whilst time-averaging the original equation (2.3) leads to the *first-moment equations* (the Reynolds equations (2.16), as we have seen), this rather more tedious procedure leads to *second-moment equations*. For the Reynolds stress $\rho\overline{u_i'u_j'}$, the second-moment equation can be written as

$$
\frac{\partial \overline{u_i'u_j'}}{\partial t} + U_k \frac{\partial \overline{u_i'u_j'}}{\partial x_k} = \underbrace{-\overline{u_k'u_j'}\frac{\partial U_i}{\partial x_k} - \overline{u_i'u_k'}\frac{\partial U_j}{\partial x_k}}_{I} \quad \underbrace{-\overline{\frac{p'}{\rho}\left(\frac{\partial u_i'}{\partial x_j} + \frac{\partial u_j'}{\partial x_i}\right)}}_{II}
$$

$$
\underbrace{-\frac{\partial}{\partial x_k}\left(\overline{u_i'u_j'u_k'} + \frac{\overline{p'u_j'}}{\rho}\delta_{ik} + \frac{\overline{p'u_i'}}{\rho}\delta_{jk} - \nu\frac{\partial \overline{u_i'u_j'}}{\partial x_k}\right)}_{III} - \underbrace{2\nu\overline{\frac{\partial u_i'}{\partial x_k}\frac{\partial u_j'}{\partial x_k}}}_{IV} \tag{2.27}
$$

The left-hand side is just the total rate of change of the Reynolds stress; the first of the two terms is zero for a steady flow, of course; the second term is sometimes called 'advection' (in preference to 'convection', which has a completely different connotation in buoyancy-affected flows). All the terms on the right-hand side relate to processes that alter the stress as a fluid element follows the mean streamline. Terms *I* denote rates of production generated by the effects of the mean rates of strain, and can be positive or negative, so they can increase or decrease the shear stresses. Term *II* is usually called the 'pressure-strain-rate' tensor and it acts essentially to redistribute energy between the three normal components, $\overline{u_i'^2}$, without changing the total energy. (Note that its trace – putting $i = j$ – is zero by continuity and we will see below that it therefore does not appear in the kinetic energy transport equation.) Terms *III* are often collectively called the diffusive transport; this is strictly only an accurate name for the viscous contribution (the fourth term). More precisely, the first (triple-product) contributions are turbulent transport terms; they can be loosely interpreted as describing the mean transport of the instantaneous stresses $\rho u_i'u_j'$ in the k-coordinate direction. The additional terms involving the fluctuating pressure demonstrate, as does term *II*, the global nature of the pressure field. These terms are all additional unknowns so, along with the ten triple-product terms, they generate significant difficulties for turbulence modellers. Note finally that term *IV* in equation (2.27) represents the viscous dissipation of the stresses. In many cases, this acts to reduce the Reynolds stresses and, unlike the viscous term in *III* which, via a straightforward scale analysis, can be shown to be negligible in high-Reynolds-number flows, it can never be ignored.

2.6 The turbulent kinetic energy equation

A more common equation used for modelling is the transport equation for turbulent kinetic energy (TKE) $k = \frac{1}{2}\overline{u_i'u_i'}$, which is a little simpler to interpret physically. To obtain the TKE transport equation, it is only necessary to set $i = j$ in equation (2.27) (implicitly summing over all i) and then divide by two. The result is

$$
\frac{Dk}{Dt} \equiv \frac{\partial k}{\partial t} + \underbrace{U_k \frac{\partial k}{\partial x_k}}_{I} = \underbrace{-\overline{u_i'u_k'}\frac{\partial U_i}{\partial x_k}}_{II} - \underbrace{\frac{\partial}{\partial x_k}\overline{u_k'\left(k + \frac{p'}{\rho}\right)}}_{IIIa+IIIb} - \underbrace{\nu\overline{\frac{\partial u_i'}{\partial x_j}\left(\frac{\partial u_i'}{\partial x_j} + \frac{\partial u_j'}{\partial x_i}\right)}}_{IV=\bar{\varepsilon}}. \tag{2.28}
$$

In writing this equation we have ignored the viscous term corresponding to the fourth term in *III* of equation (2.27) – i.e. $\nu \partial^2 k/(\partial x_k \partial x_k)$ – since, as noted above, it is generally negligible compared with all the other terms. In fact, for homogeneous turbulence – a flow in which all the fluctuating statistics are independent of the spatial position – the term is exactly zero (as are the other terms, see chapter 5). The two terms on the left-hand side – constituents of the total derivative of k – denote the rate of change of turbulent energy at a fixed point (zero in a steady flow) and, term *I*, advection of turbulent energy by the mean flow.

Consider now the three sets of terms on the right-hand side of equation (2.28). Term *II* is the rate of production of turbulence energy, extracted from the mean flow by the action of Reynolds stresses on the mean shear. It is possible to write down an equation for the mean flow kinetic energy, and this contains a term equal and opposite in sign to *II*, emphasising that it is the mean flow that provides the turbulence energy. Term *III*, the transport term, integrates to zero over the whole flow by Gauss's theorem. Just as in the Reynolds stress transport equation, it acts simply to move energy around within the flow through the action both of the turbulent velocities themselves (*IIIa*) and the fluctuating pressure (*IIIb*). It does not create or remove energy globally. (Note that, as mentioned above, there is no counterpart in the TKE equation to the pressure strain term in equation (2.27).) The final term (*IV*) can be shown to equal the viscous dissipation of turbulence energy, and it is usually given the symbol $\bar{\varepsilon}$. This is always necessarily positive (although it it important to realise that the instantaneous dissipation at a particular time and location may not be; see chapter 3), so the term represents a sink of turbulence energy, with that energy being transformed into heat. It is possible to show that the consequent rise in temperature is almost always completely negligible. There is an analogous term in the mean flow energy equation, which contains the gradients of the mean flow velocities, but since such gradients are always (except close to solid surfaces) very much smaller than the gradients of the fluctuating velocities, nearly all of the total dissipation in turbulent flows is provided by $\bar{\varepsilon}$. Figure 2.2 is a sketch of the processes taking place for a small control volume in the turbulent flow.

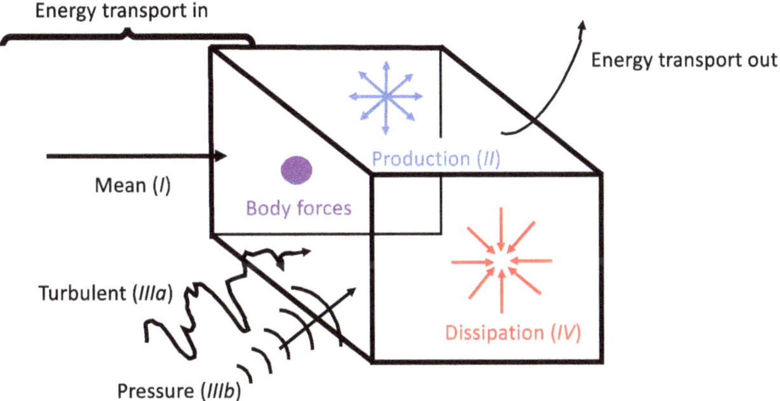

Figure 2.2. Turbulence energy transfers. The notation follows that given in equation (2.28). (Adapted from [3]).

The energy transfers in and out correspond to terms *I* and *III*, plus the viscous diffusive term, which, as noted above, is usually negligible, so is not shown. Production (term *II*) and dissipation (term *IV*) occur throughout the volume, as do any body forces which may exist (not included in equation (2.28)) and could provide net positive or negative generation of turbulence energy. Under steady flow conditions, we may write the TKE transport equation in words simply as

$$\text{Advection} = \text{Production} + \text{Transport} - \text{Dissipation,}$$

which emphasises the essential physical processes. Note that since a global integral of the transport term must be zero, then if k is not to change with time, there must be a global balance between the rates of production and dissipation. If these are not equal, then the flow cannot be steady – $\partial k/\partial t$ must be non-zero. For *homogeneous* turbulence in steady conditions, which explicitly implies that the advection and transport terms in equation (2.27) are all zero everywhere, there must be a balance between $\overline{u_i'u_k'}\partial U_i/\partial x_k$ and $-\bar{\varepsilon}$ everywhere; any imbalance would force the turbulent motions to decay either in time or space. Thus, only the presence of mean flow shear can prevent such decay. An additional simplification that can be considered is that of *isotropy*. This requires that all the statistics of the fluctuating quantities are independent of coordinate system rotations or reflections, so that turbulent shear stresses are all zero and the normal stresses are all equal. In this situation (which necessarily requires the turbulence to be homogeneous) and if the flow is steady, i.e. $\partial/\partial t = 0$, the turbulence must decay in some coordinate direction. Because of their importance in analysing turbulence theoretically, and thus understanding more of the important physics, in chapter 5 we explore the classical flows in which the turbulence is isotropic (and therefore homogeneous) or just homogeneous (but not isotropic).

A few more preliminary comments about the dissipation term (*IV*) are in order here. First, it is straightforward to show that this term can be written as

$$\bar{\varepsilon} = \nu\overline{\frac{\partial u_i'}{\partial x_j}\frac{\partial u_i'}{\partial x_j}} + \nu\frac{\partial^2 \overline{u_i'u_j'}}{\partial x_i \partial x_j}. \tag{2.29}$$

A similar expression involving the fluctuating vorticity (ω_i') can also be found and is given by

$$\bar{\varepsilon} = \nu\overline{\omega_i'\omega_i'} + 2\nu\frac{\partial^2 \overline{u_i'u_j'}}{\partial x_i \partial x_j}. \tag{2.30}$$

Consequently, in homogeneous turbulence, since the gradients of $\overline{u_i'u_j'}$ are zero,

$$\bar{\varepsilon} = \nu\overline{\omega_i'\omega_i'}. \tag{2.31}$$

This is sometimes called 'pseudo dissipation' and given a different symbol, e.g. $\tilde{\varepsilon}$ [5]. Even in more general turbulence, this result is valid if the Reynolds number is high, because mean-squared (fluctuating-) velocity derivatives are then dominated by the small scales; in contrast, derivatives of $\overline{u_i'u_j'}$ scale according to the overall size of the flow and are thus much smaller.

Secondly, one can develop a more quantitative formulation of the above argument by considering the various length scales of the turbulence. For now, we merely emphasise that the overall turbulence dissipation rate is dominated by the smallest scale motions when the Reynolds number is sufficiently large and, away from the inevitable viscous dominated regions very close to walls, the mean flow contribution can be ignored. Energy dissipation, which occurs at the smallest scales of motion, is the result of a cascade of energy, first from the mean flow into the largest scales of the turbulence, and then consecutively through smaller and smaller scales. All this is explored in detail in chapter 3.

2.7 The enstrophy transport equation

It is straightforward to produce a transport equation for the mean-squared vorticity, usually called the enstrophy, $\overline{\omega_i'\omega_i'}$ (the second term of the total enstrophy, $\frac{1}{2}\Omega_i\Omega_i + \frac{1}{2}\overline{\omega_i'\omega_i'}$). This quantity has a relationship to the mean vorticity Ω that is very similar to the relationship of kinetic energy to the mean velocity (expressed by the k transport equation (2.28)). Recall, too, its intimate relationship with the viscous dissipation of kinetic energy given by equation (2.30) and, at high Reynolds number, by the almost exact result $\bar{\varepsilon} = \nu\overline{\omega_i'\omega_i'}$ (exact in homogeneous turbulence). The enstrophy transport equation, in full, is a rather long equation, but it can be shown that its various terms are of very disparate orders of magnitude. (The mean enstrophy, for example, is negligible compared with the fluctuating part.) Retaining only the dominant terms leads to

$$\frac{\partial \frac{1}{2}\overline{\omega_i'\omega_i'}}{\partial t} + U_j\frac{\frac{1}{2}\overline{\omega_i'\omega_i'}}{\partial x_j} = \underbrace{\overline{\omega_i'\omega_j'\frac{\partial u_i'}{\partial x_j}}}_{I} - \underbrace{\nu\overline{\frac{\partial \omega_i'}{\partial x_j}\frac{\partial \omega_i'}{\partial x_j}}}_{II} \tag{2.32}$$

Even in this equation, the left-hand side terms have only two small-scale velocity derivatives, compared with three and four in terms I and II, respectively, so they can be ignored. There is therefore a close balance between the production of enstrophy by turbulent vortex stretching (I) and its viscous destruction (II). I is usually positive [7]. Imagine a vortex tube distorted by random turbulent motions. Ignoring friction, the tube will become contorted and stretched while conserving its angular momentum. Its cross-sectional area will therefore decrease, while its vorticity will increase. So turbulent vortex stretching tends to concentrate high-magnitude vorticity in narrow regions located some way from each other. The fluctuating vorticity field therefore has a 'spotty' appearance; this is the 'internal intermittency' mentioned in section 1.2.5 and we say a little more about it in the following chapter (section 3.1), but note here that since enstrophy is usually closely equal to energy dissipation, equation (2.31), the dissipation field is also intermittent. It is really only since direct numerical simulations of the full Navier–Stokes equations (2.3) have become possible that this intermittency has been directly visualised – although necessarily for relatively small Reynolds numbers.

A final remark is worth making. Noticing that the enstrophy production term (I) in equation (2.32) is an averaged triple product of various velocity derivatives, it is possible to show [2] that

$$\overline{\omega_i' \omega_j' \frac{\partial u_i'}{\partial x_j}} = -\frac{35}{2} \overline{\left(\frac{\partial u_k'}{\partial x_k}\right)^3}. \tag{2.33}$$

This is an instructive result, for it directly implies that (since term I is always positive) the skewness factor of the third moment of the velocity derivative is negative, even in homogeneous, isotropic turbulence (the simplest kind), so that turbulence can *never* be a precisely Gaussian process (for which all third-order statistics are necessarily zero, see chapter 4).

2.8 Subject giant: O Reynolds

One of the first things a student of fluid mechanics learns is the importance of the Reynolds number and, subsequently perhaps, that at high enough Reynolds numbers, typical flows are expected to be turbulent. The importance of this dimensionless number was first identified by the English Professor of Engineering, Osborne Reynolds (1842–1912). Reynolds held a chair of engineering at Owen's College (which is now the University of Manchester), where he focussed on his research in fluid mechanics. It was in 1883 that Reynolds published his article describing 'the circumstances which determine whether the motion of water shall be direct or sinuous' [6]. This article presented the results of his experiments in which he used a dye filament to observe the state of flow within a clear pipe (as illustrated in figure 2.3). It was in this paper that he used dimensional analysis to identify that the critical parameter which determines the onset of turbulence is defined by $\rho VD/\mu$. The article was actually peer reviewed by two very notable figures of the time, Sir George Stokes and Lord Rayleigh, who both supported its publication in the

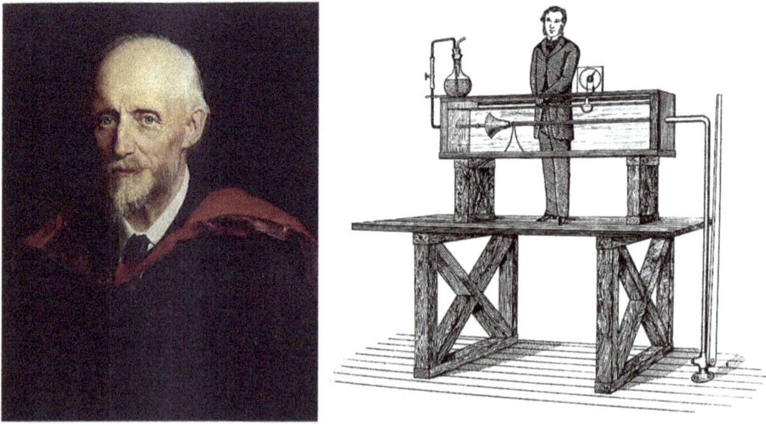

Figure 2.3. (Left) Osborne Reynolds (1842–1912) painted in 1904 by John Collier. (Right) The apparatus from Reynolds' famous experiment to determine the onset of turbulent flow in a pipe. Reprinted from [6] with permission from The Royal Society.

Philosophical Transactions of the Royal Society. Interestingly, it was not Reynolds who put his name to this dimensionless number, but rather it was only several years after his retirement that German colleagues started calling it by that name: Sommerfield in 1908, and later Prandtl in 1910.

The Reynolds number was not Reynolds' only substantial contribution to fluid mechanics; two more important concepts also bear his name. The first of these is the idea that the signal of a turbulent flow can, at any point, be dissected into a 'time-averaged' component and a fluctuating component (e.g. $u = U + u'$). This concept was first published by Reynolds in 1895 and is now known as *Reynolds decomposition*; it is explored in more detail in section 2.4 of this chapter. The second noteworthy contribution made by Reynolds was the connection between momentum and heat transfer, known as the *Reynolds analogy*. The idea is that, in turbulent flows, the momentum flux is analogous to the heat flux and that velocity and temperature profiles will have similar shapes because both processes are dominated by transfer by turbulent eddies. Although of less importance than his other two contributions, this idea has provided significant insight for many engineering applications.

Reynolds' contributions to the field of fluid mechanics and turbulence have shaped the direction of research ever since. J. J. Thomson, who in 1906 received the Nobel Prize in physics for demonstrating the existence of the electron, is quoted as remarking from personal experience as one of Reynolds' students that 'As a teacher, he did not spoon-feed his students and many did not pass his course, but the best students enjoyed his lectures and found them stimulating' [1]. In 1888, Reynolds was awarded the Royal Society's Royal Medal 'for his investigations in mathematical and experimental physics, and on the application of scientific theory to engineering'. The University of Manchester still has the apparatus from his 1883 experiment on display and the European Research Community on Flow, Turbulence and Combustion (ERCTOFAC) has, for many years, been celebrating his achievements with an annual Osborne Reynolds Day, comprising a competition for young researchers studying applied fluid dynamics in the UK.

Sample exercises

2.1 Derive the Reynolds equation (2.16) from the basic Navier–Stokes equation (2.4) using the Reynolds decomposition and anything else needed.

2.2 Derive equation (2.22) from (2.21) and show that if the turbulence is isotropic – i.e. if all statistics of the fluctuating velocity u_i' are not affected by coordinate system rotations or translations, so that the normal stresses are equal – then the shear stress is necessarily zero.

2.3 Derive the Reynolds-averaged vorticity equation (2.23) from equation (2.12).

2.4 Derive the Reynolds stress transport equation (2.27).

2.5 Show that the full viscous term in the Reynolds stress equation (2.27) (i.e. the viscous transport and the viscous dissipation) can also be written in the following forms:

$$\frac{\partial}{\partial x_k}\left(\nu\frac{\partial \overline{u_i' u_j'}}{\partial x_k}\right) - 2\nu\overline{\frac{\partial u_i'}{\partial x_k}\frac{\partial u_j'}{\partial x_k}} = \overline{\nu u_i'\frac{\partial^2 u_j'}{\partial x_k \partial x_k}} + \overline{\nu u_j'\frac{\partial^2 u_i'}{\partial x_k \partial x_k}}$$

2.6 Use the Reynolds stress transport equation (2.27) to derive the full TKE transport equation (2.28).

2.7 Show that the dissipation $\bar{\varepsilon}$ as it appears in equation (2.28),

$$\bar{\varepsilon} = \nu\overline{\frac{\partial u_i'}{\partial x_j}\left(\frac{\partial u_i'}{\partial x_j} + \frac{\partial u_j'}{\partial x_i}\right)},$$

can also be written equivalently as

$$\bar{\varepsilon} = \frac{1}{2}\nu\overline{\left(\frac{\partial u_i'}{\partial x_j} + \frac{\partial u_j'}{\partial x_i}\right)\left(\frac{\partial u_i'}{\partial x_j} + \frac{\partial u_j'}{\partial x_i}\right)}$$

or

$$\bar{\varepsilon} = \nu\overline{\frac{\partial u_i'}{\partial x_j}\frac{\partial u_i'}{\partial x_j}} + \nu\frac{\partial^2 \overline{u_i' u_j'}}{\partial x_i \partial x_j}.$$

Why is the second term on the right-hand side of this last relation identically zero in homogeneous turbulence?

References

[1] Anderson J D 1997 *A History of Aerodynamics* (Cambridge: Cambridge University Press)

[2] Batchelor G K 1953 *The Theory of Homogeneous Turbulence* (Cambridge: Cambridge University Press)

[3] Bradshaw P 1971 *An Introduction to Turbulence and Its Measurement* (Oxford: Pergamon)

[4] Libby P A 1996 *Introduction to Turbulence* (London: Taylor and Francis)

[5] Pope S B 2000 *Turbulent Flows* (Cambridge: Cambridge University Press)

[6] Reynolds O 1883 An experimental investigation of the circumstances which determine whether the motion of water shall be direct or sinuous and the law of resistance in parallel channels *Philos. Trans. R. Soc.* **174** 935–82

[7] Taylor G I 1938 The spectrum of turbulence *Proc. R. Soc. Lond.* A **164**(919) 476–90

[8] Wilcox D C 2006 *Turbulence Modelling for CFD* 3rd edn (DCW Industries)

IOP Publishing

Turbulent Flows: an Introduction

Ian P Castro and Christina Vanderwel

Chapter 3

The scales of motion

In this chapter we unpack the notion, already introduced in chapter 1, that turbulent flows always contain motions that have a wide range of spatial and temporal scales. After some initial remarks to set the scene, section 3.2 outlines an order-of-magnitude analysis that considers the production and dissipation of turbulence kinetic energy, concluding with the important result that the separation between the largest and smallest scales becomes ever larger as the Reynolds number increases. Some relationships for the smallest scales are derived in the following section and then, in section 3.4, we discuss the crucial issue of how energy is transferred between the different scales of motion. This naturally leads to an introduction to the energy spectrum in section 3.5. For this introductory book, this chapter is necessarily brief; there are texts that explore the topic in great detail and contain significant and sometimes heavy mathematics (e.g. [2, 3]); readers who wish to delve further should consult those.

3.1 Initial remarks

It is important to understand from the outset at least some of the fundamental characteristics of the various scales of motion and to develop a quantitative picture of their relative sizes in order to gain insights into numerous phenomena in general turbulent flows. The general idea that the turbulence energy cascades from the largest down to the smallest scales was first expounded by Richardson in his 1922 book on numerical weather prediction [10]. He suggested that energy enters the turbulence (from the mean flow) through the production process at the largest scales and that this energy is then transferred (by essentially inviscid, nonlinear processes) to smaller and smaller scales. At each stage, viscosity plays no part in this process because the Reynolds number appropriate for the eddies is sufficiently large; the process is predominantly driven by inertial forces. Eventually, however, as the cascade proceeds, the Reynolds number for the smallest eddies reaches the order of unity, at which point the viscous forces become significant and dissipation begins to

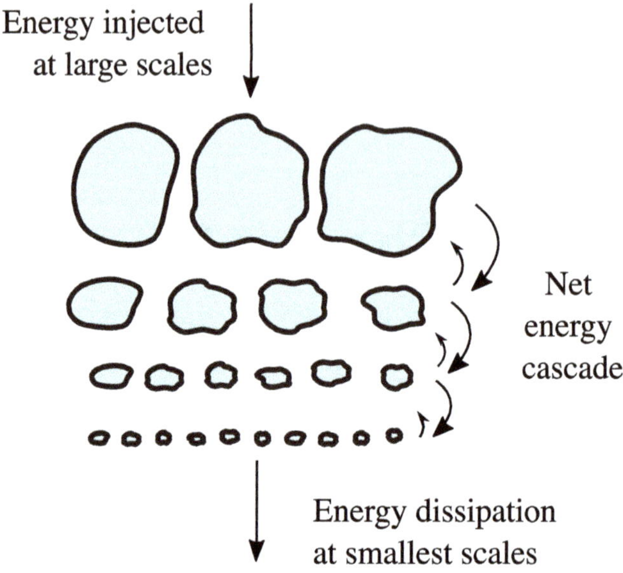

Figure 3.1. Illustration of the turbulent energy cascade imagined by Richardson (1922)[1].

be important. A well-known humorous rhyme (a parody of Jonathan Swift's poem about fleas) appeared in Richardson's chapter on eddy motions:

Big whorls have little whorls
that feed on their velocity.
Little whorls have lesser whorls
and so on to viscosity
—in the molecular sense.

This qualitative picture of the energy cascade of turbulence is illustrated in figure 3.1, showing that the largest eddies, which continuously extract energy from the mean flow, break up into smaller and smaller eddies. In the following sections, we look more closely at the production of energy that occurs at the large scales and the dissipation of energy at the smallest scales. The reason that the range of scales is an important feature of turbulence will be explored and the physical mechanisms of energy transfer between the different scales will be briefly outlined. Over the past century, since the proposal of this energy cascade, it has become clear that this one-way description of energy transfer is not entirely correct; eddies can sometimes combine together, creating an inverse flow of energy to larger scales, see section 3.4.

With reference to figure 3.3, it is difficult to illustrate in two-dimensional pictures the fact that although the largest eddies tend to fill the domain in at least one

---

[1] Redrawn with permission based on the illustration by Frisch, U., Sulem, P., and Nelkin, M. (1978). A simple dynamical model of intermittent fully developed turbulence. *J. Fluid Mech.*, 87(4), 719–736, copyright Cambridge University Press.

coordinate direction, the smallest scales are far from space filling. Because vortex stretching (as described in section 3.4) leads to individual vortex filaments becoming smaller, the vorticity field becomes less and less space filling as one moves towards the smallest scales. This is what accounts for the intermittency (called 'spottiness' in section 1.2.5) in the dissipation field. The fact of this spatial intermittency of the smallest scales leads to inevitable questions about the adequacy of some aspects of Kolmogorov's classical ideas, first introduced in section 3.3, as he assumed that the number of eddies per unit volume grows as their size reduces, so that the small eddies are as space filling as the large ones. We do not address this issue further in this book. There is a substantial literature on this small-scale intermittency and its implications; interested readers with more of a 'physics' bent could well start with Frisch's book [3].

3.2 Order-of-magnitude analysis

Consider a typical kind of turbulent flow, generated by the confluence of two streams of fluid that have different velocities. This is illustrated by figure 3.2, in which, for convenience, one of the streams has zero velocity. One can define U and x as the global velocity and length scales, respectively. The large-scale eddy motions have length and velocity scales of L and $\hat{u}$, respectively, where the latter is just some appropriate velocity that represents the turbulent fluctuations (typically, it could be the root mean square of u', u'_{rms}) and we recognise that the large scales cannot be larger than the width of the shear layer. It is intuitively clear that the rate of change of the total (volume integral) turbulence energy must be the difference between production and dissipation, i.e.

$$\frac{d}{dt}\int_V k\,dV = \int_V (P_k - \varepsilon)\,dV,$$

with the integrals taken over the volume V and assuming that $u_i = 0$ at the boundaries. For equilibrium, it follows that the right-hand size must be zero. This overall, global balance must hold for any kind of turbulent flow. If the turbulence is homogeneous, the balance must also hold *locally*, so that the turbulence energy transport equation (2.28) becomes

$$P_k \equiv -\overline{u'_i u'_k}\frac{\partial U_i}{\partial x_k} = \nu\,\overline{\frac{\partial u'_j}{\partial x_j}\left(\frac{\partial u'_i}{\partial x_j} + \frac{\partial u'_j}{\partial x_i}\right)} \equiv \varepsilon, \qquad (3.1)$$

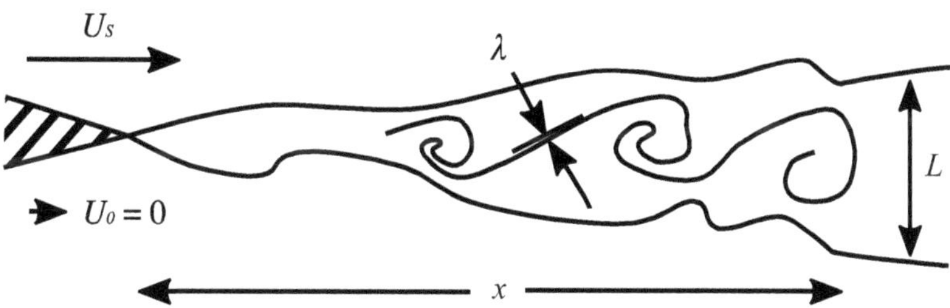

Figure 3.2. A simple turbulent shear flow.

where for convenience hereafter we have dropped the overbar for ε (see equation (2.28)), although realising that it is a time-averaged quantity. The left-hand side (the production, P_k) can be scaled using L and $\hat{u}$. Thus, $P_k \sim \hat{u}^3/L$ and so the energy dissipation per unit mass is

$$\varepsilon \sim \frac{\hat{u}^3}{L}. \tag{3.2}$$

This is a cornerstone of turbulence theory. But how should the right-hand side be scaled? If we were to use $\hat{u}$ and L again, we would have $\dfrac{\hat{u}^3}{L} \sim \nu\dfrac{\hat{u}^2}{L^2}$, which implies that $\nu \sim \hat{u}L$; this must clearly be wrong, since $\nu \ll \hat{u}L$ normally. So let us use a length scale, λ, rather than L, chosen such that

$$\frac{\hat{u}^3}{L} \sim \nu\frac{\hat{u}^2}{\lambda^2}.$$

It follows that

$$\frac{\lambda}{L} \sim \mathrm{Re}_L^{-1/2}, \ \mathrm{Re}_L = \frac{\hat{u}L}{\nu}, \ \mathrm{Re}_\lambda = \frac{\hat{u}\lambda}{\nu}. \tag{3.3}$$

Re_L is clearly a Reynolds number for the large-scale motions; Re_λ is called the Taylor scale Reynolds number. Here, λ is called the dissipation, or Taylor, microscale and represents the range of scales over which dissipation occurs. It is *not*, however, representative of the smallest possible scales, but rather a measure of the order of magnitude of velocity derivatives (i.e. anticipating equation (4.54))

$$\frac{\overline{u_i' u_j'}}{\lambda^2} = \mathcal{O}\overline{\left(\frac{\partial u_i'}{\partial x_j}\right)^2}. \tag{3.4}$$

Equation (3.3) clearly shows that the larger the Reynolds number of the flow, the greater the separation of scales (see also section 3.5). This demonstrates one of the biggest challenges in simulating and measuring turbulent flows: in most engineering problems of interest, one needs the facility to resolve many orders of magnitude of flow features. This is why there may never be the computer power to completely resolve the flow around an entire airplane nor the laboratory techniques to capture every detail of the flow in a high-Reynolds-number experiment. It is why large-scale engineering or environmental problems are often broken down into simplified, more manageable flows which can be more easily addressed.

3.3 The small scales

For the very smallest scales, whatever the Reynolds number of the large scales, there must be a range of eddy scales for which the appropriate Reynolds number is of the order of unity (so that energy dissipation occurs via viscous effects). We define these scales as v_η and η, such that

$$\frac{\hat{u}^3}{L} \sim \varepsilon \sim \nu\frac{v_\eta^2}{\eta^2} \text{ and } \frac{\eta v_\eta}{\nu} = 1. \tag{3.5}$$

Solving for η and v_η yields

$$\eta \sim (\nu^3/\varepsilon)^{1/4}, \; v_\eta = (\nu\varepsilon)^{1/4}. \tag{3.6}$$

Effectively, therefore, once the large scales $\hat{u}$ and L are fixed, multiple instabilities occur until there are fluctuations at the scales of η and v_η, such that $P_k \sim \varepsilon$. The results expressed by equation (3.5) for the smallest scales were first famously deduced by Kolmogorov [7], on the basis of his **first similarity hypothesis**:

'In every turbulent flow at sufficiently high Reynolds number the small-scale motions must have a universal form, uniquely determined by ε and ν only.'

This leads directly to equation (3.5) simply on dimensional grounds; the fact that $\eta v_\eta/\nu = 1$ follows immediately. η and v_η are known as the Kolmogorov scales. This *universal equilibrium range*, whose form depends only on ε and η, must hold for motions with length scales l satisfying $l < l_{EI}$, say. Its existence naturally relies on the smaller scales of the cascade hierarchy being sufficiently far removed from the largest scales, which extract energy from the mean flow, that they are independent of that extraction process. Kolmogorov therefore suggested the **hypothesis of local isotropy**:

'At sufficiently high Re, the small-scale motions are statistically isotropic (say for length scales $l < l_{EI}$)'.

Using the relations expressed by equations (3.2) and (3.5), we can estimate some useful ratios of scales:

$$\frac{\eta}{L} \sim \left(\frac{\nu^3}{L^4\varepsilon}\right)^{1/4} = \left(\frac{\nu^3 L}{L^4\hat{u}^3}\right) = \mathrm{Re}_L^{-3/4} \tag{3.7}$$

and

$$\frac{v_\eta}{\hat{u}} \sim \left(\frac{\nu\varepsilon}{\hat{u}}\right)^{1/4} = \left(\frac{\nu\hat{u}^3}{L\hat{u}^4}\right)^{1/4} = \mathrm{Re}_L^{-1/4}. \tag{3.8}$$

Kolmogorov [7] also used a scale analysis to argue that there must be an intermediate range of scales across which energy is transferred without being either produced or dissipated and that, under equilibrium conditions, the energy flux through this region must equal the rate at which energy is dissipated at the smallest scales, i.e. ε. Kolmogorov's **second similarity hypothesis** was thus:

'In every turbulent flow at sufficiently large Reynolds number the statistics of the motions of scale l in the range $\eta \ll l \ll L$ have a universal form uniquely determined by ε'.

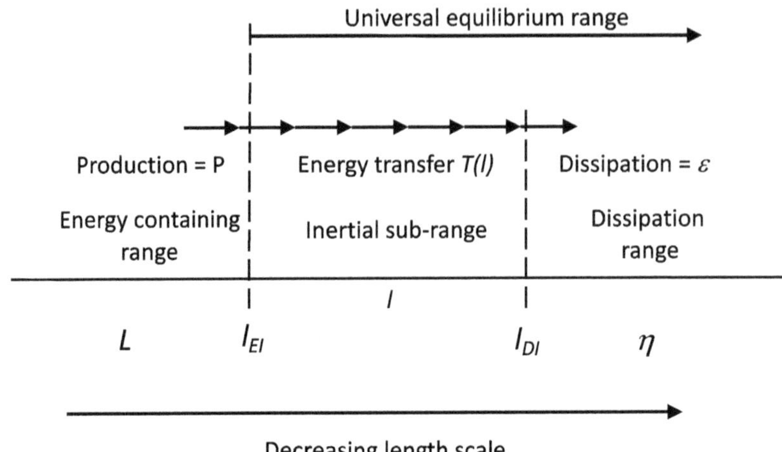

Figure 3.3. The turbulence processes as the eddy scale reduces.

Given a length scale l in this *inertial subrange*, a corresponding velocity $u(l)$, and a time $\tau(l)$, scales can be expressed as

$$u(l) = (\varepsilon l)^{1/3} = v_{\eta}(l/\eta)^{1/3} \sim \hat{u}(l/L)^{1/3} \tag{3.9}$$

and

$$\tau(l) = (l^2/\varepsilon)^{1/3} = \tau_{\eta}(l/L)^{2/3} \sim \tau_o(l/L)^{2/3}, \tag{3.10}$$

where τ_{η} is the timescale of the smallest eddies – η/v_{η} – and τ_o is the timescale of the large eddies – $L/\hat{u}$.

Figure 3.3 illustrates these various ideas, emphasising that turbulence energy is produced at the large scales, $l > l_{EI}$, (from the mean flow) and it cascades progressively through and into smaller and smaller scales, until finally the scales are small enough for viscosity to become important, so that energy is dissipated (into heat). L_{DI} in the figure delineates the lower wavelength end of the inertial sublayer. Note the dual role played by ε: it is both the energy dissipation rate (per unit mass) and it is the rate at which energy cascades across the inertial subrange.

3.4 Mechanisms of interscale energy transfer

These mechanisms for energy transfer between the different scales of eddies in turbulent flows can also be explained by imagining the transformation that individual eddies must undergo as a result of the turbulent field in which they exist. The straining field of the turbulent flow has a crucial role to play.

The large-scale eddies in a turbulent flow are often imagined as large three-dimensional vortex filaments. Imagine such a vortex tube inside a chaotic turbulent flow whose characteristic length scale is represented by its diameter, as in figure 3.4. As a result of the local strain, the vortex filament might either stretch, becoming longer but thinner, or it might undergo compression, becoming shorter but thicker. Assuming the

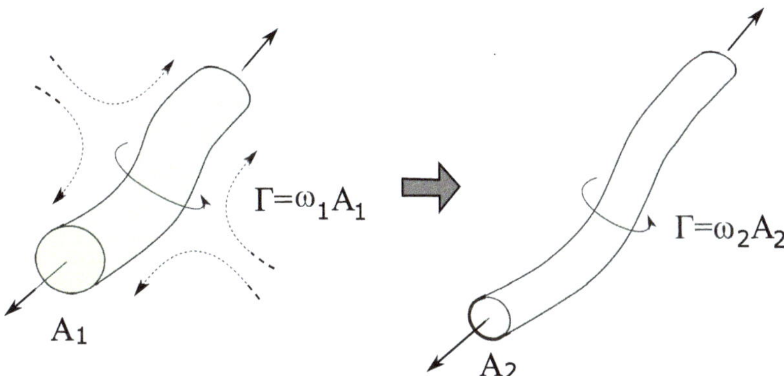

Figure 3.4. Illustration of the mechanism of energy transfer to smaller scales through vortex stretching.

vortex filament retains a constant circulation, $\Gamma = \omega A$, as it increases in length and therefore reduces in area, it will gain vorticity and energy; likewise, a vortex filament undergoing compression will lose energy. In this way, energy can actually pass to either larger or smaller scales, but, on average, vortex stretching wins out and energy is transferred from large scales to smaller scales. More recently, it has been suggested that the mechanism of vortex stretching cannot be disentangled from the fact that strain rates also steepen via nonlinear self-advection, further amplifying interscale energy transfer [6].

The assumption that the structure of turbulence consists solely of a tangle of vortex filaments like spaghetti is not entirely correct. Kuo and Corrsin [8] described the shapes of the varying eddy motions observed in turbulence as 'blobs', 'rods', 'slabs', and 'strips'. However, they did observe that at the very small scale, the eddies are predominantly rod-shaped, consistent with this idea of vortex stretching being a dominant route to dissipation in the locally isotropic universal equilibrium range described above. In that range, the term $\overline{\omega_j' \frac{\partial u_i'}{\partial x_j}}$ (term *II* from the mean vorticity balance equation (2.23)) represents the mean vorticity reorientation and stretching by turbulent fluctuations, as described in figure 3.4, and is crucial to the cascade process in which energy is transferred from larger to smaller scales.

3.5 The energy spectrum

It is natural to think about the various energy transfers illustrated in figure 3.3 by considering how the turbulence energy is distributed across the different scales. This requires us to introduce the idea of an energy spectrum. Chapter 4 will discuss this more fully, but for the moment let us simply point out first that if $E(\kappa)$ is defined as the energy contained in the small wavenumber range $d\kappa$ where κ, the wavenumber, is just $2\pi/l$ (where l is an eddy length scale in conformity with our discussion in section 3.2), then the total energy within the turbulence must be given by

$$k = \int_0^\infty E(\kappa)d\kappa. \tag{3.11}$$

Using ε and κ to non-dimensionalise the spectrum, we have

$$E(\kappa) = \varepsilon^{2/3}\kappa^{-5/3}\Psi(\kappa\eta),\tag{3.12}$$

where $\Psi(\kappa\eta)$ is usually called the (non-dimensional) 'compensated Kolmogorov spectrum function'. In the inertial subrange, Kolmogorov's second hypothesis implies that for $\kappa\eta \ll 1$, ε is the only important parameter, so that $\Psi(\kappa\eta)$ must become a constant (since it contains η). Thus

$$E(\kappa) = C\varepsilon^{2/3}\kappa^{-5/3}\tag{3.13}$$

which is the famous $-5/3$ spectrum. Alternatively, we could non-dimensionalise $E(\kappa)$ using ε and ν, to yield

$$E(\kappa) = (\varepsilon\nu^5)^{1/4}\psi(\kappa\eta)\tag{3.14}$$

where $\psi(\kappa\eta)$ is the 'Kolmogorov spectrum function'. So $\psi(\kappa\eta) = E(\kappa)/(\varepsilon\nu^5)^{1/4}$ is a universal (non-dimensional) function within the universal range of wavenumbers, provided the Reynolds number of the flow is sufficiently high. We consider how high in due course.

It is important at this point to clarify that the various spectral functions introduced above, principally $E(\kappa)$, are three-dimensional. This should be obvious, since k, the turbulence energy on the left-hand side of equation (3.11), includes fluctuations in all three coordinate directions. Both the scale l and the wavenumber κ are therefore actually vector quantities. To obtain $E(\kappa)$ experimentally is thus not straightforward, but it is much easier (with certain necessary assumptions) to obtain the one-dimensional spectra from single-point time traces of velocity. We can denote these functions by $E_{11}(\kappa)$, $E_{22}(\kappa)$, or $E_{33}(\kappa)$, respectively, for the u_1', u_2' and u_3' velocity components (i.e. u', v' and w' for the typical Cartesian coordinate system). It can be shown that the constants corresponding to C in equation (3.13) are related via $C_{11} = \frac{18}{55}C$ and $C_{22} = C_{33} = \frac{24}{55}C$; the wavenumbers in the various related expressions are then simply the appropriate elements of the full vector.

Kolmogorov's development of the energy cascade ideas originally proposed by Richardson, his clarification of the multiscale nature of turbulence, and his identification of the energy transfer rate, ε, as the crucial parameter that controls the inertial subrange (leading, for example, to equation (3.13)) have been foundational for all subsequent developments in turbulence theory. Although there are aspects of his theories which have, in more recent decades, been shown to be inadequate (as we explore briefly in chapter 5), it can be confidently claimed that 'everything we know today about turbulent flows rests on his papers of 1941', as Jiménez states [5]. The same author also correctly points out that 'the name of the most widely used engineering turbulence model ($k - \varepsilon$) clearly shows how much engineering practice is indebted to him'. We emphasise that results such as equation (3.13) were based solely on theoretical ideas. Their experimental verification did not begin until a decade or more later.

As an example of sets of experimental measurements of E_{11}, figure 3.5 shows an adjusted version of data compiled by Saddoughi and Veeravalli [11]. They are

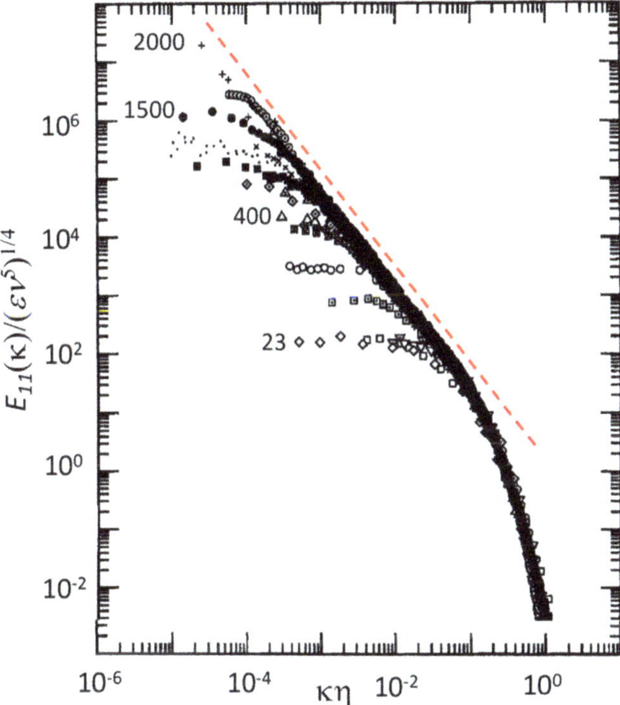

Figure 3.5. Experimental E_{11} spectra in flows at numerous Reynolds numbers (Re_λ)[2]. The red dashed line has a slope of $-5/3$ and the numbers on the figure give the approximate value of Re_λ for the data set closest to them.

plotted using Kolmogorov's universal scaling as expressed in equation (3.14) and comprise data from numerous authors for different flows and at different Reynolds numbers. In terms of the Taylor scale Reynolds numbers, Re_λ, these range from very low to very high; values for some of the data sets are indicated on the figure. The highest are the famous data of Grant *et al* [4], which were obtained in a tidal channel (the Discovery Channel on the west coast of Canada). This was the first set of data obtained at such a high Reynolds number, and the particularly high value of Re_λ ($\sim$2000) has rarely been exceeded since. The data of Saddoughi and Veeravalli [11] included $\mathrm{Re}_\lambda \sim 1500$, obtained in a boundary layer in the largest wind tunnel at the NASA Ames Research Centre. Only a handful of wind tunnel facilities in the world can achieve such high values of Re_λ, although, more recently, some specialist facilities (e.g. using helium) have reached values in excess of 3000.

There are a number of noteworthy points in figure 3.5. First, it is clear that the range of wavelengths over which the inertial subrange holds increases with Re_λ, as expected. The data of Grant *et al* [4], with an Re_λ value of around 2000, yielded about three decades of $-5/3$ slope, but for the lowest values there is almost certainly

---

[2] Data collected by Saddoughi and Veeravalli [11]; reproduced with permission, copyright Cambridge University Press.

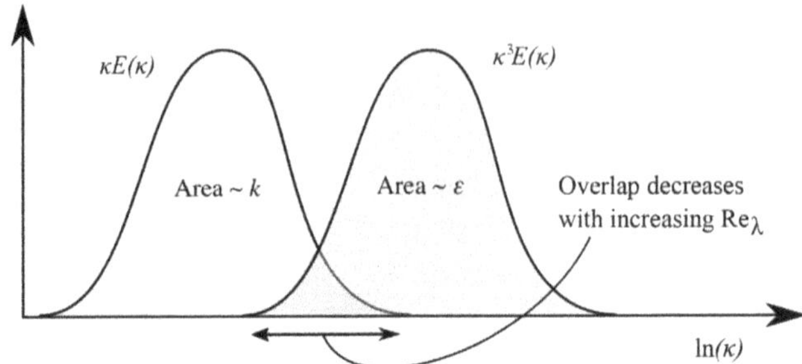

Figure 3.6. $\kappa E(\kappa)$ and $\kappa^3 E(\kappa)$ versus $\ln \kappa$ for a typical flow.

no inertial subrange at all. We return to this point below. Second, universality seems to hold beyond the inertial subrange right down into the dissipation range, also as expected, beginning at around $\kappa\eta = 10^{-1}$. Third, for each set of data, E_{11} 'peels over' towards a constant value at the low wavenumber end. This is inevitable because, as we will see later, $E(k)$ can be expressed as an integral of the velocity spatial correlation function (see chapter 4) which must tend to zero as the spatial separation becomes large, or κ in wavenumber space becomes very small. For the data sets with lower Reynolds numbers, there is simply not enough separation between the largest scales (with $\kappa \to 0$) and the smallest dissipative scales (η); i.e. $\mathrm{Re}_{L_x}^{3/4}$ (see equation (3.7)), is simply too small. It can be difficult to determine exactly how much of a subrange is present in any particular set of data, not least because the spectrum must inevitably have an asymptote of a $-5/3$ slope somewhere between the lowest and highest wavenumbers. Whether or not $E(\kappa) \sim \kappa^{-5/3}$ over any appreciable range of κ is perhaps best determined by plotting $\kappa^{5/3}E(\kappa)$ versus κ – often called a compensated spectrum – since that will have a constant plateau over some range of κ if an inertial subrange exists. A typical example is shown in section 4.5.

It is possible to show that the rate of energy dissipation, ε, is related to the energy spectrum, $E(\kappa)$ by

$$\varepsilon = \int_0^\infty \kappa^2 E(\kappa)d\kappa = \int_0^\infty \kappa^3 E(\kappa)d(\ln \kappa). \tag{3.15}$$

Since we can write k similarly, as $\int_0^\infty \kappa E(\kappa)d(\ln \kappa)$, this means that plots of $\kappa E(\kappa)$ and $\kappa^3 E(\kappa)$ versus $\ln \kappa$ will give curves whose integrated areas are k and ε, respectively. The sketch of figure 3.6 illustrates this and shows the small area of overlap between the curves. This overlap decreases in size with increasing Re_λ, emphasising that it is the latter which determines, in any turbulent flow, the extent of the inertial subrange. We discuss spectra more fully in the following chapter, as part of the exploration of the various fundamental functions which can be used to explore the nature of turbulent flows.

3.6 Subject giants: A Kolmogorov and G K Batchelor

Andrei Kolmogorov (1903–1987) was a Professor of Mathematics at the Moscow State University. Early in his career, he had the opportunity to travel to Germany

Figure 3.7. (Left) Photograph of Andrei N. Kolmogorov (1903–1987), courtesy of Konrad Jacobs (CC BY-SA 2.0). (Right) Portrait of George K. Batchelor (1920–2000) by Rupert Shephard from 1984; this portrait hangs in DAMTP, Cambridge, the Department founded under Batchelor's leadership in 1959[3].

and France in 1930–1931 and was influenced by other great mathematicians of the time, including Richard Courant, Hermann Weyl, Maurice René Fréchet, and Paul Levy. Although his travel was limited thereafter due to the politics of the time, it did not stop Kolmogorov from establishing a productive 'Russian school of turbulence' with notable peers including Alexander Obukhov, Andrei Monin, and Akiva Yaglom, who are also big names in turbulence.

Kolmogorov studied probability theory and stochastic processes before turning his attention to turbulence, where he applied his mathematical viewpoint to come up with his famous similarity hypotheses of turbulent flows. We summarised these in section 3.3 and ever since they were published in 1941 a significant research effort has attempted to prove or disprove the universality of the hypotheses.

One major criticism was made in 1944 by the Russian physicist Lev Davidovich Landau, who argued that because turbulence is often very intermittent, the dissipation rate of energy must vary both in space and time, even at the smallest scales of the flow. Therefore, a universal law cannot be based on the average dissipation rate of a flow because it is so inhomogeneous. Kolmogorov acknowledged this limitation of his original hypotheses and responded with his refined similarity hypotheses in 1962, replacing the globally averaged dissipation rate in his original statements with a locally averaged dissipation rate. This local dissipation

---

[3] Image reproduced with permission from Davidson, Kaneda, Moffatt, and Sreenivasan, (ed). (2011). *A Voyage Through Turbulence* (Cambridge: Cambridge University Press), copyright Cambridge University Press.

rate is averaged in space over the volume of a sphere of radius equal to the scales of eddies of interest and in this way it becomes a function of space and time, dependent on the flow. The task of proving (or disproving) this refined hypothesis is particularly challenging, as it requires fully time-resolved, three-dimensional knowledge of the flow, and is still an active topic of research. Frisch [3], for example, includes a discussion of this topic.

George Keith Batchelor (1920–2000) was best known as the founder Editor of the Journal of Fluid Mechanics, the first Head of the Department of Applied Mathematics and Theoretical Physics (DAMTP) at the University of Cambridge, and the co-Founder and first Chairman of EUROMECH. Originally from Melbourne, Australia, Batchelor moved to Cambridge after the Second World War to work with the famous Professor G I Taylor, and continued his research in turbulence [9].

One of Batchelor's most significant contributions to the study of turbulence is his 1953 book on *The Theory of Homogeneous Turbulence* [2], which summarised the research on this topic by himself and his colleague Alan Townsend at Cambridge. This was published after his paper that did so much to introduce Kolmogorov's turbulence work to the Western world [1]. Although homogeneous turbulence is an idealised situation, in which one assumes that turbulence fluctuations are independent of direction and boundary conditions, it is a powerful theoretical concept that allows significant simplification of the transport equations, which then generates many insights into the behaviour of turbulence. Sections 5.1 and 5.2 in chapter 5 build on these ideas and the lessons learned by considering such canonical flows.

Sample exercises

3.1. Estimate the range of scales involved in the flow in the boundary layer over an airplane wing, assuming an appropriate Reynolds number and that the largest scales are of the order of the boundary layer thickness.

3.2. Identify the units of the variables in equations (3.12) and (3.14), to prove that $\Psi(\kappa\eta)$ and $\psi(\kappa\eta)$ are non-dimensional.

References

[1] Batchelor G K 1947 Kolmogoroff's theory of locally isotropic turbulence *Proc. Camb. Phil. Soc.* **43** 533–59

[2] Batchelor G K 1953 *The Theory of Homogeneous Turbulence* (Cambridge: Cambridge University Press)

[3] Frisch U 2002 *Turbulence* (Cambridge: Cambridge University Press)

[4] Grant H L, Stewart R W and Moilliet A 1962 Turbulence spectra from a tidal channel *J Fluid Mech.* **12** 241–68

[5] Jiménez J 2004 The contributions of AN Kolmogorov to the theory of turbulence *Arbor* **178** 589–606

[6] Johnson P L 2020 Energy transfer from large to small scales in turbulence by multiscale nonlinear strain and vorticity interactions *Phys. Rev. Lett.* **124** 104501

[7] Kolmogorov A N 1941 The local structure of turbulence in an incompressible viscous fluid for very large Reynolds numbers *Proc. USSR Acad. Sci.* **30** 301–5 Doklady Akademiia Nauk SSSR

[8] Kuo A Y-S and Corrsin S 1972 Experiment on the geometry of the fine-structure regions in fully turbulent fluid *J. Fluid Mech.* **56** 447–79

[9] Moffatt H 2011 George Batchelor: the post-war renaissance of research in turbulence *A Voyage Through Turbulence* ed P A Davidson, Y Kaneda, K Moffatt and K R Sreenivasan (Cambridge: Cambridge University Press) pp 276–304

[10] Richardson L F 1922 *Weather Prediction by Numerical Processes* (Cambridge: Cambridge University Press)

[11] Saddoughi S G and Veeravalli S V 1994 Local isotropy in turbulent boundary layers at high Reynolds number *J. Fluid Mech.* **268** 333–72

Turbulent Flows: an Introduction

Ian P Castro and Christina Vanderwel

Chapter 4

Statistical functions and tools

This chapter first discusses, in sections 4.2 and 4.3, some of the most basic amplitude-domain and time-domain functions which can be used to describe features of random (or deterministic) signals of any kind. To illustrate them, use is made of some simulated signals with specific properties, as well as a typical turbulence signal (in fact, the one shown in figure 1.2 of chapter 1). The general approach is intentionally made as straightforward as possible. The functions are then generalised into the approach that is most appropriate for exploring any kind of turbulent flow. Section 4.4 introduces the spectrum functions, and this is followed by a section discussing Taylor's hypothesis – often crucial for estimating spatial information from time records. Section 4.6 explores how best to obtain these various functions from digital records of the velocity field. Codes used to perform the analyses, and the signals themselves, are available to readers so that they can explore how factors such as the sampling rate and sampling time, for example, can affect the accuracy of measurements. Some of the sample exercises in the final section are designed to allow exploration of these issues, in order to generate confidence when analysing whatever turbulent flow signals are of interest.

4.1 Initial remarks

We have already discussed some statistical features of turbulence – e.g. the time-mean quantities such as the mean velocity U_i and the mean shear stress $\overline{u_i'u_j'}$ – and emphasised, back in section 1.2, that significant progress in understanding turbulent flows can be made by considering their statistical properties. But what exactly *are* the various statistical quantities that might be of interest in a particular turbulent flow and *how* does one obtain them? Although, in this book, we do not discuss specific computational or laboratory techniques, which are developing all the time, it is important to consider the statistical tools available for dealing with the random signals which are typically produced by those techniques. These tools can be considered to comprise those that seek to determine either the amplitude statistics,

doi:10.1088/978-0-7503-3619-2ch4   4-1

such as the mean, root mean square and, more fundamentally, the probability density function, or those that lead to the estimation of time- (or frequency-) domain quantities, i.e. correlations and the energy spectra. We will discuss them separately, whilst recognising that there are inevitable links between the two sets of statistics, which we will also explore. These topics all fall in the general area of signal processing, which, these days, is almost exclusively *digital* signal processing, since the sequences of data produced by laboratory experiments are usually digital – inevitably so in the case of numerical computation of the full Navier–Stokes equations. There are numerous books which discuss probability theory and its ramifications for measuring amplitude statistics, e.g. [3, 4]. Likewise, there are many books which discuss time-domain issues, including specific texts on Fourier transform theory and its application, e.g. [7, 10]. Nearly all of these books contain significantly more information than is necessary to understand the fundamental matters needed for analysing turbulent flow processes. It is these matters, presented relatively simply, which we discuss in the following sections. We also explore some of the fundamental implications about turbulence that arise from a basic consideration of the various statistics.

4.2 Amplitude-domain statistics

4.2.1 Stationarity

We must first distinguish between ensemble averaging and time averaging, introducing the concept of stationarity. Consider two realisations of all possible time histories of some property of a random (or turbulent) process. As examples, we could use the two realisations of mean velocity shown in figure 1.2, which we repeat here as figure 4.1 with a different annotation. The two realisations could have been obtained by repeating the experiment and selecting equal-length records from each. The collection of all possible realisations is the 'random process'. Each realisation, $x_n(t)$, yields a specific value of x_n at time t_1 from the start, $x_n(t_1)$. The mean value of these samples from all the possible realisations is just

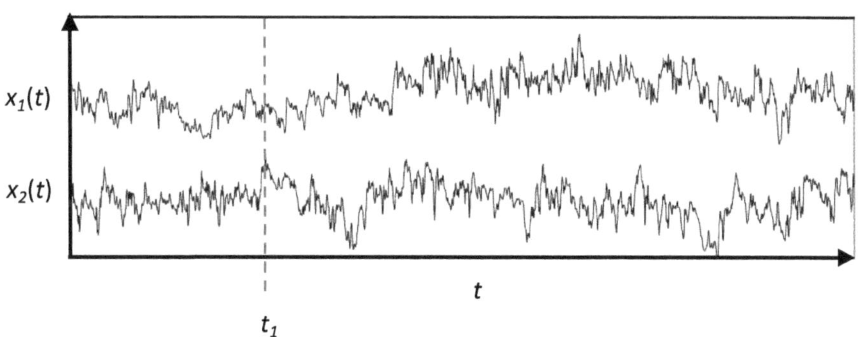

Figure 4.1. Two equal-period records of a random signal, obtained at widely separated times.

$$\overline{x(t_1)} = \lim_{N \to \infty} \frac{1}{N} \sum_{n=1}^{\infty} x_n(t_1) \qquad (4.1)$$

and similarly for all powers of $x(t)$ (such as the mean square). This is called an ensemble average. If these ensemble averages do not vary with t_1 the process is *stationary* (invariant to time translations). It is also possible to describe the properties of $x(t)$ by computing time averages over specific realisations. Consider the nth realisation, $x_n(t)$. It has a mean value given by

$$\overline{x_n} = \lim_{T \to \infty} \frac{1}{T} \int_0^T x_n(t)dt. \qquad (4.2)$$

If $x(t)$ is stationary and $\overline{x_n}$ does not depend on n (the particular realisation chosen), then the whole process is said to be *ergodic*. Thus, for ergodic processes, the time-averaged mean values are equal to the corresponding ensemble means – i.e. $\overline{x(t)} = \overline{x_n}$. Note that only stationary processes can be ergodic. It is a fortunate fact that, in practice, random data representing stationary physical phenomena are generally ergodic, as this means that we can analyse them properly on the basis of a single observed time history, provided it is long enough. For non-stationary processes, typified by those that contain a regular periodic component superposed on the turbulent components of the flow such as, for example, an in-cylinder engine flow, then the only way of obtaining the turbulence properties separately from the periodic component is via ensemble averaging.

Note that the mean values defined by equations (4.1) and (4.2) are only true means if N and T are infinite. In practice, of course, this is never the case, and the values are just averages over finite N and T. Furthermore, for nearly all purposes in modern turbulence research, one will be dealing with digital signals, i.e. a series of N consecutive numbers (samples), obtained at a specific rate (e.g. Δt apart) for a total time $T = (N - 1)\Delta t$, for which a digitally summed equivalent of equation (4.2) for the finite time period, i.e.

$$\overline{x(t)} = \frac{1}{N} \sum_{n=0}^{N-1} x(n\Delta t),$$

is, in practice, the most appropriate expression to use in order to obtain the time average. (In earlier days, analogue devices were routinely used to obtain time-averaged quantities from the analogue signals obtained in the laboratory, but the digitisation of analogue signals is now ubiquitous.) It can be shown that the average value converges rather slowly towards the true mean – according to $1/\sqrt{N}$, in fact – so one normally requires N to be very large if accurate measurements of the mean value are required. It is also necessary for the process to be sampled for a sufficiently long time, T; one can imagine that if the sampling rate were extremely high, a very large number of samples would be obtained from a very short period of the signal, which might well be too short to capture all the longer-timescale variations. Just how long is necessary for a specified accuracy will be considered in due course.

4.2.2 The probability density function and its moments

The mean value is just one of the quantities defining the amplitude statistics of the signal. The most fundamental property, from which this and all others can be deduced, is the probability density function (PDF). Consider again our stationary random signal $x(t)$. At any particular instant, there exists a specific probability that the signal amplitude lies between x and $x + \Delta x$. If T_x is the amount of time, summed throughout the total signal time T, during which $x(t)$ lies within this range, then this specific probability is just T_x/T. As Δx tends to zero, so must T_x/T (there is an infinitesimally small likelihood that $x(t)$ has a specific value, except in the rather uninteresting case of a constant amplitude signal). But a probability *density* function $p(x)$ can be defined for small Δx by

$$p(x) = \lim_{\Delta x \to 0} \left(\frac{T_x}{\Delta x T} \right), \tag{4.3}$$

which, we emphasise, is not a probability. Unless the signal is band-limited (in amplitude), T_x/T must lie in the range $0 \leqslant T_x/T \leqslant 1$, whatever particular values of x and Δx are chosen. Real signals are almost always limited to a certain amplitude range, so that $|x(t)| \leqslant a$, say. The probability corresponding to $|x| > a$ must be zero and $p(x)$ is not strictly defined for $|x| > a$ – although, since T_x is always zero in that range, it is natural to set $p(x) = 0$ also. If $x(t)$ is a random signal, $p(x)$ will tend to some limiting value for each x as $T \to \infty$.

It is important to note that equation (4.3) contains no reference as to how the signal is arranged in time – i.e. its time-domain statistics. One could imagine snipping out some section of the signal and inserting it somewhere else in the record; this would have no influence at all on $p(x)$, although it is obvious that it would generally change the time structure of the signal significantly. Furthermore, given our observational period T, $p(x)$ exactly describes the amplitude-domain statistics within that period, but it may have little resemblance to the $p(x)$ that would have been obtained for a different observational period, whether of the same duration or not. This can be simply illustrated by considering a sine-wave signal. It is straightforward to show that for $x(t) = a \sin(2\pi f t)$, an observation period T equal to any integral multiple of the period $1/f$ of the signal has a probability density function given by

$$p(x) = \begin{cases} \dfrac{1}{\pi}(a^2 - x^2)^{1/2} & \text{if } |x| \leqslant a, \\ 0 & \text{if } |x| > a. \end{cases} \tag{4.4}$$

This is the usual result quoted for the PDF of a sine wave, but it is obvious that if T were a non-integral multiple of $1/f$, a very different result would be obtained unless $T \to \infty$. Recall too that changing the time structure of the sine wave by arbitrarily cutting part of it and reinserting that portion elsewhere would not change $p(x)$, but would change the signal's time structure. Note the importance of recognising that $p(x)$ is *not* a probability; it is clearly not correct to state that the probability that x

takes the value of, for example, $0.99a$ is given by $(1/\pi)(1 - 0.99^2)^{1/2} = 2.26$! The probability is actually given by

$$\lim_{\varepsilon \to 0} \int_{0.99a-\varepsilon}^{0.99a+\varepsilon} p(x)dx = 0.$$

This expression introduces the idea of the *cumulative* probability density function, $P(x)$, defined by

$$P(x) = \int_{-\infty}^{x} p(z)dz, \tag{4.5}$$

which must tend to unity as $x \to \infty$ (not surprisingly – all the signal must lie in the amplitude range $-\infty$ to $+\infty$). $P(x)$ is a genuine probability (a number between zero and unity), unlike the PDF, $p(x)$.

All the individual amplitude quantities of the signal can be written in terms of $p(x)$ which, from now on, we will assume to have been generated by a velocity signal, $u(t)$. The mean value must equal the sum of 'all the individual amplitude values u times their frequency (i.e. probability) of occurrence $p(u)du$', i.e.

$$U = \int_{-\infty}^{\infty} up(u)du. \tag{4.6}$$

More commonly, of course, the mean is found directly using a finite-period equivalent of equation (4.2), i.e.

$$U = \frac{1}{T} \int_{0}^{T} u(t)dt, \tag{4.7}$$

noting, by comparison with equation (4.2) as written, that T will, in practice, be finite. Equations (4.6) and (4.7) are only equivalent and thus yield identical values for U, if the $p(u)$ in the former is measured over the same time period T as that used in the latter.

Similarly, the mean square of the signal must be the sum of 'all the squared amplitudes times their frequency of occurrence', i.e.

$$\overline{u^2} = \int_{-\infty}^{\infty} u^2 p(u)du \tag{4.8}$$

with an expression such as equation (4.7) for the equivalent time integration of u^2. The Reynolds decomposition introduced in chapter 2 implies that $\overline{u^2} - U^2 = \overline{u'^2} = \sigma^2$, the signal's variance in the usual parlance – i.e. the energy in the fluctuating part of the signal. Its square root, σ, is the standard deviation, which, for turbulent flows, we usually denote by u'_{rms}.

Higher-order moments of the PDF can also be deduced either from the PDF or by the more common time integration. Thus, generally, for the nth moment, we may write

$$\overline{u^n} = \int_{-\infty}^{\infty} u^n p(u)du. \tag{4.9}$$

It is more common to define 'central moments' in terms of the fluctuating quantity, i.e. $(u(t) - U)$, and they can be normalised by the appropriate power of the standard deviation. So,

$$\gamma_n = \frac{\overline{u'^n}}{\sigma^n} = \frac{\int_{-\infty}^{\infty}[u(t) - U]^n p(u)\,du}{\left(\int_{-\infty}^{\infty}[u(t) - U]^2 p(u)\,du\right)^{n/2}}. \tag{4.10}$$

Note that when $n = 1$, the central moment (the numerator in equation (4.10)) is zero, as expected. The third and fourth normalised central moments ($n = 3$ and 4) provide the skewness and the flatness factor (or kurtosis), respectively.

Some examples of PDFs of typical signals, which differ from a genuine turbulent signal to a greater or lesser extent, are discussed in section 4.4.1 and illustrated in figure 4.6.

4.3 Time-domain characteristics

It was noted earlier that the PDF of a signal does not unambiguously provide all the information the signal contains and, in particular, it tells us nothing about the time structure. This can be seen by considering the two signals illustrated in figure 4.2. Neither of these are turbulence signals, but for convenience we retain the use of $u(t)$ to denote each signal. Both have been constructed to have exactly the same PDF – the well-known Gaussian (or normal) distribution specified by

$$p(u) = \frac{1}{\sigma\sqrt{2\pi}}e^{-u^2/2\sigma^2}, \tag{4.11}$$

where we have assumed a zero mean value. Each of the two signals shown has a variance of unity. It is obvious that they have very different time structures.

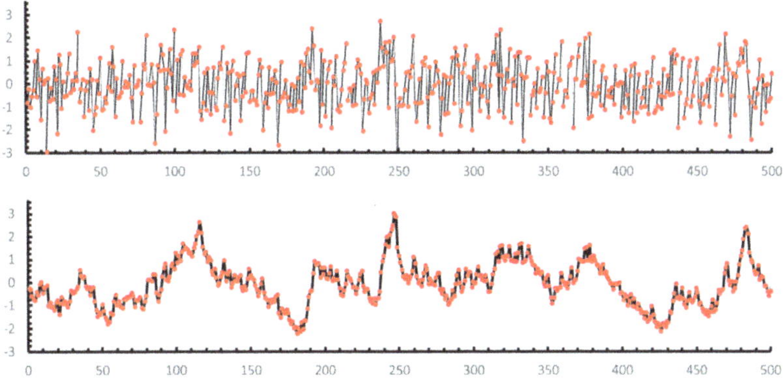

Figure 4.2. Equal-period records of different Gaussian signals; each record comprises the first 500 consecutive samples (in red) of the roughly 130000 available; they can be thought of as digitised samples from analogue signals, so the x-axis is essentially time. The samples are joined by smoothed lines. The upper signal is white noise and the lower one is (first-order) pink noise, defined later.

To explore the time structure, we need to consider other measures; these are discussed in the following sections.

4.3.1 Correlations of a time signal

First we discuss how to quantify the dependence of the signal value at one time, say, t with that at another time, separated from t by, say, τ. This dependence, or *autocorrelation*, is the simplest two-time statistic for a single signal. It is defined by the autocovariance

$$C(\tau) = \lim_{T\to\infty} \frac{1}{T} \int_0^T u(t)u(t+\tau)dt,$$

or, after subtracting the mean value, by

$$R(\tau) = \lim_{T\to\infty} \frac{1}{T} \int_0^T u'(t)u'(t+\tau)dt \qquad (4.12)$$

so that $R(\tau) = C(\tau) - U^2$. For turbulence signals, we expect no correlation between $u(t)$ and $u(t+\tau)$ for large enough time lags, τ, i.e. $R(\tau) \to 0$ as $\tau \to \infty$. In other words, for sufficiently large τ, the signal has no 'memory' of what it was like much earlier. In that case, it follows from the definitions above that $C(\infty) = U^2$ and so the mean value is given by $U = \sqrt{C(\infty)}$. It is important to note that this cannot be true for genuinely deterministic signals – like a sine wave, for example, which clearly has 'perfect memory', for it is always possible to predict future values exactly at *any* time on the basis of current and past values. This is never true for turbulence signals. In all the following, we will assume that the mean value is zero, so that the autocovariance and autocorrelation functions are identical. It is clear from equation (4.12) that for zero time lag, $R(0)$ is simply the variance of the signal, i.e.

$$R(0) = \overline{u'^2}. \qquad (4.13)$$

It is worth noting at this point that, as expected simply by looking at figure 4.2, $R(\tau)$ is very different for the two signals. In fact, noting that the signals have unit variance, for the white noise signal it is given by $R(\tau) = \delta(\tau)$, where $\delta(\tau)$ is the delta function defined as unity for $\tau = 0$ and zero otherwise. Therefore, there is no correlation between successive values of $u'(t)$, however short the time lag, except for the perfect correlation at $\tau = 0$. Along with a Gaussian PDF, this, in fact, defines white noise. On the other hand, pink noise is defined as a signal that has a decaying autocorrelation given simply by $R(\tau) = e^{-\lambda\tau}$, where λ is the decay rate, along with a Gaussian PDF. Many natural phenomena lead to a signal behaviour similar to this, which makes pink noise a particularly relevant analytic generalisation for studying, for example, probable measurement errors in real situations. The reader can explore the whole signal, and others, in exercise 4.4. Figure 4.3 sketches the normalised $R(\tau)$ for white noise and pink noise signals.

A measure of the typical time during which a signal value has significant memory of its past is given by the integral timescale defined by

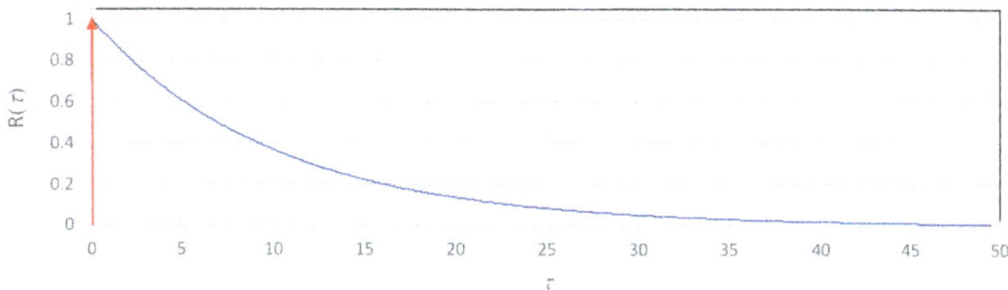

Figure 4.3. The autocorrelation $R(\tau)$ (normalised by the signal's mean square) for a white noise signal (red) and a pink noise signal (blue). The latter has $R(\tau) = e^{-\lambda\tau}$ with $\lambda = 0.1$.

$$T = \frac{1}{\overline{u'^2}} \int_0^\infty R(\tau)d\tau; \qquad (4.14)$$

note that this is not the only possible measure. (For example, one could use an integral whose upper limit coincides with the time when $R(\tau)$ first becomes negative, if it does.) For the white noise in figure 4.2, $T = 0$, whereas it is easily shown to be just $1/\lambda$ for the pink noise. We return to a fuller discussion of timescales later in the next section 4.3.2, linking them to the discussion in chapter 3.

In some standard texts, the autocorrelation function is defined in non-dimensional terms by dividing $R(\tau)$ by the signal variance $(\overline{u'^2} \equiv \sigma^2)$. This makes the autocorrelation unity for $\tau = 0$. We use $R(\tau)$ in both ways in what follows; it will always be evident from the context whether the normalised or unnormalised function is meant. In either case, it is straightforward to show that $R(\tau) = R(-\tau)$ – i.e. $R(\tau)$ is an even function. Note that the prefix 'auto' is used because these functions describe the time structure of a *single* signal, rather than the relationships between two or more signals, at the same or different times. The latter are normally termed *cross*-correlations.

4.3.2 Spatial correlation and the implications

We will now generalise the definitions of $p(u)$ and $R(\tau)$ to include recognition of the fact that turbulence has a three-dimensional spatial structure as well as a structure in time. To understand a particular flow, it is therefore commonly necessary to consider, say, a velocity signal captured at one point in the field $(\mathbf{x})$ along with that captured simultaneously (or, perhaps, after a specific delay) at some other point that has a spatial separation of $\mathbf{r}$. Such considerations require the exploration of both correlations and, equivalently, spectra. Often, in an experiment on a turbulent flow, only a single one-point signal of, say, u_1 is available as a function of time, so we also discuss (in section 4.5) how one might obtain spatial information from such a signal, and under what specific circumstances the necessary assumptions might be valid.

Writing, for convenience, $\mathbf{x}' = \mathbf{x} + \mathbf{r}$, we define a spatial (cross, or two-point) correlation by

$$R(\mathbf{r}, \mathbf{x}, t) = \overline{u_i'(\mathbf{x}, t)u_j'(\mathbf{x} + \mathbf{r}, t)}, \qquad (4.15)$$

which emphasises that (i) the velocity field is three-dimensional (the two positional references $\mathbf{x}$ and $\mathbf{r}$ are vectors) and (ii) the correlation is a second-order tensor and thus there are nine individual correlation functions, each dependent on which velocity components are chosen. This is the simplest statistic that contains information about the spatial structure of the velocity field. It is easiest to discuss it in the context of homogeneous turbulence, which implies that R_{ij} is not changed by an equal shift in the vectors $\mathbf{x}$ and $\mathbf{x}'$ so that the correlation is dependent only on the separation $\mathbf{r}$. On the understanding that u_i' and u_j' are obtained at the same time, t, we can drop this parameter and, recognising that for homogeneous turbulence there is no dependence on $\mathbf{x}$, we can write for conciseness

$$R_{ij}(\mathbf{r}) = \overline{u_i'(\mathbf{x})u_j'(\mathbf{x}')}. \tag{4.16}$$

Note that although, in homogeneous turbulence, R_{ij} is necessarily an even function – i.e. $R_{ij}(\mathbf{r}) = R_{ij}(-\mathbf{r})$ – this is *not* generally the case if the turbulence is inhomogeneous. It is always true, however, that for sufficiently large separations there is no correlation between the two velocity signals, so that $R_{ij}(\mathbf{r}) \to 0$ as $\mathbf{r} \to \infty$, mirroring the behaviour of the one-point time (auto) correlation, $R(\tau)$, discussed in section 4.3.1. For zero separation ($\mathbf{r} = 0$), the correlation clearly reduces to the local Reynolds stress at $\mathbf{x}$, simply written as

$$R_{ij}(0) = \overline{u_i'u_j'}. \tag{4.17}$$

As mentioned for the autocorrelation (in the previous section 4.3.1), these correlations are sometimes defined in normalised form, thereby becoming correlation *coefficients*, by dividing the right-hand side of equation (4.16) by $(\overline{u_i'(\mathbf{x})^2})^{1/2}(\overline{u_j'(\mathbf{x}')^2})^{1/2}$. A typical shape for one of the R_{ij} components, R_{11}, with separations in the x-direction, is shown in figure 4.4 for a particular turbulent flow. We discuss in section 4.5 exactly how this particular $R(r)$ was obtained.

In chapter 3, we introduced the various important scales of motion, but now, given these definitions of spatial correlations, some more precise scale relationships can be deduced and these are explored next.

Consequences for integral scales
The separation, $\mathbf{r}$, required for the correlation to fall significantly, can naturally be characterised by a spatial integral scale. A common definition is that for the integral scale based on velocities in the direction parallel to the main flow direction (i.e. $i = j = 1$) – the axial direction x, say – with separations r_1 in that direction. This can be written as:

$$L_x \equiv L_{11}^{[1]} = \frac{1}{\overline{u_1'^2}} \int_0^\infty R_{11} dr_1. \tag{4.18}$$

L_x is a scale characteristic of the larger (most energetic) eddies in the flow and was used in the discussions in chapter 3. However, it is clearly only one of the various possible integral scales that can be defined using an integral of equation (4.16), whose relative sizes will depend on the particular structure of the turbulent flow in

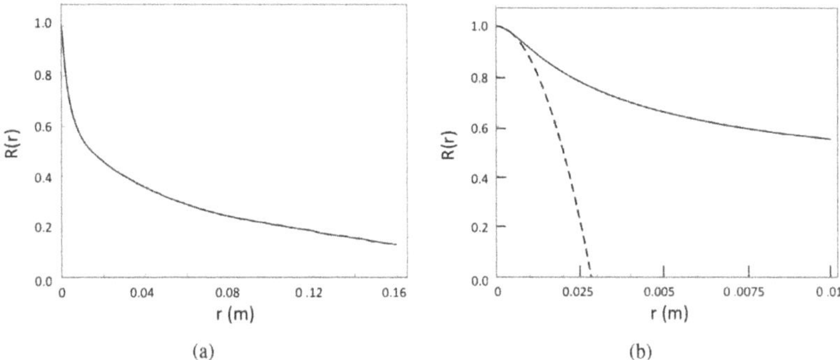

(a)   (b)

Figure 4.4. An example of a normalised spatial correlation curve; $R(r)$ has been normalised by the variance, $\overline{u'^2}$. (a) measured up to a separation of 160mm; (b) a magnified portion of the region near $r = 0$. The dashed line is a parabola fitted to the region very near $r = 0$, which has the same second derivative at $r = 0$ as the correlation curve.

question, in particular, the structure of the large eddy motions. The corresponding scale deduced from the velocity components in the normal direction can be defined by

$$L_y \equiv L_{22}^{[1]} = \frac{1}{\overline{u_2'^2}} \int_0^\infty R_{22} dr_1 \tag{4.19}$$

and these two scales – commonly termed the longitudinal and transverse integral scales – can, in homogeneous isotropic turbulence (for which $\overline{u_1'^2}$ and $\overline{u_2'^2}$ are equal), be shown to be simply related by $L_{22}^{[1]} = \frac{1}{2} L_{11}^{[1]}$.

Other commonly used scales are those formed from the axial or vertical velocity simultaneously measured at two points separated in one of the two directions orthogonal to x. So, for example, and noting that these L_y are not the same as those defined by equation (4.19),

$$L_y \equiv L_{11}^{[2]} = \frac{1}{\overline{u_1'^2}} \int_0^\infty R_{11} dr_2, \text{ or } L_y \equiv L_{22}^{[2]} = \frac{1}{\overline{v_1'^2}} \int_0^\infty R_{22} dr_2. \tag{4.20}$$

These two expressions yield different values for L_y, unless the turbulence is both homogeneous and isotropic (the subject of section 5.1 in chapter 5).

Consequences for Taylor microscales

In addition to the relations between R_{ij} and the integral scales, it is also possible to relate the cross-correlation R_{ij} to the Taylor microscale introduced in chapter 3 by considering the second derivatives of R_{ij} at $\mathbf{r} = 0$, which are directly related to mean-squared velocity gradients. For example, on differentiating equation (4.16) twice with respect to r_m and r_n, choosing the $i = j = 1$ diagonal component, and setting $\mathbf{x} = \mathbf{x}'$, we have

$$-\frac{\partial^2 R_{11}}{\partial r_m \partial r_n}\bigg|_{\mathbf{r}=0} = \overline{\left(\frac{\partial u_1'}{\partial x_1}\right)^2}. \tag{4.21}$$

For very small separations ($\mathbf{r}$), the diagonal components of R_{ij} can be expanded as a Taylor series in $\mathbf{r}$, so that for this $i = j = 1$ component and setting the normalising scale to λ,

$$\frac{R_{11}}{\overline{u_1'^2}} = 1 - \frac{r_1^2}{\lambda^2} + \cdots \tag{4.22}$$

Using equation (4.21), this leads directly to

$$\frac{2\overline{u_1'^2}}{\lambda^2} = \overline{\left(\frac{\partial u_1'}{\partial x_1}\right)^2}, \tag{4.23}$$

which relates the Taylor microscale, λ, to the square of an appropriate velocity derivative. The parabolic behaviour of R_{ij} at small separations is illustrated in figure 4.4(b) and the curve expressed by equation (4.22) has been fitted to the correlation, ensuring that the second derivative of the two curves is the same at $r = 0$. In this case, the resulting microscale is about 2.8 mm (where the curve crosses the r axis), which is a factor of about 17 smaller than the integral scale (47 mm). Note, however, that these results and figure 4.4 were actually derived from a one-point time record along with the use of Taylor's hypothesis, which is explored further in section 4.5.

Consequences for Kolmogorov scales
The Taylor microscale is always much smaller than the integral scale, L_x, as we will see. It is much larger, however, than the Kolmogorov scale, which is appropriate to the smallest eddies. Before considering the latter, note incidentally that there are nine different velocity derivatives and thus a corresponding number of different microscales. It is therefore best to think of λ, the Taylor microscale, in terms of an order of magnitude of the velocity derivatives (as indicated in figure 4.4), i.e.

$$\frac{\partial u_i'}{\partial x_j} = \mathcal{O}\left(\frac{u_{\mathrm{rms}}'}{\lambda}\right). \tag{4.24}$$

Recall now that the kinetic energy dissipation, ε, can be written as

$$\varepsilon = \nu \overline{\frac{\partial u_i'}{\partial x_j} \frac{\partial u_i'}{\partial x_j}} \tag{4.25}$$

(equation (2.29) with the second term zero for homogeneous turbulence); thus, using equation (4.24), we can express a similar order-of-magnitude expression for ε in terms of λ, i.e.

$$\varepsilon = \mathcal{O}\left(\frac{\nu u_{\mathrm{rms}}'^2}{\lambda^2}\right). \tag{4.26}$$

The Kolmogorov scale, η, can be estimated by considering velocity derivatives, because at these scales, the velocity field is smooth – i.e. the magnification of, say, a velocity–time record over those scales does not reveal further structure (the change in velocity with time is a straight line). Denoting the velocity difference between two points by Δu_η, the velocity derivative can be written as

$$\frac{\partial u_i'}{\partial x_j} = \mathcal{O}\left(\frac{\Delta u_\eta}{\eta}\right). \tag{4.27}$$

By comparison with equation (4.24), this implies that

$$\frac{\lambda}{\eta} = \mathcal{O}\left(\frac{u_{\mathrm{rms}}'}{\Delta u_\eta}\right). \tag{4.28}$$

Since the velocity differences for small separations are very small compared with those appropriate to the large-scale motions (i.e. $\Delta u_L = \mathcal{O}(u_{\mathrm{rms}}')$, say), this implies that $\lambda \gg \eta$. It also follows that $\lambda \ll L$ (an integral scale), so that

$$\eta \ll \lambda \ll L,$$

which is entirely consistent with the expressions given in chapter 3 – equations (3.3) and (3.7) – for the (order-of-magnitude) sizes of the microscale and the Kolmogorov scale, respectively.

Comments on R_{ij} in isotropic turbulence
There are some instructive consequences for R_{ij} in the special case of homogeneous isotropic turbulence (explored more fully in chapter 5), one of which is worth mentioning here. It follows from the fact that R_{ij} must then be an isotropic tensor. Since the only two second-order tensors that can be formed from the vector $\mathbf{r}$ are δ_{ij} and $r_i r_j$, it can be expressed in terms of two scalar functions $f(r)$ and $g(r)$:

$$R_{ij}(\mathbf{r}) = \overline{u'^2}\left(g(r)\delta_{ij} + \left[f(r) - g(r)\right]\frac{r_i r_j}{r^2}\right); \tag{4.29}$$

note that $\overline{u'^2}$ is the variance of all three velocity fluctuations. This allows one to deduce, for the various components, that

$$R_{11} = \overline{u'^2}f(r), \quad R_{22} = \overline{u'^2}g(r), \quad R_{33} = R_{22} \text{ and } R_{ij} = 0 \text{ for } i \neq j. \tag{4.30}$$

We see, therefore, that f and g are the longitudinal and transverse correlations and that the corresponding integral scales introduced above as L_x and L_y in equations (4.18) and (4.19) can be written in terms of f and g, respectively – i.e. $L_x = \int_0^\infty f(r)dr$ and $L_y = \int_0^\infty g(r)dr$. Also, since $\partial R_{ij}/\partial r_j = 0$ by continuity, equation (4.29) implies that

$$g(r) = f(r) + \frac{1}{2}r\frac{\partial}{\partial r}f(r). \tag{4.31}$$

In isotropic turbulence, it is thus evident that the two-point correlation $R_{ij}(\mathbf{r})$ is completely determined by the longitudinal correlation function $f(r)$. The relations above can be used to prove the relation $L_y = \frac{1}{2}L_x$, mentioned earlier (for isotropic turbulence).

4.3.3 The joint probability function and its implications

It was stated earlier that there are inevitable links between the amplitude- and time-domain statistics. This is most simply seen by considering the autocorrelation in terms of probability functions. The probability that $u(t)$ lies in the range $u_1 < u(t) < u_1 + \Delta u_1$ while, simultaneously, $u(t + \tau)$ lies in the range $u_2 < u(t + \tau) < u_2 + \Delta u_2$ is given by

$$\lim_{T \to \infty} \left[\frac{T_x(u_1, u_2)}{T} \right]$$

where $T_x(u_1, u_2)$ is the total amount of time for which $u(t)$ and $u(t + \tau)$ lie in these ranges simultaneously during the observation period T. The joint probability density function $p(u_1, u_2)$ is then defined by

$$p(u_1, u_2) = \lim_{\Delta u_1, \Delta u_2 \to 0} \lim_{T \to \infty} \frac{1}{T} \left[\frac{T_x(u_1, u_2)}{\Delta u_1 \Delta u_2} \right]. \tag{4.32}$$

Denoting $u(t)$ by u_1 and $u(t + \tau)$ by u_2 the autocovariance is given by

$$C(\tau) = \int_{-\infty}^{\infty} \int_{-\infty}^{\infty} u_1 u_2 p(u_1, u_2) du_1 du_2. \tag{4.33}$$

If the mean value of the signal is zero, then this is also $R(\tau)$. It must be emphasised that whilst, for a sufficiently long observation time, neither $p(u_1)$ nor $p(u_2)$ will depend in any way on the time structure of the signal, $p(u_1, u_2)$ certainly will. For stationary random signals, note that $p(u_1)$ and $p(u_2)$ will normally be identical.

The joint probability density function can also relate two *separate* signals measured either at the same point or at different points in space and at the same or different times. The joint probability density function $p(u, v)$, which relates two specified velocity records (in this case, arbitrarily named $u(t)$ and $v(t)$), is defined by

$$p(u, v) = \lim_{\Delta u, \Delta v \to 0} \lim_{T \to \infty} \frac{1}{T} \left[\frac{T_x(u, v)}{\Delta u \Delta v} \right]. \tag{4.34}$$

For example, consider an axial velocity $u_1(\mathbf{x}, t)$ and a transverse velocity $v_1(\mathbf{x} + \mathbf{r}, t + \tau)$, obtained at different points in space and at different times from the records $u(t)$ and $v(t)$. The joint probability that $u_1 \leqslant u(t) \leqslant u_1 + du_1$ and $v_1 \leqslant v(t) \leqslant v_1 + dv_1$ are satisfied simultaneously is $p(u_1, v_1) du_1 dv_1$.

Assuming that the flow is homogeneous, so that there is no dependence on $\mathbf{x}$, the time-averaged product of the two velocities is then simply (and by comparison with equation (4.33))

$$R_{12}(\mathbf{r}, \tau) = \overline{u_1 v_1} = \iint u_1 v_1 p(u_1, v_1) du_1 dv_1, \tag{4.35}$$

in which the integrals are taken over all possible amplitudes of u_1 and v_1, we assume (for convenience) that the mean values are zero, and that the two-time records are sufficiently long to yield an accurate measure of $p(u_1, v_1)$. As always for a correlation, it is a measure of the interdependence between the two random variables and it must normally decay to zero as $\tau \to \infty$ and $\mathbf{r} \to \infty$.

We can generalise the correlation yet further by obtaining the two velocity contributions at *different* times, t and $t + \tau$, hence writing

$$R_{ij}(\mathbf{r}, t, \tau) = \overline{u_i'(\mathbf{x}, t)u_j'(\mathbf{x} + \mathbf{r}, t + \tau)}, \qquad (4.36)$$

which is a two-point, two-time cross-correlation. Appropriate integrals can yield length and time scales that identify numerous additional correlation scales for $u(t, \mathbf{x})$, but these complications (although straightforward) are not fully pursued in this book. Nonetheless, we will need to bear this two-point, two-time correlation in mind in the following sections.

Clearly, depending on the two velocities measured and the choice of time lag and spatial separation, this expression can yield any of the usual Reynolds stresses (e.g. $\overline{u'^2}$, $\overline{u'v'}$, etc., for $\tau = \mathbf{r} = 0$), or spatial correlations with or without a specific time delay for a specific separation. Normally, of course, such quantities would be determined directly from the velocity signals (i.e. without bothering to generate $p(u_1, v_1)$ for each specific τ and $\mathbf{r}$), but there are some practical circumstances in which probability concepts can be used to great effect. For example, near the edges of many turbulent flows (e.g. jets, wakes, and boundary layers) the *intermittency* is often defined as the probability that the flow at a particular point is turbulent rather than laminar. Determining this probability makes it possible to obtain weighted averages (of, say, the mean axial velocity or Reynolds stresses) within the turbulent flow only, rather than measured over all time, as discussed a little more in section 6.6.2. The reader can find more rigorous discussion of the statistics of joint random variables in the turbulence context in books such as, for example, [12].

4.4 The spectral density function

4.4.1 One-dimensional spectra

The spectral density function describes the frequency composition of the signal – how much energy there is in different frequency ranges. Let $u'(t, f, \Delta f)$ be that part of the fluctuating signal $u'(t)$ which lies in the frequency band Δf centred on f. The mean square value within this band is

$$S^2(f) = \lim_{T \to \infty} \frac{1}{T} \int_0^T u'^2(t, f, \Delta f)\,dt \qquad (4.37)$$

and the (1D) power spectral density can then be defined by

$$E'(f) = \lim_{\Delta f \to 0}\left[\frac{S^2(f)}{\Delta f \overline{u'^2}}\right]. \qquad (4.38)$$

Note the dimension of $E'(f)$ (time) and that, since all the energy must be contained in the range $0 < f < \infty$, $E'(f)$ must satisfy

$$\int_0^\infty E'(f)df = 1.$$

The alternative, unnormalised definition for the spectral density, $E(f)$, would not include the variance in the denominator, so that

$$\int_0^\infty E(f)df = \overline{u'^2}. \tag{4.39}$$

Comparing equations (4.13) and (4.39), it is evident that there must be some relationship between $R(\tau)$ and $E(f)$, because these imply that $R(0) = \int_0^\infty E(f)df$. In fact, the two functions form an exact Fourier transform pair, i.e.:

$$E(f) = 2\int_{-\infty}^\infty R(\tau)e^{-2\pi f\tau}d\tau = 4\int_0^\infty R(\tau)\cos(2\pi f\tau)d\tau \tag{4.40}$$

and

$$R(\tau) = \frac{1}{2}\int_{-\infty}^\infty E(f)e^{-2\pi f\tau}df = \int_0^\infty E(f)\cos(2\pi f\tau)df; \tag{4.41}$$

the second of each of these relationships follows because, for real data, $R(\tau)$ and $E(f)$ are even functions, as noted earlier. Knowledge of either $R(\tau)$ or $E(f)$ therefore leads to the other via a straightforward manipulation. There is no essential information contained within $R(\tau)$ that is not available in $E(f)$ and vice versa; the former simply provides information in the time domain, whereas the latter provides it in the frequency domain.

There is an entirely equivalent but alternative way of defining the spectral density function, which is (these days) more frequently used in practice. We assume that the velocity signal $u'(t)$ has a convergent Fourier integral (requiring that $u'(t) = 0$ outside some finite time range); this can be written as:

$$C(f) = \int_0^\infty u'(t)e^{-2\pi i ft}dt. \tag{4.42}$$

It can then be shown that

$$E(f) = \lim_{T\to\infty}\frac{1}{T}\left[\left|C(f)\right|^2\right], \tag{4.43}$$

where $|C(f)|^2$ is the sum of the squares of the amplitudes of the real and imaginary parts of the Fourier integral (i.e. the modulus, in the usual sense). Strictly, this is only valid if the numerator is $\text{ave}_n[|C(f)|^2] - $ i.e. an average of $|C(f)|^2$ over n independent ensembles, with $n \to \infty$. This has significant implications for the determination of the energy spectrum, which we consider in section 4.6.

As examples, consider again the two signals illustrated in figure 4.2. For the white noise signal, $E(f)$ is simply a constant over all frequencies. In practice, of course, the data for $R(\tau)$ only extend, at most, to the time lag corresponding to the available signal time, say, T, so the upper limit of the integrals in equation (4.40) and the corresponding equation (4.43) is T. Correspondingly, the lower limit of the integrals in equations (4.41) is the minimum available frequency, $1/T$. (In section 4.6, we consider how the upper limit is set in practice.) In the case of white noise $E(f)$ is then simply $\overline{u'^2}/T$. For the pink noise signal, the spectrum can easily be shown to be given by

$$E(f) = 4T\overline{u'^2}/(1 + 4\pi^2 f^2/\lambda^2), \tag{4.44}$$

so that at a high enough frequency, $E(f) \sim f^{-2}$. This illustrates immediately that the signal in figure 4.2, despite looking rather like genuine turbulence (as in figure 1.2), cannot be turbulence, since we know that at high frequencies in the inertial subrange, $E(f) \sim f^{-5/3}$ in any turbulent flow (see section 3.5). We also mention here a typical example of a deterministic signal that contains a noticeable periodicity, that of second-order pink noise. This is defined as having a Gaussian PDF (like first-order pink noise), but an autocorrelation given by

$$R(\tau) = e^{-\lambda\tau} \cos(b\tau), \tag{4.45}$$

where b is the dominant frequency.

As a final example, consider the signal shown in figure 4.5. This is part of something known as a telegraph signal. The random times between the changes of state (i.e. between $+1$ and -1) are, by definition, specified according to a Poisson distribution. It can be shown that the autocorrelation and spectrum functions of this signal are exactly the same as for the pink noise shown in figure 4.2 (i.e. $R(\tau) = e^{-\lambda\tau}$). On that basis alone, the signal is indistinguishable from pink noise. It is starkly obvious, however, that the probability density function is, unlike that of pink noise, certainly not Gaussian. It consists simply of two delta functions at $u = -1$ and $u = +1$, as can be anticipated by simple inspection of the signal. This emphasises the earlier point that a signal's one-time PDF provides no information about its time structure and, correspondingly, signals may have the same spectrum but completely different PDFs. Other examples which make this point are given in Pope's book [12], in which random processes are discussed in more detail than is appropriate here.

To illustrate the various points made above, the PDFs and spectra for the three simulated signals are illustrated in figure 4.6, where they are compared with the corresponding functions obtained from a genuine turbulence signal. A number of points should be noted:

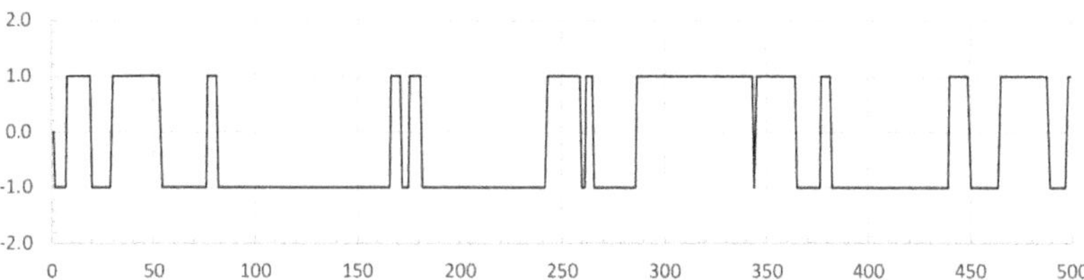

Figure 4.5. The first 500 samples of a 65000-sample long telegraph signal, which has a zero mean and a unit rms.

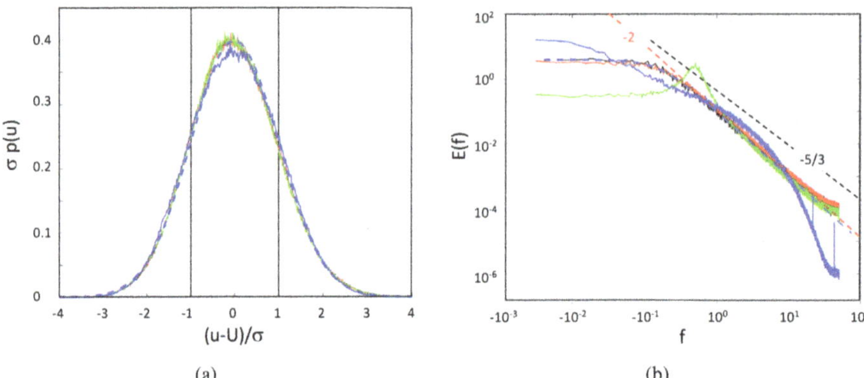

Figure 4.6. PDFs (a) and spectra (b) of various signals. The black and red, almost coincident, lines represent telegraph and pink noise signals, respectively; the green line denotes second-order pink noise and the heavier blue line depicts a real turbulence signal. The blue dashed lines correspond to the exact Gaussian PDF and the exact spectrum for pink noise and telegraph signals. The dashed black and red lines in (b) have slopes of $-\frac{5}{3}$ and -2, respectively. All signals are normalised to have a zero mean and a unit variance, and the axes in (b) have arbitrary units, but such that $\int_0^\infty E(f)df = 1$ (the variance).

1. The PDF for the telegraph signal consists of delta functions at $u = -1$ and $+1$, whereas the other simulated signals – first- and second-order pink noise – have Gaussian PDFs and are thus essentially coincident with the ideal Gaussian PDF given by (for a zero mean and a unit variance) $p(u) = (1/\sqrt{2\pi})e^{-u^2/2}$.

2. Although the PDF of the turbulence signal appears to be closely Gaussian, it is measurably not so, because the skewness and flatness factor for that particular signal are about 0.05 and 2.7, respectively, rather than the true Gaussian values of exactly zero and 3.0. (Incidentally, a nonzero skewness is, in fact, an essential part of the dynamics of turbulence – recall equation (2.33) in section 2.7.)

3. The spectra of all three simulated signals decay like f^{-2} at high frequencies, as is evident in figure 4.6(b), and the second-order pink noise contains the anticipated spectral peak.

4. The telegraph and pink noise spectra essentially collapse over all frequencies onto the anticipated theoretical spectrum (the blue dashed line), given by equation (4.44).

5. The turbulence signal has a spectrum which contains (at best) only a very short region with the expected $-5/3$ slope – short, because the Reynolds number is not very large (see section 3.5). At higher frequencies, $E(f)$ falls off more rapidly towards the dissipation range.

6. All the spectra shown in figure 4.6(b) except, naturally, the theoretical one for pink noise (the dashed blue line), appear to rise slightly above what might be expected as the lowest $E(f)$ values are approached at the highest available frequency. This is a result of a phenomenon known as aliasing, which will be discussed in section 4.6; it arises because the sampling rate (i.e. the time lag between the individual samples of data constituting each signal) is inevitably nonzero; digital signals are *not* continuous in time.

We emphasise, finally, that knowledge of the spectrum (or autocorrelation) and the one-time PDF discussed in section 4.2.2 is not sufficient to completely define a general random process, unless that process is a statistically stationary Gaussian one. In that case, it *is* fully defined by $p(u)$ and $E(f)$ (or $R(\tau)$). In contrast to the three simulated signals introduced above, the turbulence signal is *not* completely characterised by $E(f)$ and $p(u)$. We have already seen that turbulence, even when statistically stationary, is *not* Gaussian. Complete information therefore requires all the 'N-time' PDFs – i.e., for turbulence, the phase information between eddy structures at different locations within the flow.

4.4.2 Three-dimensional spectra and their implications

Autocorrelations and the spectra of *time* series are intimately related, so it is now natural to explore the spectral equivalent of the *spatial* cross-correlations ($R_{ij}(\mathbf{r}, t)$) discussed above. Such spectra are normally presented in *wavenumber* form, using the wavenumber vector $\boldsymbol{\kappa}$, which contains three components (κ_1, κ_2, κ_3), just as $\mathbf{r}$ has the three components (r_1, r_2, r_3). A characteristic turbulent eddy with a length of l has a characteristic wavenumber of $\kappa = 2\pi/l$. We start by writing the Fourier transform of the one-time, two-point spatial correlation defined by equation (4.16):

$$\boldsymbol{\Phi}_{ij}(\boldsymbol{\kappa}, t) = \frac{1}{(2\pi)^3} \int_{-\infty}^{+\infty} \int_{-\infty}^{+\infty} \int_{-\infty}^{+\infty} R_{ij}(\mathbf{r}, t)e^{-i\boldsymbol{\kappa}\mathbf{r}}d\mathbf{r}, \qquad (4.46)$$

which is known as the three-dimensional wavenumber spectrum (with the volume integral performed over all $\mathbf{r}$) and, like R_{ij}, is a tensor. Note that defining the spectrum in these terms requires the turbulence to be homogeneous, since, otherwise, the correlation R_{ij} would depend on $\mathbf{x}$ as well as the separation $\mathbf{r}$. The inverse transform is

$$R_{ij}(\mathbf{r}, t) = \int_{-\infty}^{+\infty} \int_{-\infty}^{+\infty} \int_{-\infty}^{+\infty} \boldsymbol{\Phi}_{ij}(\boldsymbol{\kappa}, t)e^{i\boldsymbol{\kappa}\mathbf{r}}d\boldsymbol{\kappa}, \qquad (4.47)$$

where the volume integral is here performed over all wavenumber space. In these terms, and as an example, we can write the one-dimensional scalar wavenumber spectrum for, say, the u-component correlation with a separation of r_1 in the x-direction, $R_{11}(r_1)$, as

$$E_{11}^{[1]}(\kappa_1) = \frac{1}{2\pi} \int_{-\infty}^{+\infty} R_{11}(r_1)e^{-i\kappa_1 r_1}dr_1. \qquad (4.48)$$

By comparison with equation (4.18), defining the corresponding integral scale, it is clear that

$$L_{11}^{[1]} = \frac{\pi}{u_1'^2}E_{11}^{[1]}(\kappa_1 = 0), \qquad (4.49)$$

where we have implicitly recognised that for homogeneous turbulence, the R_{ij}s (and the corresponding spectral functions) are even. Similar relations exist that link the other length scales to one-dimensional spectral functions. We mention, incidentally, that the three-dimensional *scalar* energy spectrum, $E(\kappa)$, introduced in section 3.5, can be obtained from $\Phi_{ij}(\kappa)$ by integrating the trace, $\Phi_{ii}(\kappa)$, over the surface (A) of a sphere of radius κ:

$$E(\kappa) = \frac{1}{2} \int \Phi_{ii}(\kappa)dA. \tag{4.50}$$

Recall from section 3.5 (equation (3.11)) that the integral of $E(\kappa)$ over all κ equals the total turbulence kinetic energy, k.

Just as for R_{ij}, there are specific consequences for $\Phi_{ij}(\kappa)$ when the turbulence is not only homogeneous but also isotropic. For example, Batchelor [2] showed that the transverse spectra $E_{22}(\kappa_1)$ and $E_{33}(\kappa_1)$ are uniquely determined by the longitudinal spectrum $E_{11}(\kappa_1)$, according to

$$E_{22}(\kappa_1) = E_{33}(\kappa_1) = \frac{1}{2}\left(1 - \kappa_1\frac{\partial}{\partial\kappa_1}\right)E_{11}(\kappa_1), \tag{4.51}$$

where κ_1 is the longitudinal wavenumber and the three components yield the three velocity variances, via

$$\int_0^\infty E_{11}(\kappa_1)d\kappa_1 = \overline{u_1'^2}, \ \int_0^\infty E_{22}(\kappa_1)d\kappa_1 = \overline{u_2'^2}, \text{ and } \int_0^\infty E_{33}(\kappa_1)d\kappa_1 = \overline{u_3'^2}. \tag{4.52}$$

These relationships mirror those for R_{ij} in homogeneous isotropic turbulence, given in section 4.3.2.

Since a good deal of the theoretical research into turbulence has concentrated on this special case, a number of texts are available which explore it all in considerable detail (of which Batchelor's [2] is the classic), but we do not pursue it further here. There is, however, a crucial question concerning the relation between the wavenumber spectra and the frequency spectra discussed in section 4.4.1 above, not least because laboratory measurement of the former, especially the full three-dimensional spectrum, is very difficult. Even the one-dimensional scalar spectrum – e.g. as expressed for the u–component velocity in equation (4.48) – requires the use of at least two probes in the flow. Classically, nearly all measurements of spectra have been obtained from time records obtained at one point, followed by the use of a concept called Taylor's hypothesis to deduce the spatial structure within the flow. Nowadays, the widespread use of numerical simulation of the full Navier–Stokes equations has meant that direct deductions of wavenumber spectra can be obtained rather more easily but, nonetheless, the use of Taylor's hypothesis remains ubiquitous. We point out, for example, that all of the wavenumber spectra shown in figure 3.5 were deduced from time records obtained at a single point in the various flows. It is therefore important to discuss this hypothesis; this is the topic of the next section.

4.5 Taylor's frozen turbulence hypothesis

Taylor [14] pointed out that if the turbulence level is sufficiently low, the time variations in the velocity u' observed at a fixed point are approximately the same as those due to the convection with a mean velocity U of an unchanging spatial pattern past the point. If this is true, we can write $u'(r, t) \approx u'(r - Ut, t)$, where r and t represent the distance measured downstream in the mean flow direction and time, respectively. Taylor showed that this abstraction of turbulent flows as fields of frozen eddies convected by the mean flow – which is, strictly speaking, merely an approximation rather than a hypothesis – is indeed approximately true for the case of turbulence behind a grid in a wind tunnel (a flow we consider in chapter 5).

The importance of the Taylor hypothesis is that because it can be written in terms of velocity gradients, it allows spatial information to be estimated from time records. Take, for example, the x-component of the fluctuating velocity in a mean flow field of velocity U, when the Taylor hypothesis assumes that:

$$\frac{\partial u'}{\partial t} = -U_j \frac{\partial u'}{\partial x_j} \approx -U_c \frac{\partial u'}{\partial x}, \tag{4.53}$$

where U_c is some convection velocity (often taken simply as U). Until the advent of full Navier–Stokes simulations, which yield fluctuating velocities at all points in the flow field and at all times, subject to the numerical limitations imposed by the spatial grid and the time step, obtaining spatial correlations and (in particular) spatial spectra was very difficult, as indicated earlier. A common way to obtain such information is to transform one-point time records. So, for example, to measure the Taylor microscale, λ, one can assume Taylor's hypothesis and write equation (4.23) as

$$\frac{2\overline{u_1'^2}}{\lambda^2} = \frac{1}{U_c^2} \overline{\left(\frac{\partial u_1'}{\partial t}\right)^2}. \tag{4.54}$$

In a similar manner, the spatial correlation $R(r)$, with a separation r in the x-direction, for example, can be approximated by $R(U_c\tau)$. In fact, this is exactly the way that figure 4.4 was constructed, using a time record of the axial velocity measured in a turbulent shear flow. The mean velocity of the time record used was about 10 m s^{-1}, so time lags τ used in the autocorrelation computations from that time record were simply multiplied by 10 to obtain the x–axis in meter units for the (approximate) spatial separations r. The microscale deduced from equation (4.54) turns out to be about 2.8 mm. Likewise, the integral length scale defined by equation (4.18) can be estimated, using the Taylor-approximated transformation of that equation:

$$L_x \approx U_c T_x = \frac{1}{\overline{u_1'^2}} \int_0^\infty R(\tau)d(U_c\tau), \tag{4.55}$$

which, in this particular case, yields an L_x of about 47 mm.

Taylor's hypothesis is also widely used to obtain spatial spectra, which is essentially performed by writing the wavenumber κ as $2\pi f / U_c$, where f is frequency. Wavenumber spectra can thus be obtained from frequency spectra. It is straightforward to transform equation (4.48) (for the one-dimensional scalar wavenumber spectrum of the u-component correlation with a separation of r_1 in the x-direction) into an equation for the $E(f)$ in equation (4.40), allowing measurement of the latter to be used to deduce the former. This is what all the authors of the data shown in figure 3.5 did to obtain $E_{11}(\kappa_1)$ as a function of κ_1 from their single-point time records of u_1', yielding $E(f)$ as a function of f. However, this process implicitly assumes not only that the turbulence intensities are small but also that the convection velocity is the same for all wavenumbers; even if the former is true, the latter is clearly unlikely. Although the large-scale structures (which have small wavenumbers) might well not change much as they are swept across a certain (not-too-large) separation, so that $U_c = U$ may be appropriate, the (small) eddies corresponding to large wavenumbers are much more likely to change over the same separations, making $U_c = U$ a very unlikely approximation.

Many authors have sought ways to estimate how the convection velocity varies with wavenumber, in order to make appropriate corrections to Taylor's hypothesis for particular flows; see, for example, the work of Del Alámo and Jiménez [1]. The problem is quite a difficult one and it remains common for the hypothesis to be used fairly freely (and often uncritically!) in situations in which the turbulence intensities are low (typically below, say, 10%), which is the lowest-level requirement for the approximation to be reasonable. It is now possible, given full simulations of the Navier–Stokes equations, to assess the adequacy of Taylor's hypothesis in much more detail than is possible using laboratory measurements, but such study is inevitably limited to relatively low Reynolds numbers.

As an example of the use of Taylor's hypothesis, consider the record of axial velocity $u(t)$, two separate portions of which were initially shown in figure 1.2. We have already shown the spatial correlation $R_{ij}(r)$ deduced from this record using Taylor's hypothesis, see figure 4.4. As mentioned above, the separation distances, r, were taken as $U\tau$, where U is the mean velocity of the entire record, which led to the values for L_x and λ given above. Writing $\kappa = 2\pi f / U$ and using equation (4.43) the resulting spectrum is shown in figure 4.7, plotted as $E(\kappa)$ versus κ. As noted in section 4.4.1, the extent of the expected inertial subrange region that has a slope of $-\frac{5}{3}$ is small. A better way to check that such a region exists is to plot the premultiplied spectrum, which is included in the figure, since any inertial region would appear with a constant value of $\kappa^{5/3}E(\kappa)$ which, according to the Kolmogorov result (equation (3.13) in section 3.4), would be equal to $C\varepsilon^{2/3}$. For this case, such a region is clearly very limited. The Taylor microscale Reynolds number, $Re_\lambda = \lambda u_{rms}'/\nu$, is, for this case, only about 310. Larger values would lead to a greater separation of the integral and Kolmogorov scales and a wider inertial subrange (see figure 3.5). Incidentally, given an accepted value for C (see chapter 5), using the value of $C\varepsilon^{2/3}$ in the inertial subrange of the compensated spectrum provides a common and convenient way to estimate the dissipation rate without having to differentiate the velocity–time record

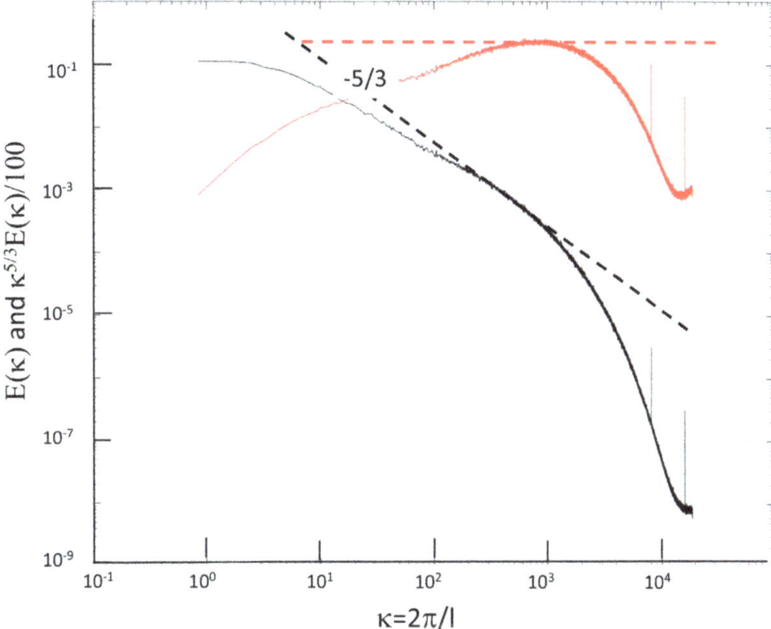

Figure 4.7. The wavenumber spectrum (black line), obtained using Taylor's hypothesis and the Fourier transform of a turbulence velocity signal $u(t)$, with $\kappa = 2\pi f / U$. The red line shows the spectrum premultiplied by $\kappa^{5/3}$.

in order to use equation (2.29) (after applying Taylor's hypothesis and assuming homogeneity). This latter route for obtaining ε requires a high time-resolution velocity record because the derivatives are dominated by small-scale motions, as mentioned in section 2.6.

We can immediately deduce some of the limitations of Taylor's approximation by considering the rate of change equation for any general quantity, e.g. the turbulence kinetic energy k, which can be written as

$$\frac{Dk}{Dt} \equiv \frac{\partial k}{\partial t} + U_j \frac{\partial k}{\partial x_j} = \text{Transport} + \text{Source} + \text{Sink}, \qquad (4.56)$$

which is just another way of writing equation (2.28), examined in section 2.6. It is obvious that the time and spatial derivatives are well correlated only if the terms on the right-hand side are all zero (i.e. Equation (4.53) holds). Using the full Navier–Stokes equations, Lin [11] showed long ago that the approximation is valid only if the turbulence levels are low, the viscous forces are negligible, and the mean shear is small. It would therefore be unreasonable to expect the hypothesis to apply throughout the whole of, say, a wall boundary layer or a jet. Nonetheless, it remains useful to pursue the idea that slowly distorting eddies are convected downstream by the mean flow at some steady velocity, even though the latter may not be equal to the mean velocity and even though, in fact, differently sized eddies may be convected at different velocities.

4.6 Issues in the digital processing of turbulence signals

In days gone by, many turbulence signals obtained in the laboratory were analysed using analogue electronics only. A major question that then faced the experimentalist – in addition to those concerning instrument amplitude and frequency bandwidth – was that of the required integration time. This clearly had to be sufficiently long to average the lowest-frequency content of the signal. Nowadays, such signals are almost always digitised, usually 'online' – i.e. as they are being collected by the instrument and before any analysis is carried out, although the latter often also happens 'online'. One then has a sequence of samples and the questions become:

1. What digitiser (amplitude) resolution is required?
2. How rapidly must the signal be digitised (i.e. what sampling rate is necessary)?
3. How many samples are required?

The first of these is a direct result of replacing an analogue signal, which has an infinite number of amplitude values, by a digital signal which can only take a finite number of values (the amplitude discretisation). This leads to inevitable 'quantisation errors', which can usually be made insignificant. Sixteen-bit analogue-to-digital converters (ADCs) are normally the minimum used, but higher resolutions (e.g. 24-bit) are common, and the only concern the experimentalist may have is to ensure that the analogue signal spans a large enough proportion of the ADC's amplitude bandwidth. Similar considerations are necessary for optical methods of data collection, e.g. laser Doppler anemometry or particle image velocimetry (PIV), but further considerations are needed in those cases, and there are numerous texts and papers covering the various issues involved. It should be recognised, however, that the effective amplitude resolution of such methods is often considerably less than is usual with the use of ADCs for single or multiple analogue signals generated by, for example, hot-wire anemometry. Quantisation issues therefore often require careful consideration.

The second and third questions are linked. Obtaining a sufficiently long time period can obviously be done in very many ways, since the product of the sampling rate and the number of samples collected gives the sampling period. It should be obvious that the answers to these two questions will depend on the particular measurement envisaged. If one merely requires quantities such as the mean or mean square, or indeed any characteristic of the amplitude statistics, one can make the sampling rate sufficiently low to ensure that consecutive samples are essentially uncorrelated and then obtain a sufficient quantity of them to (i) capture a long enough period to ensure adequate averaging of the lowest-frequency content of the signal and (ii) minimise the errors which arise from having only a finite number of samples. On the other hand, obtaining the autocorrelation or spectrum obviously requires much more rapid sampling but, again, for a long enough period to ensure adequate accuracy for the lowest-frequency components of the signal. Consider figure 4.3, for example. Using a sampling rate of 1/20 (i.e. choosing the time lag τ

between consecutive samples to be 20) when digitising the pink noise signal would yield almost uncorrelated samples and, in that sense, would be relatively efficient for generating amplitude statistics. However, the resulting sequence of samples could not be used to obtain $R(\tau)$ for $\tau \leqslant 20$ which, in fact, is the most interesting part of the signal as far as time- or frequency-domain information is required.

The question of how many independent (i.e. uncorrelated) samples are needed for adequate measurement of amplitude statistics is a straightforward one and only a brief discussion is necessary. Fuller details can be found in standard statistical texts; a classical one is that of Kendall and Stuart [8]. We start by reiterating equation (4.1) and its equivalent for the variance, in terms of a set of N samples of a single signal, $u(t)$:

$$\hat{U} = \frac{1}{N} \sum_{i=1}^{N} u_i, \quad \hat{\sigma}^2 = \frac{1}{N} \sum_{i=1}^{N} (u_i - \hat{U}_i)^2, \tag{4.57}$$

where $\hat{U}$ and $\hat{\sigma}^2$ are just 'estimators' (in common probability theory parlance) of the true values (U and σ^2) that would be obtained if $N = \infty$. $\hat{U}$ and $\hat{\sigma}^2$ are thus, themselves, random variables. One of the basic sampling theorems states that a random sample of size N taken from a (large) population of mean U and variance σ^2 yields a random variable $\hat{U}$ whose mean value is U and whose variance is σ^2/N. Likewise, the variance of $\hat{\sigma}^2$ can be shown to be $\sigma^4/(N/2)$. The *standard error*, ε_U say, in the estimate of U is thus $\sigma/\sqrt{N}$ and the *standard error* in the estimate of the variance, ε_{σ^2} say, is $\sigma^2\sqrt{(2/N)}$. These errors will clearly be small if N is large. Sometimes, one wants to be able to make probabilistic confidence statements about the likely differences between U and $\hat{U}$ and σ^2 and $\hat{\sigma}^2$; standard texts should be consulted for a discussion of this.

However, simple statements about the standard errors are often all that are needed. As an example, consider the boundary layer turbulence signal already used in earlier sections of this chapter, for which the mean is about 10 m s^{-1} and $\sigma^2 = \overline{u'^2}$ is about $2.76 \text{ m}^2 \text{ s}^{-2}$. The turbulence intensity defined by u'_{rms}/U is thus about 16%, which suggests that the signal was obtained deep in the layer – fairly close to the wall (see chapter 8). According to equation (4.57), 1000 uncorrelated samples of this signal would give standard errors of about 0.05 and 0.123 in the mean and variance, respectively; when normalised by the mean value and the variance, respectively, these are relative standard errors of 0.5% and 4.5%, respectively. For some measurement systems, $N = 1000$ could normally be very significantly exceeded, even by orders of magnitude, but for other techniques (e.g. PIV), suitably large sample numbers are more demanding. We emphasise that these expressions for standard errors rely on the samples being statistically independent (uncorrelated). Recalling the earlier discussion, they would not be valid if the sampling rate gave time lags of less than, say, 20 units (see figure 4.3)

The estimation of the full probability density function requires the use of the digital equivalent of equation (4.3),

$$\hat{p}(u) = \frac{N_x}{N \Delta u}, \tag{4.58}$$

where N_x is the number of samples out of the total number N which lie in the amplitude range $u - \Delta u/2$ to $u + \Delta u/2$. Thus, decisions need to be made not only about the number of samples to collect but also about the amplitude range to be covered and the number of $p(u)$ estimates, M say, required in that range. (If M estimates are required in a u-range of $\pm R$, say, then $\Delta u = 2R/(M - 1)$ for a constant 'slot' width.) Theoretical deductions about the sampling distribution of $\hat{p}(u)$ are very difficult to obtain, but it is obvious that, as usual, the standard error decreases as N increases. One can also anticipate that any set of sampling parameters that leads to a higher N_x/N in a given slot will lead to smaller variability in that particular $\hat{p}(u)$ estimate. Choosing a smaller Δu gives higher resolution across the amplitude range, but without concomitant increases in N, this would increase the standard error of $\hat{p}(u)$. Reasonable error estimates can be obtained using the approximate result for $\hat{p}(u)$ variability given by

$$\text{var}[\hat{p}(u)] \approx \frac{p^2(u)}{N \Delta u \hat{p}(u)}, \tag{4.59}$$

so that a standard error normalised by the peak $p(u)$ (which is often near, if not at, $p(0)$) can be approximately expressed as

$$\varepsilon' = \varepsilon_{p(u)} \Big/ p(0) = \frac{1}{p(0)} \sqrt{\frac{p(u)}{N \Delta u}}. \tag{4.60}$$

It is straightforward to show that for a Gaussian PDF, if one were satisfied with a 5% standard error, $M = 32$ would require over 5000 samples. Reducing ε' to 2% or increasing M to 200 would increase the required N to about 33000, illustrating the point that large sample sizes are required to obtain accurate PDF estimates, particularly if fine amplitude resolution is also required.

All this discussion has taken place on the basis of uncorrelated samples. However, as discussed earlier, the estimation of time-domain quantities requires much more rapid sampling, so that consecutive samples are highly correlated. In these circumstances, the variability of amplitude-domain quantities is essentially determined by the total sampling time and the spectral nature of the signal. Standard texts derive expressions for the variability in these circumstances – see, for example, Bendat and Piersol [3]. However, theoretical results that can be obtained for simple random signals (such as white noise, for example) are not very useful for signals with more complex spectral content, as in turbulence, for which no such results can be obtained. The sensible approach in such cases is simply to ensure that the effective averaging time is significantly longer than the period of the lowest-frequency fluctuations in the signal. This inevitably requires some element of trial and error in the experimental (or numerical) procedures.

For the measurement of time- or frequency-domain statistics, it should be clear from the above discussion that, assuming amplitude quantisation is adequate, errors will arise largely because of the signal's quantisation in time or an inadequate sampling period. If the sampled values are separated too far apart in time, they could represent either low or high frequencies in the original signal. This becomes obvious if one considers the digitisation of a simple sinusoidal signal. If sampling were to occur just once in every cycle, a set of samples of identical amplitude would be obtained, representing energy at a zero frequency rather then the frequency of the original signal. This phenomenon is known as *aliasing*, which we now explore briefly.

Consider the sampling of a continuous signal at a rate of $1/\Delta t$, so that individual samples are Δt apart in time. The maximum frequency which can be unambiguously recovered from the sample values can be shown to be $1/(2\Delta t)$. Any energy in the original signal at frequencies higher than this is 'folded back' and appears as lower-frequency components in the energy spectrum. This arises essentially because of the circular nature of the Fourier transform process – equation (4.42). Recall that $\cos(2\pi f \Delta t) = \cos[2\pi(f \pm n/\Delta t)\Delta t]$, where n is an integer. If we define $f_a = 1/(2\Delta t)$, all data at frequencies of $f \pm 2nf_a$ have the same cosine function as the data at frequency f and are therefore indistinguishable from the latter. Consequently, measurements of the signal's energy content at frequencies less than f_a are contaminated by all energies at frequencies $2nf_a \pm f$. Here, f_a is termed the *aliasing frequency*.

We illustrate aliasing by considering the simulated pink noise signal used earlier. This was created with an effective sampling rate of 100 Hz. Calculating the spectrum using that rate leads to a spectrum with an aliasing frequency of 50 Hz, as shown in figure 4.8. Using only every 8th or every 64th available sample reduces the aliasing frequency by factors of 8 and 64, respectively; the resulting spectra are included in the figure. It is straightforward to show that provided the number of samples is large (which it is for these cases), the estimated spectral density at the aliasing frequency f_a is too high by a factor of $\pi^2/4$, independent of f_a. In every case, the additional total energy above that given by the ideal f^{-2} line is equal to all the energy in the complete signal at all frequencies above f_a, and the noticeable distortion in the spectrum begins at about $f_a/3$. The amplitude and extent of the distortion depend on the particular spectrum shape; for the turbulence signal shown in figure 4.7, it is significantly smaller than that seen in figure 4.8, because the rate of fall of $E(\kappa)$ in the dissipation range is faster and there is very little energy in the analogue signal above the aliasing frequency, so the slow down in the fall of $E(\kappa)$ towards the aliasing frequency is only slight. For laboratory measurements, the usual ways of surmounting the aliasing effect are either to filter the signal before digitising, removing all signal components at frequencies above f_a, or to make the sampling rate sufficiently high that the energy levels for frequencies above f_a are negligible. This latter is practically the case for the turbulence signal, since it was sampled at 60 kHz.

To obtain spectra, it is common (as was done here) to use the digital equivalent of equation (4.43); this can be expressed as

$$\hat{E}(k\Delta f) = \frac{2}{N_b T}\left(\sum_{j=1}^{N_b} |C_j(k\Delta f)|^2\right)(\Delta t)^2, \quad k = 1, 2, 3, \ldots, N \tag{4.61}$$

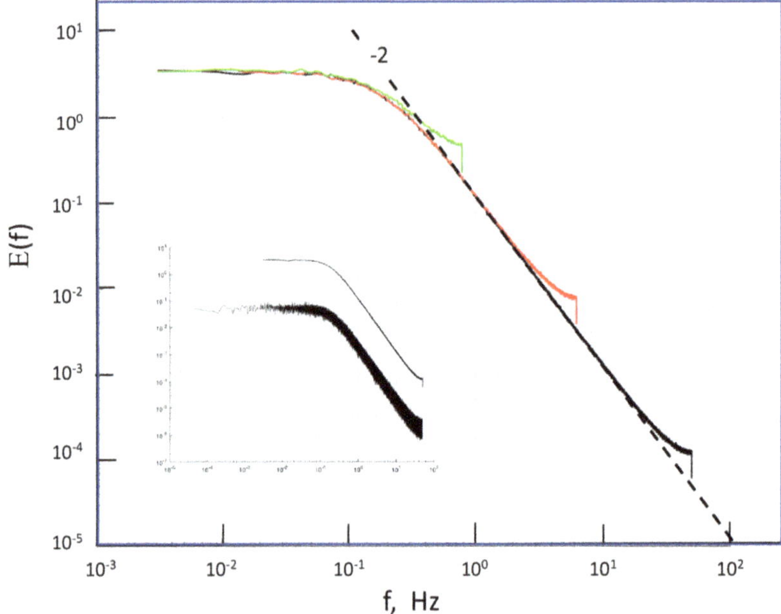

Figure 4.8. Frequency spectra for a simulated pink noise signal. Sampling rates: 100 Hz, black; 12.5 Hz, red; 1.56 Hz, green. The dashed black line shows the theoretical f^{-2} behaviour for high frequencies. The insert shows a repeat of the black line result along with a (arbitrarily shifted) spectrum obtained using far fewer (and thus shorter) segments of the original signal.

with

$$C(k\Delta f) = \sum_{n=1}^{N} u'(n)e^{-2\pi ikn/N}, \quad k = 1, 2, 3, \ldots, N \tag{4.62}$$

which is the discrete Fourier transform (DFT) with $\Delta f = 1/(N\Delta t)$. In equation (4.61), N_b is the number of blocks (or segments) of the complete digitised signal $u'(n)$ chosen for each transform, with N consecutive samples in each block ($N_{\text{tot}} = NN_b$ is thus the total number of samples available – i.e. the maximum value of n – assuming that the N_b segments are contiguous). The spectral estimates at each frequency are averaged over the N_b blocks, and T is the total time for each segment – i.e. $T = N\Delta t$, where Δt is the time between consecutive samples (the inverse of the sampling frequency). Note that the mean value of each block of data must be subtracted before performing the DFT.

Two other points about obtaining spectra are pertinent. First, the lowest frequency for which $E(f)$ can be obtained, $f_{\min}$ say, is naturally $1/T = 1/(N\Delta t)$. For the simulated pink noise, T was about 330 s for the spectra in figure 4.8, so $f_{\min} \approx 0.003$ Hz. Second, there is an important point to make about the accuracy of the spectral estimates. It is common to obtain the spectrum by obtaining a number of independent spectra, N_b using the above notation, by partitioning the complete data set into (typically) N_b equal-length segments, calculating the spectrum for each segment and then averaging all the spectral estimates, as shown by equation (4.61).

The accuracy of the resulting estimates is dependent solely on N_b and naturally reduces as N_b increases. This is illustrated by the insert in figure 4.8, which shows the original spectrum (at the maximum sampling frequency), which was obtained using 1024 segments (of length 330 s), along with one obtained using only 16 segments, which were consequently each 16 times longer – so f_{min} was correspondingly smaller. In some circumstances, it would be useful to obtain spectral estimates down to much lower frequencies, but the disadvantage is clearly that there is much more scatter (i.e. uncertainty) in the individual spectral estimates. It can be shown that the variance of the spectral estimates obtained using equation (4.61) is given by

$$\mathrm{var}[\hat{E}(f)] = \varepsilon^2 = \sigma^2/N_b, \qquad (4.63)$$

so halving the standard error in the estimates requires quadrupling the number of segments used. We emphasise that the accuracy of the spectral estimates in each block is independent of the number of samples in the block. Provided only that the basic N spectral estimates are obtained (as in equation (4.62)), these are essentially independent of one another (exactly so for Gaussian signals), but note that if, as always in turbulence, $u'(n)$ is real, the second $N/2$ estimates are a repeat of the first $N/2$ and are reflected symmetrically about the aliasing frequency. They are therefore not shown in any of the spectra illustrated above.

It is, of course, equally possible to obtain the spectrum from the autocorrelation (i.e. by using the digital equivalent of equation (4.40)). One would thus first obtain the latter, using

$$\hat{R}(m\Delta\tau) = \frac{1}{N_{\mathrm{tot}} - m} \sum_{i=0}^{M_{\mathrm{tot}}-m} u'(i\Delta\tau)u'(i\Delta\tau + m\Delta\tau), \ m = 0, 1, 2, \ldots, M - 1, \quad (4.64)$$

where N_{tot} is the total number of samples available in the digitised record. Note that although for the mth lag there are $N_{\mathrm{tot}} - m + 1$ cross products, using a divisor of $N_{\mathrm{tot}} - m$ ensures that the consecutive estimates are unbiased (in the statistical sense). Of course, the number of cross products that contributes to each estimate falls as m increases, so that unless M were chosen to be much smaller than the number of samples, N_{tot}, the variability of the estimates for the longer time lags would be significantly higher than those for short lags. It is therefore usual to ensure that $M \ll N_{\mathrm{tot}}$. This is equivalent to requiring that the total signal time is long enough to ensure adequate averaging of the longest lag components of $R(\tau)$.

After obtaining the autocorrelation via equation (4.64), the digital equivalent of equation (4.40) can be used to obtain the spectrum – i.e. by obtaining the real part of the Fourier transform of $R(\tau)$. Given that $M \ll N_{\mathrm{tot}}$, the resulting variance in the spectral estimates obtained at the same frequency intervals as those in equation (4.61) can be shown (for a Gaussian signal) to be proportional to $1/(NN_b)$, which is obviously considerably lower than the $1/N_b$ of equation (4.61). Naturally, if *all* the possible $R(\tau)$ estimates were obtained (NN_b of them) then the spectral estimates' variabilities would be identical. Given the ease, these days, of collecting and storing large quantities of rapidly sampled digitised data, and the relative simplicity of performing fast digital Fourier transforms (FFTs), it is perhaps more common to obtain spectra directly using equations (4.61) and (4.62), as explained above. We therefore do not explore obtaining them via the autocorrelations any further. There are, however, some practical subtleties in using the above expressions for spectra, which are worth mentioning briefly.

In the codes used for the spectral calculations discussed above (available to readers for exploration in the solution of exercise 4.7 posted online) a standard FFT algorithm was used. As normal, this FFT requires that N (the number of samples to be transformed) is an integral power of two; often, the chosen record lengths have to be padded with zeros to satisfy that condition. To minimise the accuracy issues this causes, the N_b segments are frequently overlapped (and are thus not contiguous) and a Hamming window function is also used on each segment. These devices are designed to reduce the variability in the spectral estimates – they are equivalent in some ways to digital smoothing. There is a significant literature on the design of FFTs (e.g. [5, 13]). The latter reference specifically explores the way MATLAB constructs FFTs.

4.7 Subject giant: J L Lumley

John L Lumley (1930–2015) was an American Professor of Aeronautics who pioneered several statistical tools during his research into turbulence modelling, geophysical fluid mechanics, and laboratory experiments of turbulent flows [9]. He began his career as a Harvard graduate and then went on to pursue his graduate studies at Johns Hopkins University under the guidance of Stanley Corrsin. In 1959, he joined the Pennsylvania State University and in 1977, he moved to Cornell University, where he built the Cornell turbulence group. He authored several books, including 'A First Course in

Figure 4.9. John L Lumley (1930–2015) from the film Deformation of continuous media[1], copyright 1963 by the Education Development Center, Inc. (EDC). Used with permission and with all other rights reserved.

---

[1] http://web.mit.edu/hml/ncfmf.html

Turbulence', co-authored with Henk Tennekes in 1973 [15], which was aimed at final year undergraduates or first-year graduate students interested in the topic of turbulence, much like this book. He is also famous for the introduction of the method of proper orthogonal decomposition (POD) as a tool with which to identify coherent structures, which is detailed in his book *Turbulence, Coherent Structures, Dynamical Systems and Symmetry* [6]. Although beyond the scope of this introductory text, POD analysis has become commonplace as a tool for isolating the most energetic repeating structures in bounded turbulent flows as a basis for model reduction.

Sample exercises

4.1. Show that for $x(t) = a\sin(2\pi ft)$, an observation period T equal to any integral multiple of the period $1/f$ yields a probability density function of $x(t)$ given by equation (4.4).

4.2. Show that $\overline{u'^2} = \int_{-\infty}^{\infty}(u - U)^2 p(u)du$.

4.3. The file 'TurbulenceSample.txt'[2] contains data obtained by hot-wire anemometry in a wind tunnel boundary layer. The streamwise velocity signal was sampled at 60 kHz for a total time of 30 s. Estimate the probability density function of the data, as defined by equation (4.3), by plotting the histogram of the data. Choose appropriate bin sizes Δx.

4.4. Obtain the autocorrelation $R(\tau)$ for the turbulence signal provided by 'TurbulenceSample.txt'[3] and compare it with those for the white and pink noise signals in 'NoiseSample.txt'[4]. Assuming Taylor's frozen flow hypothesis, apply equations (4.54) and (4.55) to calculate the integral length scale L_x and the Taylor microscale λ of the turbulence.

4.5. The file 'TurbulenceSample2.txt'[5] contains the streamwise u, vertical v, and spanwise w components of velocity near the bottom wall of a channel flow from DNS.

    (a) Plot the joint probability density function of u/u_{rms} and v/v_{rms}, choosing appropriate bin sizes. A nice way to visualise this is to superimpose contours of the joint pdf on top of a scatter plot of u/u_{rms} and v/v_{rms}. Does such a plot imply a positive or negative correlation between the two components? The slope of the line of best fit should equal the correlation coefficient.

    (b) Determine the Reynolds shear stress (i) from the joint pdf as in equation (4.35) and (ii) as the time average of $u'v'$ from the data records.

4.6. Show that $E(f)$ for the first-order pink noise signal (i.e. the signal for which $R(\tau) = e^{-\lambda\tau}$) is given by equation (4.44). In addition, find the

---

[2] https://github.com/cvanderwel/TurbulentFlows/blob/main/data/TurbulenceSample.txt

[3] https://github.com/cvanderwel/TurbulentFlows/blob/main/data/TurbulenceSample.txt

[4] https://github.com/cvanderwel/TurbulentFlows/blob/main/data/NoiseSample.txt

[5] https://github.com/cvanderwel/TurbulentFlows/blob/main/data/TurbulenceSample2.txt

spectrum function for the second-order pink noise signal whose autocorrelation is given by equation (4.45).

4.7. Plot the one-dimensional energy spectrum of the velocity signal from the file 'TurbulenceSample.txt'[6]. Explore how (i) the period and (ii) the sampling rate of the given turbulence signal affect the location of the first spectral estimate (i.e. $E(1)$) and the aliasing frequency. In addition, explore how splitting the total available signal into shorter batches can be used to improve the accuracy of spectral estimates. Compare this spectrum of a real turbulence signal with those computed for the white noise and pink noise signals provided in 'NoiseSample.txt'[7]

References

[1] Álamo J C D and Jiménez J 2009 Estimation of turbulent convection velocities and corrections to Taylor's approximation *J. Fluid Mech.* **640** 5–26

[2] Batchelor G K 1953 *The Theory of Homogeneous Turbulence* (Cambridge: Cambridge University Press)

[3] Bendat J S and Piersol A G 1966 *Measurement and Analysis of Random Data* (New York: Wiley)

[4] Bendat J S and Piersol A G 1986 *Random Data: Analysis and Measurement Procedures – Revised and Expanded* (New York: Wiley-Interscience)

[5] Harris F J 1978 On the use of windows for harmonic analysis with the discrete Fourier transform *Proc. IEEE* **66** 51–83

[6] Holmes P, Lumley J L and Berkooz G 1996 *Turbulence, Coherent Structures, Dynamical Systems and Symmetry* 1st edn (Cambridge: Cambridge University Press)

[7] James J F 2011 *A Student's Guide to Fourier Transforms: With Applications in Physics and Engineering* 3rd edn (Cambridge: Cambridge University Press)

[8] Kendall M G and Stuart A 1977 *The Advanced Theory of Statistics* vols 1-3 4th edn (London: Charles Griffen)

[9] Leibovich S and Warhaft Z 2018 John Leask Lumley: Whither Turbulence? *Annu. Rev. Fluid Mech.* **50** 1–23

[10] Lighthill M J 1958 An Introduction to Fourier Analysis and Generalised Functions *Cambridge Monographs on Mechanics and Applied Mathematics* (Cambridge: Cambridge University Press)

[11] Lin C C 1953 On Taylor's hypothesis and the acceleration terms in the Navier-Stokes equations *Quart. Appl. Math.* **10** 295–306

[12] Pope S B 2000 *Turbulent Flows* (Cambridge: Cambridge University Press)

[13] Schmid H 2012 *How to use the FFT and Matlab's pwelch function for signal and noise simulations and measurements* Institute of Microelectronics, University of Applied Sciences, Northwestern Switzerland

[14] Taylor G I 1938 The spectrum of turbulence *Proc. R. Soc. Lond* A **164** 476–90

[15] Tennekes H and Lumley J L 1972 *A First Course in Turbulence* (Cambridge, MA: MIT Press)

---

[6] https://github.com/cvanderwel/TurbulentFlows/blob/main/data/TurbulenceSample.txt
[7] https://github.com/cvanderwel/TurbulentFlows/blob/main/data/NoiseSample.txt

IOP Publishing

Turbulent Flows: an Introduction

Ian P Castro and Christina Vanderwel

Chapter 5

Canonical turbulent flows

There are a few relatively simple turbulent flows which can be viewed as canonical, in the sense that they have, for more than three-quarters of a century, provided the basis for fundamental explorations of turbulence which are not possible in the context of the more general (non-idealised) turbulent flows usually encountered in practice. The primary case is that of homogeneous isotropic turbulence (HIT). Indeed, as noted in section 3.6, the classical monograph by Batchelor [4] is devoted to this ideal case. The addition of a mean flow shear adds a degree of complication, but the study of homogeneous turbulent shear flow also reveals important and fundamental aspects of more general turbulent flows. These two flows are the subject of this chapter. Another relatively simple case is that of two-dimensional turbulent flow in a channel. This case provides the start of the discussion in chapter 7, since it is the simplest turbulent flow which involves the influence of a wall.

5.1 Homogeneous isotropic turbulence

5.1.1 Basics

It was noted in section 2.6 that, in the absence of mean flow shear, there must be a balance between the production and dissipation of turbulence energy if the flow is to be steady and, further, that if the turbulence is isotropic, thus ensuring zero energy production, then the energy *must* decay (i.e. the flow must be unsteady) because only the dissipation term remains nonzero on the right-hand side of the turbulent kinetic energy equation (2.28). This equation then becomes simply:

$$\frac{dk}{dt} = -\varepsilon. \tag{5.1}$$

The classical context in which this flow was first studied is its approximate experimental realisation provided by the flow behind a planar grid in a wind tunnel. Figure 5.1(a) is a sketch of such a grid, comprising a set of horizontal and vertical bars forming a square

doi:10.1088/978-0-7503-3619-2ch5

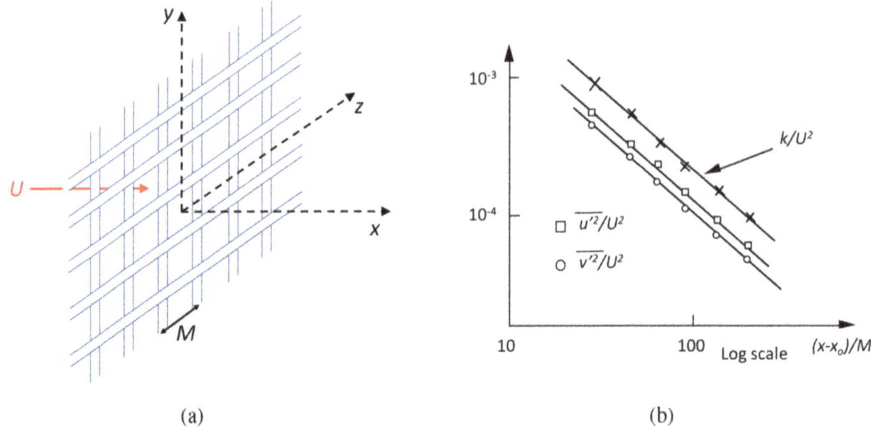

Figure 5.1. (a) Sketch of a biplanar grid, mounted normal to a uniform axial flow, U. (b) Measurements of the normalised variance of the axial and transverse velocity components. Turbulence energy k estimated as $\frac{1}{2}(\overline{u'^2} + 2\overline{v'^2})$.

grid of mesh size M. The grid spans the wind tunnel's working section (in the y and z directions), and ideally M is chosen so that there are an exactly integral number of meshes in each direction, with those near the walls spaced a distance $M/2$ from them. This ensures minimal wall effects further downstream and helps to establish a fairly early development of a flow which is closely homogeneous in the two cross-stream directions. In the laboratory frame the flow is statistically stationary, and away from significant wall influences the statistics only vary in the downstream (x) direction once x/M is sufficiently large – typically $x \geqslant 30M$. In a frame moving with the mean velocity the turbulence is approximately homogeneous, evolving with time ($t = x/U$).

Just behind the grid, however, the flow consists of a discrete set of vortices generated by the bars, and the flow obviously cannot be spatially homogeneous in the y and z directions. Nonetheless, at sufficiently high Reynolds numbers, the vortices rapidly develop turbulent cores and then mingle and interact so that a state of fully developed turbulence is reached and variations in the statistics across the y and z directions become minimal – i.e. the flow becomes homogeneous in those directions. The full range of scales is present, from the integral scale (essentially set by the parameters of the mesh – M and the bar width) down to the Kolmogorov microscale. Turbulence energy is concentrated in the former, while enstrophy, and hence dissipation, is largely confined to the small eddies with sizes of around η. This is a state of freely decaying turbulence, in which there is significant energy dissipation. As the turbulence decays (i.e. as k decreases with x) the appropriate Reynolds number, $\mathrm{Re} = \hat{u}L_x/\nu$, for example, where we again use $\hat{u}$ as the rms of the velocity fluctuations, decreases (see below), so that in due course viscosity effects dominate. The small-scale motions decay more rapidly than the large-scale ones, so eventually, only the larger scales remain and the flow slowly reverts, once Re is around unity, to become a collection of essentially laminar eddies, although this state is seldom reached in a wind tunnel. This is generally known as the final period of decay, in which inertia is relatively unimportant. Very early measurements behind such a grid were described by Batchelor [2] in a paper which provided perhaps the first introduction to Western scientists of Kolmogorov's (and other Russian authors') theories. A more complete comparison with experiments followed [5].

Measurements of the axial and transverse energies as functions of x within the freely decaying range are shown in figure 5.1(b), obtained in a typical laboratory wind tunnel at a Reynolds number given by UM/ν of around 100000. (For convenience, the three velocity components, u_i', $i = 1 - 3$, are here denoted as u', the axial velocity, and v', w', the two transverse components in the wind tunnel axes x, y, z, respectively.) The figure also includes the total turbulence energy, $k = \frac{1}{2}(\overline{u'^2} + \overline{v'^2} + \overline{w'^2})$, deduced by taking $\overline{w'^2} = \overline{v'^2}$. Notice first that the isotropy is not quite exact – the transverse variance (shown here as $\overline{v'^2}$ but it could equally be $\overline{w'^2}$ in a well-designed experiment) is rather lower than the axial variance. In one of the early experiments on this flow, it was shown that placing the grid at a suitable location within an upstream contraction of the working section led to much closer isotropy [7]. Second, note that the x axis necessarily includes a 'virtual origin' x_0, which is a result of the inhomogeneous behaviour closer to the grid. Obtaining the value of x_0 (typically a few mesh lengths) is, in some ways, not straightforward, but it is not appropriate for our purposes to discuss the issue here. (For interested readers, the discussion in section 4.2 of [12] provides a suitable introduction.) It is evident that the variances and thus, in particular, k, decay according to a power law, i.e.:

$$\frac{k}{U^2} = A\left(\frac{x - x_0}{M}\right)^{-n}. \tag{5.2}$$

In the moving frame, we can thus write k and ε as functions of t, using equation (5.1):

$$k = k_o\left(\frac{t}{t_o}\right)^{-n} \quad \text{and} \quad \varepsilon = \varepsilon_o\left(\frac{t}{t_o}\right)^{-(n+1)}. \tag{5.3}$$

Here, t_o is an arbitrary reference time, k_o is the value of k at that time, and $\varepsilon_o = nk_o/t_o$. In wind-tunnel experiments (where $t \equiv x/U$), n is typically found to lie around 1.3, although there is considerable variability in the literature. It is thus evident that the integral scale defined by $L_x = \hat{u}^3/\varepsilon$ (see section 3.2) varies according to $t^{1-n/2}$. With $n = 1.3$ or thereabouts, this means that although the turbulence energy decays with time (for example, according to t^{-n}), the increase in the integral scale with time is insufficient to compensate, so that the associated Reynolds number, $\text{Re}_x = \hat{u}L_x/\nu$, decreases with time – like t^{1-n} (typically, $t^{-0.3}$). As noted above, this eventually leads to a final stage of decay.

Since $d\hat{u}^2/dt$ is clearly a measure of the rate of destruction of turbulence energy, a characteristic timescale for the turbulence decay is $\hat{u}^2/(d\hat{u}'^2/dt)$, whereas a characteristic time for the energy containing eddies is $L_x/\hat{u}$. However, the overall decay can only occur as these large eddies break up into smaller ones via the cascade process, so these two timescales must be comparable. We can thus write

$$-A\frac{\hat{u}^2}{\frac{d}{dt}(\hat{u}^2)} = \frac{L_x}{\hat{u}}, \tag{5.4}$$

where, according to experiment, the value of A is around unity, so that

$$\frac{d\hat{u}^2}{dt} = -A\frac{\hat{u}^3}{L_x}. \tag{5.5}$$

In HIT, equation (5.1) with (5.5) lead to

$$\varepsilon = -\frac{dk}{dt} \equiv -\frac{3}{2}\frac{d\hat{u}^2}{dt} = \frac{3}{2}A\frac{\hat{u}^3}{L_x} = \sqrt{\frac{2}{3}}A\frac{k^{3/2}}{L_x} \tag{5.6}$$

(since in HIT, $k = \frac{3}{2}\hat{u}'^2$). Now recall equation (4.25), which expresses energy dissipation in terms of velocity gradients:

$$\varepsilon = \nu\overline{\frac{\partial u_i'}{\partial x_j}\frac{\partial u_i'}{\partial x_j}}.$$

It can be shown (for example, by following the route suggested in example (5.28) of [16]) that in HIT, this relation can be expressed more simply by

$$\varepsilon = 15\nu\overline{\left(\frac{\partial u_1'}{\partial x_1}\right)^2}. \tag{5.7}$$

We could, of course, equally use the derivative of either of the normal velocity fluctuations (u_2' or u_3') in the corresponding normal directions (x_2 or x_3, respectively), since the turbulence is isotropic. But note that these derivatives are *not* equal to the derivates of a velocity component in a direction normal to it. In fact, in HIT, it can be shown that, for example,

$$\overline{\left(\frac{\partial u_1'}{\partial x_1}\right)^2} = \frac{1}{2}\overline{\left(\frac{\partial u_2'}{\partial x_1}\right)^2} = \frac{1}{2}\overline{\left(\frac{\partial u_3'}{\partial x_1}\right)^2}. \tag{5.8}$$

We can also express the dissipation for HIT in general terms as

$$\varepsilon\delta_{ij} = 3\nu\overline{\frac{\partial u_i'}{\partial x_k}\frac{\partial u_j'}{\partial x_k}}; \tag{5.9}$$

it can clearly be partitioned (for $i = j$) into three equal components,

$$\varepsilon_{11} = \varepsilon_{22} = \varepsilon_{33} = \frac{1}{3}\varepsilon \tag{5.10}$$

(and is zero for $i \neq j$). The gradient term on the right-hand side of equation (5.7) can be replaced by $2\hat{u}^2/\lambda^2$ (equation (4.23)), from which it follows that

$$\varepsilon = \frac{30\nu\hat{u}^2}{\lambda_f^2}.$$

(5.11)

(The reason for now using the subscript f will become apparent below.)

5.1.2 R_{ij} in homogeneous isotropic turbulence

It is now appropriate to point out the consequences of HIT on the spatial correlation, R_{ij}, defined for homogeneous turbulence in equation (4.29). When the turbulence is also isotropic, R_{ij} must naturally be an isotropic tensor. Since the only two second-order tensors that can be formed from the vector $\mathbf{r}$ are δ_{ij} and $r_i r_j$, it can be expressed in terms of two scalar functions $f(r)$ and $g(r)$:

$$R_{ij}(\mathbf{r}) = \hat{u}^2\left(g(r)\delta_{ij} + [f(r) - g(r)]\frac{r_i r_j}{r^2}\right),$$

(5.12)

in which we note that $\hat{u}^2$ is the variance of all three velocity fluctuations. This allows one to deduce, for the various components, that

$$R_{11} = \hat{u}^2 f(r), \quad R_{22} = \hat{u}^2 g(r), \quad R_{33} = R_{22}, \quad \text{and} \quad R_{ij} = 0 \quad \text{for} \quad i \neq j. \quad (5.13)$$

We see, therefore, that f and g are the longitudinal and transverse correlations and that the corresponding integral scales introduced as L_x and L_y in equations (4.18) and (4.19) can be written in terms of f and g, respectively – i.e.

$$L_x = \int_0^\infty f(r)dr \quad \text{and} \quad L_y = \int_0^\infty g(r)dr.$$

(5.14)

Also, since $\partial R_{ij}/\partial r_j = 0$ by continuity, equation (5.9) implies that

$$g(r) = f(r) + \frac{1}{2}r\frac{\partial}{\partial r}f(r).$$

(5.15)

In HIT, it is thus evident that the two-point correlation $R_{ij}(\mathbf{r})$ is completely determined by the longitudinal correlation function $f(r)$. This latter equation can clearly be written as

$$g(r) = \frac{1}{2}\left(f(r) + \frac{\partial}{\partial r}[rf(r)]\right),$$

(5.16)

from which it is straightforward to show that $L_y \equiv L_{22}^{[1]} = \frac{1}{2}L_{11}^{[1]} \equiv L_x$ (as mentioned in section 4.3.2). Now recall that the Taylor microscale introduced conceptually by equation (3.3) was later defined in the context of R_{11} in section 4.3.2. For convenience, we call this microscale λ_f – hence the subscript in equation (5.11). Similarly, with the analogous definition of the transverse microscale, λ_g, it can be shown that $\lambda_g^2 = \frac{1}{2}\lambda_f^2$. Rather than the order-of-magnitude arguments used in sections 3.3 and 3.4, we can now be rather more specific about the relative magnitudes of the scales, by considering the turbulence Reynolds number based on the integral scale, Re_{L_x}. For example, taking $A = 1$, we find from the above relations that

$$\frac{L_x}{\lambda_g} = \left(\frac{\mathrm{Re}_{L_x}}{10}\right)^{1/2}, \qquad \frac{L_x}{\lambda_f} = \left(\frac{\mathrm{Re}_{L_x}}{20}\right)^{1/2}. \tag{5.17}$$

We emphasise again the increasing separation of the large and small scales of motion as the Reynolds number rises, seen directly in these two relationships. It is the Taylor scale, defined by λ_g, that is commonly used to characterise grid turbulence via the 'Taylor-scale Reynolds number', $\mathrm{Re}_\lambda = \hat{u}\lambda_g/\nu$ and it is evident that this varies with the integral-scale Reynolds number according to

$$\mathrm{Re}_\lambda = \left(\frac{20}{3}\mathrm{Re}_{L_x}\right)^{1/2}. \tag{5.18}$$

Incidentally, it was Taylor [22] who originally defined the microscale, λ_g, during his development of a statistical theory for HIT. He showed that the dissipation is given by

$$\varepsilon = \frac{15\nu\hat{u}^2}{\lambda_g^2}, \tag{5.19}$$

(see, correspondingly, equation (5.11)) and it was largely he who inspired the subsequent work of Batchelor and Townsend and others in the succeeding decades in Cambridge, some of which is mentioned below.

5.1.3 Energy transfer and changes in $f(r)$

Without engaging in detailed analysis, it is instructive to consider the way in which the scalar energy density function $E(\kappa)$, first introduced in chapter 3, develops over time in the specific case of isotropic turbulence in the absence of a mean velocity field. By taking the Fourier transform of the Navier–Stokes equation and then assuming isotropy, it can be shown that (in the absence of a mean flow)

$$\frac{\partial E}{\partial t} = T - 2\nu\kappa^2 E \tag{5.20}$$

where T represents the transfer of energy to the wavenumber κ from all other wavenumbers (and is a function only of $\kappa = |\kappa|$) and the subsequent term describes viscous dissipation. It is important to realise that T does not destroy or create energy – it merely redistributes it between wavenumbers, and it does this via triple velocity product terms. This term dominates at the low wavenumber end of the spectrum; the dissipation term only becomes important at higher wavenumbers, where it first becomes comparable with T and then eventually dominates.

Given the direct relationship between $E(\kappa)$ and $R_{ij}(r)$ (recall sections 4.4.1 and 4.4.2), it is natural to ask how the latter varies with time, since that could lead to theoretical estimates of how the variance $\hat{u}^2$ varies with time, for comparison with the experimental findings given above. In isotropic turbulence, $R_{ij}(r)$ depends only on $f(r)$, as explained above, and an equation for the rate of change of $f(r)$ can be obtained by using the Navier–Stokes equations to eliminate some of the time derivatives in the equation resulting from differentiating $R_{ij}(r)$ with respect to time in isotropic turbulence. This was first done long ago [11] and typically leads to

$$\frac{\partial}{\partial t}(\hat{u}^2 f) - \frac{\hat{u}^3}{r^4}\frac{\partial}{\partial r}(r^4 K) = \frac{2\nu\hat{u}^2}{r^4}\frac{\partial}{\partial r}\left(r^4\frac{\partial f}{\partial r}\right), \tag{5.21}$$

where K in the second term is uniquely determined by the longitudinal *triple* correlation $(\overline{u_1(\mathbf{x},\,t)^2 u_1(\mathbf{x}+\mathbf{r},\,t)})/\hat{u}^3$, just as in isotropic turbulence $R_{ij}(r)$ is uniquely determined by $f(r)$. This is the Kármán–Howarth equation (KHE) and it is exact. It results only from the Navier–Stokes equations (including continuity) and the assumption of isotropy. Not surprisingly, however, it is unclosed because there are two unknowns – f and K; this is the usual and inevitable result of dealing with averaged equations, as first noted in chapter 2. For $r = 0$, it can be shown that the triple-product (K) term vanishes and then the equation reduces simply to the energy equation (5.6), i.e.:

$$\frac{d\hat{u}^2}{dt} = 10\nu\hat{u}^2\frac{\partial^2 f}{\partial r^2}\bigg|_{r=0} = -\frac{10\nu\hat{u}^2}{\lambda_f}. \tag{5.22}$$

K is the spatial equivalent of the spectral transfer term T in equation (5.20). The transfer of energy from larger to smaller scales, described in spectral terms by T, is here described by the term containing K for, according to Kolmogorov's ideas, it is essentially an inertial process. Incidentally, since it contains (only) third moments of the fluctuating velocity, K would be zero if the turbulence were truly Gaussian, so it is clear that the energy cascade depends crucially on the non-Gaussian aspects of the velocity field, as was noted at the end of section 2.3.

A detailed consideration of the KHE under certain assumptions allows the way in which the turbulence energy varies with time to be deduced. We first mention the situation at large times, when the Reynolds number becomes small. In this case, the triple-product K term can be ignored and extensive analysis leads to the result that $\hat{u}^2 \sim t^{-5/2}$ and $f \sim e^{-r^2}$. This is the final period of decay mentioned earlier, in which inertial effects are negligible – the decay occurs through uninhibited viscous dissipation. It is actually a self-preserving solution of the KHE (without the K term), in the sense that f can be replaced by a function F, say, where $F = f(r/\lambda)$, so that f is assumed to preserve its shape over all values of r. However, this final-period, $t \to \infty$ solution – although confirmed by experiment – is actually of little practical importance.

Secondly, and more significantly, the early theoreticians considered the full KHE for the conditions under which a particular integral appearing in the analysis was assumed to be constant. For interest, this integral can be written as

$$B = \hat{u}^2 \int_0^\infty r^4 f(r,\,t)dr, \tag{5.23}$$

which is commonly called the Loitsyanskii integral [13]. For B be to be constant with time, the correlation functions must be self-preserving over all r, and the scaling parameter for r is now L rather than λ; equation (5.23) then implies (for sufficiently large t) that $\hat{u}^2 L^5$ is constant, provided that the integral converges. If this is so, the empirical result expressed by equation (5.5), i.e.

$$-A\frac{\hat{u}^2}{\frac{d}{dt}(\hat{u}^2)} = \frac{L}{\hat{u}},$$

can be combined with $\hat{u}^2 L^5 = $ constant to yield

$$\hat{u}^2 \sim t^{-10/7}, \quad L \sim t^{2/7} \quad \text{and} \quad \lambda \sim t^{1/2}. \tag{5.24}$$

This solution, originally proposed by Kolmogorov, was fully explored by Batchelor [3]. The astute reader will realise that the decay rate is not too different from what is typically measured in grid turbulence (equation (5.3) with $n \approx 1.3$). On the other hand, it has since been shown, initially by Saffman [18], that the integral can in fact diverge, depending on how the turbulence was initially created. In that case, it turns out from a further detailed analysis of the KHE that $\hat{u}^2 L^3 = $ constant, yielding

$$\hat{u}^2 \sim t^{-6/5}, \quad L \sim t^{2/5}, \tag{5.25}$$

which is also not far from the empirical $n = 1.3$ mentioned earlier. Equations (5.24) and (5.25) are sometimes called 'Batchelor' turbulence and 'Saffman' turbulence, respectively, and they have different forms of the energy spectral density function.

In recent decades, there have been many computer simulations of isotropic turbulence and both types have been obtained. But which type, if either, is wind-tunnel grid turbulence? This issue has been studied by (among others) Krogstad and Davidson [12], who have demonstrated (theoretically) that in strictly homogeneous, isotropic turbulence, the Saffman decay law requires A in equation (5.5) to be genuinely constant. However, in grid turbulence this is usually not the case, probably because of weak inhomogeneities, or perhaps because of the inevitable decline in Re_λ. This changes the decay exponent somewhat. Their own grid turbulence experiments suggest that the Saffman result, $\hat{u}^2 L^3 = $ constant, approximately holds, but that the exponent n is around 1.13. They admit, nonetheless, that different grids could produce different results. The decay laws for Batchelor and Saffman turbulence are not, in fact, the only possible solutions of the KHE and there will likely be ongoing work in this whole area.

A related issue concerns the 'constant' A which appears in the expression for dissipation, $\varepsilon = \sqrt{\frac{2}{3}}Ak^{3/2}/L_x$, in equations (5.6) (recall also equation (3.2)). The different decay laws discussed above lead to different behaviour for A [12]. We can write the dissipation relation more generally as the famous 'equilibrium dissipation law',

$$\varepsilon = C_\varepsilon k^{3/2}/L, \tag{5.26}$$

(with a constant C_ε). This assumes, principally, the adequacy of the Kolmogorov hypothesis that the small-scale motions evolve much more rapidly than the large-scale motions – i.e. more rapidly than the timescale of the evolution of the whole flow – so that the energy dissipation rate is (at asymptotically large Reynolds numbers) independent of the fluid viscosity, being set by the large-scale motions. Equation (5.26) then follows on dimensional grounds and has been called 'one of the cornerstone assumptions of turbulence theory' [23]; it is intimately related to the Richardson–Kolmogorov equilibrium cascade. A related result of Kolmogorov's theoretical arguments is what is commonly known as the 'four-fifths law', which links the time average of the third-order moment of the longitudinal velocity difference, $\Delta u'(r)$, over a longitudinal distance r within the inertial range (i.e. for a scale remote from both the large-scale and the dissipative motions) directly to the dissipation via

$$\overline{\Delta u'(r)^3} = -\frac{4}{5}\varepsilon r. \tag{5.27}$$

This result is, perhaps, the only non-trivial exact result in turbulence theory and was originally derived directly from the KHE (i.e. assuming isotropy) and also assuming the existence of an inertial range of scales. For details of Kolmogorov's and other, alternative, derivations the interested reader should consult the texts of Frisch [9] or Mathieu and Scott [14], for example.

A number of experiments have suggested that stationary HIT at a high enough Reynolds number does indeed yield a constant C_ε. This has been demonstrated numerically, using direct numerical simulation (DNS) for 'forced' turbulence in a box – i.e. with a forcing term included in the NSE specifically to ensure that the turbulence does not decay in time, [20] and [10], for example. The same result has been obtained in a laboratory, where non-decaying HIT can be produced (using top and bottom stirrers in a rotating tank of water) [6]. On the other hand, various authors have shown that in decaying HIT (and in other flows too), C_ε in equation (5.26) is actually a weak function of time. Taylor [22] thought it was constant 'for geometrically similar boundaries', implying that its independence on Reynolds number was not likely to be sufficient to assure its constancy for all flows. Grid turbulence, in any case, is not strictly homogeneous (because $d\hat{u}^2/dx \neq 0$). A compilation of the available dissipation data was presented by Sreenivasan in 1984 [19], which suggested that once Re_λ exceeds about 100, C_ε is essentially independent of Re_λ for square mesh grid flows. But it has now become clear that even in HIT, whether generated by a grid in a wind tunnel or modelled computationally using DNS, the precise value of C_ε is, in fact, dependent on the initial conditions [1, 20] even at large Re_λ. This is well illustrated by the data reviewed by Sreenivasan [20] and shown in figure 5.2. Furthermore, there is some evidence that in a range of turbulent flows, C_ε varies with the *local* Reynolds number [8] (as well as the initial conditions); this implies a *non-equilibrium* dissipation law, which appears to break the link between classical cascade arguments and the $-5/3$ inertial range spectrum. Vassilicos has provided a full discussion of all this [26].

One of the features of many flows of practical importance, not recognised by Kolmogorov or, indeed, present in HIT, is the prevalence of large-scale coherent

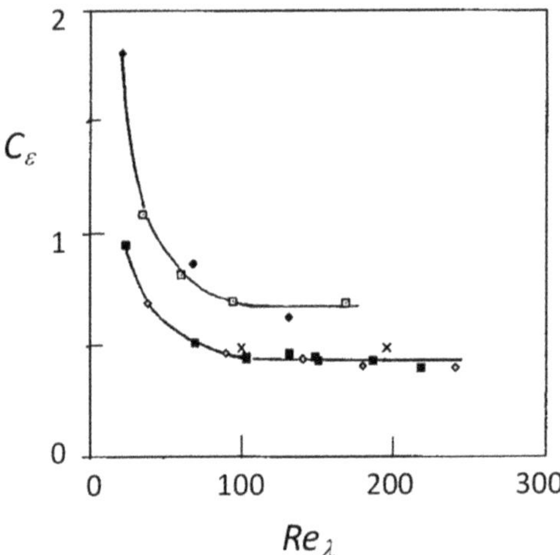

Figure 5.2. The 'constant' in equation (5.26), as determined from numerous numerical simulations of HIT; the various symbols refer to the different data sources[1].

motions. Reynolds and Hussain [17] were perhaps the first to consider the influence of these motions; they showed that the details of the energy transfer between different scales (encapsulated in the T in equation (5.20)) is significantly affected by the quasi-periodic dynamics of these coherent motions. A brief introduction to some of the subsequent efforts to identify more precisely the influence of large-scale coherent motions on the randomly fluctuating smaller-scale motions and, in particular, the cascade process, is provided in [24]. Such efforts are ongoing.

We have made all these remarks (in the previous three paragraphs) simply to indicate that although Kolmogorov's dissipation law remains central to the basic understanding of turbulence, and the $-5/3$ inertial range behaviour seems quite robust for sufficiently large Re_λ and for flows very different to HIT, there are good reasons to exercise care not to be too dogmatic about its full validity. Indeed, there are certainly some aspects of Kolmogorov's theories which are now well accepted as being invalid, to a greater or lesser extent. Explorations of the whole topic will undoubtedly continue. It should be noted, nonetheless, that equation (5.26) is a crucial relation implied in nearly all modern turbulence modelling approaches used for practically important flows; in that sense, it certainly remains 'a cornerstone in turbulence theory' and is likely to be so for quite some time.

5.2 Homogeneous shear flow turbulence (HSFT)

5.2.1 Setting the scene

Isotropic turbulence, as we have seen, cannot sustain itself and must decay in time unless it is artificially forced, which is often done in numerical simulations of HIT in

---

[1] Redrawn from [20] with the permission of AIP Publishing.

a periodic box, such as some of those discussed by Sreenivasan, producing data like those shown in figure 5.2. The simplest kind of turbulent flow that *can* sustain itself is homogeneous shear flow. If a constant velocity gradient is superimposed on initially isotropic turbulence, a shear stress is developed, so that the production term in equation (2.28) is nonzero; turbulence energy is then extracted locally from the mean flow. Figure 2.1 illustrates the situation, in which, in this ideal case, we imagine that any domain boundaries are remote. Let us suppose that the flow has only one nonzero mean velocity component, $U(y)$, in the x direction (i.e. $V = W = 0$). Now assume that the imposed mean shear, β, is defined simply by $U = \beta y$, with $\beta > 0$, and that it is constant – i.e. we have

$$\beta = \frac{dU}{dy} = \text{constant}. \tag{5.28}$$

If the turbulence is initially homogeneous, $\partial \overline{u_i' u_j'}/\partial x_j = 0$, so the turbulence does not enter the mean flow equation (2.16). The velocity field is, of course, a solution of the mean flow equation, provided the mean pressure is constant. β^{-1} is effectively a timescale associated with the mean shear, so that βt is a non-dimensional time – a measure of the cumulative mean strain.

This initial turbulence must remain homogeneous, because shifts in the coordinate origin and subsequent changes to a new inertial frame based on the mean velocity leave the equation for the fluctuating statistics unchanged. (The same is true for any flow in which the mean velocity derivatives do not vary with position.) We emphasise that, as for isotropic turbulence, the evolving turbulence field does not affect the mean flow. But how does the turbulence evolve? Consider first the turbulent kinetic energy equation (2.28), which in this simple case becomes

$$\frac{dk}{dt} = -\beta \overline{u'v'} - \varepsilon, \tag{5.29}$$

recalling that equation (2.28) assumed that the Reynolds number is sufficiently large that the viscous term in the full equation can be ignored. If the flow is initially isotropic, then it is symmetric about the plane $z = 0$ – positive and negative fluctuations in w' are equally likely to be associated with any u', v' – and this means that the two time-averaged shear stresses $\overline{u'w'}$ and $\overline{v'w'}$ are both zero. Initially, $\overline{u'v'}$ is also zero, so that the turbulence must begin to decay just as in isotropic turbulence (i.e. as in equation (5.1)). However, the mean shear will soon begin to have an 'anisotropising' effect, so that $\overline{u'v'}$ becomes nonzero and all three terms in equation (5.29) now operate. We can investigate this by considering the shear stress transport equation. This follows from the general transport equation for the Reynolds stresses (equation (2.27)) with $i = 1$, $j = 2$. Ignoring the non-dissipative viscous term, as usual, this reduces to

$$\frac{d\overline{u'v'}}{dt} = -\beta \overline{v'^2} + \frac{\overline{p'}}{\rho}\left(\frac{\partial u'}{\partial y} + \frac{\partial v'}{\partial x}\right) - 2\nu \overline{\frac{\partial u'}{\partial x_k} \frac{\partial v'}{\partial x_k}}, \tag{5.30}$$

in which the dissipative term is usually small and dominated at high Reynolds numbers by the small scales, which (in the usual Kolmogorov thinking) are isotropic, so that this viscous term is actually zero (see equation (5.9) with $i \neq j$). Experiment shows that $\overline{u'v'}$ first becomes negative, because of the first term in this equation (β is positive). This means, because of equation (5.29), that the turbulence decays less rapidly than it would in the absence of the production term, $-\beta\overline{u'v'}$. Eventually, $\overline{u'v'}$ becomes sufficiently negative that production exceeds dissipation and the turbulence energy begins to grow again. But what is the influence on this process of the pressure strain term (the second term on the right-hand side of equation (5.30)) which, it would appear, tends to offset the influence of the first term? We explore this by a brief discussion of the behaviour of p'.

5.2.2 The pressure strain term

The fluctuating pressure, p', obeys a Poisson equation, which can be obtained by taking the divergence of the Navier–Stokes equations. This leads to

$$\nabla^2 p = -\rho \frac{\partial^2 u_i u_j}{\partial x_i \partial x_j}, \tag{5.31}$$

which can be time averaged in the usual way to give

$$\nabla^2 P = -\rho \frac{\partial^2}{\partial x_i \partial x_j}\left(U_i U_j + \overline{u_i'u_j'}\right) \tag{5.32}$$

for the mean pressure. Subtracting this from the previous equation yields, for the fluctuating pressure,

$$\nabla^2 p' = -\rho \frac{\partial^2}{\partial x_i \partial x_j}\left(U_i u_j' + U_j u_i' + u_i'u_j' - \overline{u_i'u_j'}\right). \tag{5.33}$$

The solution of this equation requires the use of a Green's function, which includes a volume integral and a surface integral over the boundaries, and leads to expressions for the velocity–pressure correlations (e.g. $\overline{p'u_k'}$) whose divergence appears in the energy equation. These expressions contain double- and triple-velocity correlations at *two* points; the latter are clearly part of the closure problem and the former (i.e. the double correlations) are a result of non-locality. So, as seen also in equation (5.33), the fluctuating pressure arises through two processes, one of which is linear in the fluctuating velocity and the other of which is nonlinear. We can thus split the above expression for $\nabla^2 p'$, into two components, written as

$$\nabla^2 p'^{(1)} = -2\rho \frac{\partial U_i}{\partial x_j}\frac{\partial u_j'}{\partial x_i} \quad \text{and} \quad \nabla^2 p'^{(2)} = -\rho \frac{\partial}{\partial x_i \partial x_j}\left(u_i'u_j' - \overline{u_i'u_j'}\right). \tag{5.34}$$

The first and second relations are clearly linear and nonlinear contributions, respectively. Ignoring the viscous term, considering the equation for fluctuating pressure, and reverting to the above notation for our particular shear flow, equation (5.30) can thus be written as

$$\frac{d\overline{u'v'}}{dt} = \underbrace{-\beta\overline{v'^2}}_{\text{Prod.of}-u'v'} + \underbrace{\overline{\frac{p'^{(1)}}{\rho}\left(\frac{\partial u'}{\partial y} + \frac{\partial v'}{\partial x}\right)}}_{\text{Linear}} + \underbrace{\overline{\frac{p'^{(2)}}{\rho}\left(\frac{\partial u'}{\partial y} + \frac{\partial v'}{\partial x}\right)}}_{\text{Non-linear}}. \tag{5.35}$$

The first pressure strain term can be rewritten in a form that contains the strain β and, as indicated, is linear in the fluctuating velocities. $p'^{(1)}$ is called the 'rapid' pressure, because it responds immediately to any change in the mean velocity gradient (see the first relation in equations (5.34)). On the other hand, the second term is nonlinear, but it does not depend on the mean velocity (it is of third order in the fluctuating quantities) and it is relatively slow-acting. $p'^{(2)}$ is thus called the 'slow' pressure and in isotropic turbulence it can be shown to be identically zero. If the turbulence is initially isotropic, therefore, only the linear part of the pressure strain correlation is nonzero as soon as the mean shear is applied. It is thus this part of the pressure strain that tends to resist the growth of $-\overline{u'v'}$.

We now consider the relevant timescales. The mean shear timescale is β^{-1} and the timescale of the large scales can be characterised, as usual, by the eddy turnover time $L/\hat{u}$, which is the typical time needed for the large scales to act on themselves and decay in the absence of production. If the mean strain is large enough, so that its timescale is very small compared with this eddy turnover time (i.e. in the limit $\beta L/\hat{u} \to \infty$), the turbulence does not have time to 'act upon itself' and thus decay, so that nonlinear effects (through the slow pressure term) are negligible. The basic equations for the fluctuating velocity field can then be linearised, making their solutions mathematically more tractable. This allows the use of a particular kind of methodology to solve the equations, usually termed 'Rapid Distortion Theory' (RDT) because the equation describing the evolving turbulence contains only the rapid part, $p'^{(1)}$, of the pressure strain. The interested reader can explore some of the fundamental details of RDT by referring to the texts by, for example, Pope [14] and Mathieu and Scott [16]. Of course, if the initial turbulence is *not* isotropic, RDT does not apply (the slow pressure term is nonzero).

Nonlinear effects will, in any case, eventually become important when the timescale ratio $\beta L/\hat{u} = \mathcal{O}(1)$. The slow pressure term will then slow down the energy growth, eventually leading to an equilibrium condition. At the opposite limit, if initially $\beta L/\hat{u} \ll 1$, the mean shear is weak enough to allow initial decay of the turbulence as if there were no shear at all. After a while, however, $\hat{u}$ will have fallen enough to make $\beta L/\hat{u} = \mathcal{O}(1)$ and the turbulence will behave as described above. We summarise the effect of mean flow strain on the Reynolds shear stress in figure 5.3, for a case in which the strain is large so that the slow pressure strain term is initially negligible.

Incidentally, if, in a case for which there is an initial linear phase, the mean shear were suddenly reversed (a 'thought experiment', as this would be impossible to achieve experimentally!), the production term in the shear stress transport equation (5.35) would immediately change sign and the turbulence would return to its initial

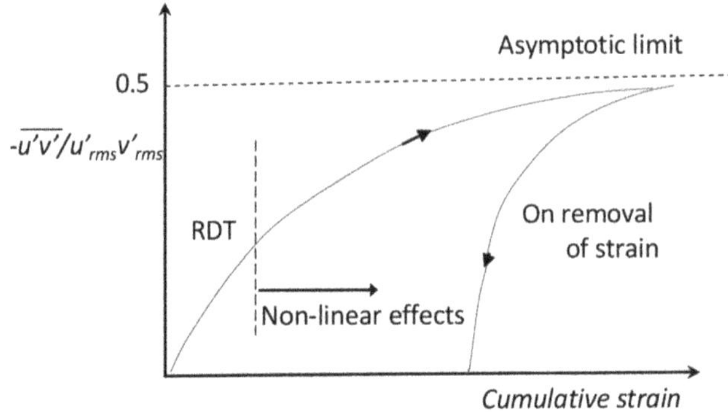

Figure 5.3. Sketch of the growth in Reynolds shear stress, normalised by the u' and v' rms velocities, for a case in which $\beta L/\hat{u}$ is initially large, so that RDT can be used. The cumulative strain is simply the integral of βdt, i.e. βt if the strain remains constant with time.

state. On the other hand, once the nonlinear processes are operative, reversing the mean shear (or turning it off entirely) will not have the same effect, since whilst the first two terms on the right-hand side of equation (5.35) change sign (or become zero) immediately, the third does not (for it does not contain the flow gradient). This term tends to reduce $-\overline{u'v'}$, so the shear stress returns rather more rapidly to zero – as indicated in figure 5.3 – giving a kind of hysteresis to the response of the turbulent field to changing mean shear.

5.2.3 The individual stress components

A consideration of the transport equations for the three normal Reynolds stresses, not fully pursued here, shows that (not unnaturally) the three stresses do not all increase at the same rate, so that even if there is an initial linear development, the turbulence becomes anisotropic. Ignoring viscous effects as before, the transport equation (2.27) for the axial stress, i.e. with $i = j = 1$, becomes

$$\frac{d\left(\frac{1}{2}\overline{u'^2}\right)}{dt} = -\beta\overline{u'v'} + \frac{\overline{p'\,\frac{\partial u'}{\partial x}}}{\rho} - \frac{1}{3}\varepsilon, \tag{5.36}$$

in which equations (5.9) and (5.10) have been used for the dissipative term. Note that this contains a (positive) energy production term, $-\beta\overline{u'v'}$; the corresponding equations for the other two normal stresses do not. Thus, once the shear stress, $-\overline{u'v'}$, begins to rise above zero as described earlier, one expects the axial stress to rise, with the other two following subsequently by the action of the pressure strain terms only. The anisotropy levels increase further as the effects of the strain increase, and the nonlinear effects (through the slow pressure strain term) become more significant. This leads to an increased slowdown in the growth of turbulence energy and eventually some kind of asymptotic state is reached. Experiments show that this equilibrium regime is characterised by constancy in the various stress component ratios. For example, in the case of the shear stress correlation coefficient, $\overline{u'v'}/(u'_{\mathrm{rms}}v'_{\mathrm{rms}})$, a ratio of around -0.5 is typically reached, as indicated in figure 5.3.

It is important to realise that although stress ratios may eventually become constant, the stresses themselves will continue to rise, as will the length scales of the turbulence. In any experimental realisation of this ideal homogeneous shear flow, such as in a wind tunnel, sooner or later there will therefore be a downstream region in which the homogeneity is destroyed because of the influence of the walls. Nonetheless, there have been numerous reasonably successful attempts to explore homogeneous shear flow experimentally. Some data from a particular set of experiments [21] are shown in figure 5.4. They are plotted against the total strain, defined by $(x/U)(dU/dy)$, where x is the distance from the initiating set of shear-generating plates and subsequent mesh, placed near the upstream end of the wind tunnel's working section. (Taylor's hypothesis can be invoked to equate this total strain to the βt notation used above.) Exercise 5.2 further explores the data from figure 5.4 and some points, in particular, are worthy of note.

First, the axial normal stress initially falls (in the region in which the shear stress $-\overline{u'v'}$ is very small), but $\beta L/\hat{u}$ rapidly increases to become $\mathcal{O}(1)$ and thereafter all the stresses rise. Secondly, note that the stresses have magnitudes in the following order: $\overline{u'^2} > \overline{w'^2} > \overline{v'^2} > |\overline{u'v'}|$. The streamwise stress receives energy directly from the mean shear via the $-\beta\overline{u'v'}$ production term in equation (5.36) and thus rises first, while the other two normal stresses receive their energy by their coupling to the streamwise stress through the pressure strain terms, as described above. These

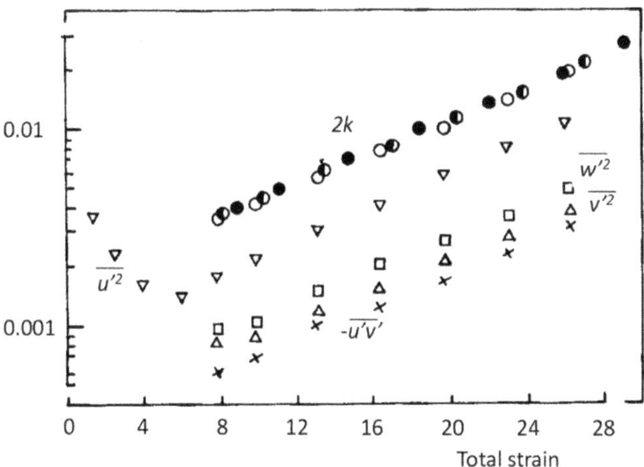

Figure 5.4. Growth of the Reynolds stresses in a wind-tunnel realisation of a homogeneous shear flow[2]. All stresses are normalised by U^2 (the mean velocity on the tunnel centreline), and the label for each one is placed near the appropriate symbol. Note that, as usual, $k = \frac{1}{2}(\overline{u'^2} + \overline{v'^2} + \overline{w'^2})$ and v' is the component of fluctuating velocity in the direction (y) of the mean shear.

---

[2] Data from [21]. Reproduced with permission, copyright Cambridge University Press.

relative sizes of the stresses are, in fact, very common in more general flows in which there is a dominant mean shear – as we will discover in later chapters. The fact that mean shear creates anisotropy in the stresses is an important feature of all practically important turbulent flows, since these all have at least some regions of mean shear in one direction or another. The degree of anisotropy can be described by the *Reynolds stress anisotropy tensor* defined (as we noted in section 2.4) by

$$a_{ij} = \overline{u_i'u_j'} - \frac{2}{3}k\delta_{ij} \tag{5.37}$$

or its normalised version,

$$b_{ij} = \frac{a_{ij}}{2k} = \frac{\overline{u_i'u_j'}}{\overline{u_m'u_m'}} - \frac{1}{3}\delta_{ij}. \tag{5.38}$$

The data suggest that the components of b_{ij} reach approximately constant values and, in particular, $b_{12} = \overline{u'v'}/2k \approx -0.15$. It is also found that in this asymptotic equilibrium state, the ratio of production to dissipation $(P_k/\varepsilon = -\beta\overline{u'v'}/\varepsilon)$ becomes constant (and of the order of unity).

Third, note from figure 5.4 that the growth of the stresses is closely exponential – straight lines on the log-linear axes used in the figure. Given that the turbulence timescale k/ε $(=\tau_t$, say) does not change significantly with time, such behaviour is entirely consistent with similarly constant values for the stress ratios and P_k/ε. It is straightforward to show that in these circumstances the turbulent kinetic energy equation (5.29) can be solved to yield

$$k = k_o e^{a\beta t} \quad \text{with} \quad a = \frac{1}{\beta\tau_t}\left(\frac{P_k}{\varepsilon} - 1\right) \tag{5.39}$$

where k_o is an initial turbulence energy at the start of the equilibrium regime (which is clearly not where the mean shear is first applied). This result suggests that if $P_k < \varepsilon$, a is negative and the turbulence would decay. However, this, with the concomitant decrease in eddy length scale, is unlikely, since a decrease of the latter to an infinitesimally small scale would imply infinite dissipation, which is not consistent with the constant production. Consequently, $a = 0$ is a lower bound and, indeed, experiments have always shown that that $P_k/\varepsilon \geqslant 1$.

One of the features of HSFT is that, because of the homogeneity, P_k is uniformly distributed in space (over constant-x planes and sufficiently far from the walls in a wind-tunnel experiment). This almost always never happens in practically important turbulent flows, where there are usually specific, localised regions of dominant production. Nonetheless, the various other features discussed above are symptomatic of many flows, which is why this flow is worthy of careful study. There are also structural (dynamical) features of the idealised HSFT which are mirrored to a greater or lesser extent in other much more complex flows. For example, coherent structures, typically the packets of horseshoe vortex and hairpin structures commonly found in wall turbulence (see chapter 8), have been shown to be present in HSFT [25], which provides a further incentive to explore this flow more fully.

5.3 Subject giant: S Corrsin

Stanley Corrsin (1920–1986) was an American researcher who studied various areas of fluid dynamics throughout his career and, in particular, turbulence, analysing a number of canonical turbulent flows. As a result of his work on turbulence, he was awarded the Fluid Dynamics Prize of 1983, given annually by the American Physical Society, whose citation mentioned his 'contributions to the understanding of turbulent transport … through ingenious experiment and physical insight'. The American Society of Civil Engineers honoured him with the Theodore von Kármán Medal for his contributions to the study of turbulence, awarded on the day after his death in 1986 at their conference given in his honour.

Corrsin was born in Philadelphia to immigrant Jewish parents from Romania. After obtaining his undergraduate degree from the University of Pennsylvania, he went to CalTech in 1940 (recruited by Clark B Millikan) for graduate studies in the Guggenheim Aeronautical Laboratory under Theodore von Kármán. After his M.S. degree he studied for his doctorate under (mainly) Hans Liepmann. His research there involved the use of hot wires to study the decay of turbulence behind a grid and the velocity and temperature fields in round jets. The publication of his thesis was delayed by the Second World War, when he was otherwise occupied teaching aerodynamics to the military at CalTech, but in 1947 he was awarded his PhD. Thereafter, he began his career at the Johns Hopkins University in Baltimore, in the newly formed Aeronautics Department led by Francis H Clauser, eventually becoming, in 1981, the Theophilus Halley Smoot Professor

Figure 5.5. Professor Stanley Corrsin explaining decaying isotropic turbulence behind a grid in a wind tunnel. Image reproduced with permission from [15], copyright Cambridge University Press.

of Engineering. He had many distinguished students, including John Lumley (see section 4.7), and continued working on unraveling some of the intricacies of turbulence by studying canonical flows. Corrsin's interests were wide, however, and he also worked on human locomotion and in the biological fluid dynamics field, even studying the aerodynamics of the albatross and of formation flying in flocks of birds, for example. But it is for his numerous contributions to the understanding of turbulence that he will mainly be remembered. To mention just two of his noteworthy contributions, it was Corrsin who discovered the characteristic intermittent region between a laminar and a turbulent flow at the edge of any turbulent flow with a free stream boundary and he was the first to apply statistical theory to turbulent, reacting flows. His contributions also provided much of the pivotal understanding of scalar transport, the fine-scale structure of passive scalars, and homogeneous turbulence.

Sample exercises

5.1. The file 'HITData.txt'[3] tracks the variation over time of the properties of decaying isotropic turbulence, from a 512^3 DNS by Wray [27]. The data include total energy, enstrophy, and integral length scale.

    (a) Plot the total kinetic energy energy versus time and determine the exponent of the best power law fit (equation (5.3)).

    (b) Determine the rate of turbulent kinetic energy dissipation using (i) $\varepsilon = -dk/dt$ (Equation (5.1)) and (ii) $\overline{\nu \omega_i' \omega_i'}$ (equation (2.31)).

    (c) Plot the development of the integral length scale in time.

    (d) Explore the validity of the equilibrium dissipation law (equation (5.26)) and estimate C_ε.

5.2. The file 'HSFData.txt'[4] contains measurements of the growth of the turbulent kinetic energy and Reynolds stresses in a wind-tunnel realisation by Tavoularis and Karnik [21] of a homogeneous shear flow with $\beta = 84$ s^{-1}.

    (a) Calculate the turbulent kinetic energy and plot it to see how it varies downstream. Does it grow or decay?

    (b) Fit a line to the plot of k vs τ using a logarithmic y-axis to determine the exponent of the exponential function $k = k_o e^{a\beta t}$ (equation (5.39)), noting that in this case $\tau \approx \beta t$. Compare this value of a with that predicted by

$$a = \frac{\varepsilon}{\beta k}\left(\frac{P_k}{\varepsilon} - 1\right),$$ noting that $P_k = -\beta \overline{u'v'}$ and $\varepsilon = P_k - \beta\frac{dk}{d\tau}$.

    (c) Determine the anisotropy coefficients defined by equation (5.38).

References

[1] Antonia R A and Pearson B R 2000 Effect of initial conditions on the mean energy dissipation rate and the scaling exponent *Phys. Rev. E* **62** 8086–90

[2] Batchelor G K 1947 Kolmogoroff's theory of locally isotropic turbulence *Proc. Camb. Phil. Soc.* **43** 533–59

---

[3] https://github.com/cvanderwel/TurbulentFlows/blob/main/data/HITData.txt
[4] https://github.com/cvanderwel/TurbulentFlows/blob/main/data/HSFData.txt

[3] Batchelor G K 1948 Energy decay and self-preserving correlation functions in isotropic turbulence *Q. J. Appl. Math.* **6** 97–115

[4] Batchelor G K 1953 *The Theory of Homogeneous Turbulence* (Cambridge: Cambridge University Press)

[5] Batchelor G K and Townsend A A 1948 Decay of isotropic turbulence in the initial period *Proc. Roy. Soc. Lond.* A **193** 539–58

[6] Cadot O, Couder Y, Douady A and Tsinober A 1997 Energy injection in closed turbulent flows: stirring through boundary layers versus inertial stirring *Phys. Rev.* E **56** 427–33

[7] Comte-Bellot G and Corrsin S 1966 The use of a contraction to improve the isotropy of grid generated turbulence *J. Fluid. Mech.* **25** 657–82

[8] Dairay T, Obligardo M and Vassilicos J C 2015 Non-equilibrium scaling laws in axisymmetric turbulent wakes *J. Fluid. Mech.* **781** 166–95

[9] Frisch U 2002 *Turbulence* (Cambridge: Cambridge University Press)

[10] Kaneda Y, Ishihara T, Yokokawa M, Itakura K and Uno A 2003 Energy dissipation rate and energy spectrum in high resolution direct numerical simulations of turbulence in a periodic box *Phys. Fluids.* **15** L21–4

[11] von Karman T and Howarth L 1938 On the statistical theory of isotropic turbulence *Proc. R. Soc. Lond.* A **164** 192–215

[12] Krogstad P-A and Davidson P A 2010 Is grid turbulence Saffman turbulence? *J. Fluid. Mech.* **642** 373–94

[13] Loitsyanskii L G 1939 [In Russian] Some basic laws for isotropic turbulent flow *Trudy Tsentr Aero-Giedrodin Inst* **440** 3–23

[14] Mathieu J and Scott J 2000 *An Introduction to Turbulent Flow* (Cambridge: Cambridge University Press)

[15] Meneveau C and Riley J R 2011 *A Voyage Through Turbulence* ed P A Davidson, Y Kaneda, K Moffatt and K R Sreenivasan (Cambridge: Cambridge University Press)

[16] Pope S B 2000 *Turbulent Flows* (Cambridge: Cambridge University Press)

[17] Reynolds W C and Hussain A K M 1972 The mechanics of wave-organised wave in turbulent shear flow. Part 3. Theoretical models and comparison with experiments *J. Fluid. Mech.* **54** 263–88

[18] Saffman P G 1967 The large-scale structure of homogeneous turbulence *J. Fluid. Mech.* **27** 581–93

[19] Sreenivasan K R 1984 On the scaling of the turbulence energy dissipation rate *Phys. Fluids.* **27** 1048–59

[20] Sreenivasan K R 1998 An update on the energy dissipation rate in isotropic turbulence *Phys. Fluids.* **10** 528–9

[21] Tavoularis S and Karnik U 1989 Further experiments on the evolution of turbulent stresses and scales in uniformly sheared turbulence *J. Fluid. Mech.* **204** 457–78

[22] Taylor G I 1935 Statistical theory of turbulence parts i–iv *Proc. R. Soc. Lond.* A **151** 421–78 Part i: 10.1098/rspa.1935.0158, part ii: 10.1098/rspa.1935.0159, part iii: 10.1098/rspa.1935.0160, part iv: 10.1098/rspa.1935.0161

[23] Tennekes H and Lumley J L 1972 *A First Course in Turbulence* (Cambridge, MA: MIT Press)

[24] Thiesset F and Danaila L 2020 The illusion of a Kolmogorov cascade *J. Fluid. Mech.* **902** 1–4

[25] Vanderwel C and Tavalouris S 2011 Coherent structures in uniformly sheared turbulent flow *J. Fluid. Mech.* **689** 434–64

[26] Vassilicos J C 2015 Dissipation in turbulent flows *Annu. Rev. Fluid. Mech.* **47** 95–114

[27] Wray A A 1997 Database of a direct numerical simulation of decaying turbulence in a triply periodic box at 512^3 resolution *A Selection of Test Cases for the Validation of Large-Eddy Simulations of Turbulent Flows* AGARD Advisory Report NO 345

IOP Publishing

Turbulent Flows: an Introduction

Ian P Castro and Christina Vanderwel

Chapter 6

Free turbulent shear flows

In the previous chapters we discussed the two, perhaps simplest, canonical flows. We now turn to free shear flows. This is a class of flow that is more complex than the homogeneous shear flow discussed in section 5.2, mainly because each flow lacks homogeneity in two directions. In some senses, however, the flows can also be considered to be canonical. The main members of the class are mixing layers, wakes, and jets, and in the general hierarchy of complexity of turbulent flows, they are the next simplest to study – largely because they are remote from walls, so that viscous effects can usually be ignored provided the Reynolds number is high enough. There are both planar and axisymmetric versions of each and they have several character- istics in common; these are listed in the following introductory section. Section 6.2 then outlines the general approach for dealing with these flows, identifying the relevant scales and using order-of-magnitude arguments to simplify the governing equations. The crucial self-similarity ideas that can be used to transform the governing partial differential equations (PDEs) into straightforward ordinary differential equa- tions (ODEs) governing each flow are outlined, and this is followed by a series of sections applying those ideas to various cases, first axisymmetric flows (section 6.3 and 6.4) and then, more briefly, their planar equivalents (in section 6.5). Analytic solutions, along with a discussion of some of the basic turbulence characteristics, are included for each flow, with reference to laboratory and numerical experiments. The chapter closes with a consideration of, among other things, features of the ubiquitous entrainment processes at the edges of each of the flows, and of the crucial assumptions regarding energy dissipation made in the analyses of the previous sections, which have come under particular scrutiny in recent decades.

6.1 Initial remarks

We begin the discussion by listing the major features common to all canonical free shear flows.

1. There is always a single dominant velocity component, which we take to be the x direction.
2. They are created through an uneven distribution (in y) of momentum flux (usually at the x origin), which creates cross-stream shear that in turn generates turbulence.
3. The mean flow shear is not uniform across the flow (unlike the simpler case of homogeneous shear flow in section 5.2), so the flows are necessarily neither isotropic nor homogeneous. However, they are taken to be two-dimensional – i.e. there is no variation in the mean statistical properties in the third direction (spanwise in a planar flow or circumferentially in its axisymmetric equivalents).
4. The flows are unrestricted in space but bounded on at least one side by an ambient ('free stream') fluid, which is often nonturbulent and usually irrotational. As they evolve downstream, the momentum mainly diffuses across (far more than along) the flow.
5. The highly contorted boundary between the turbulent motions and the ambient irrotational fluid (sometimes called the *intermittency surface*) leads to entrainment of the latter into the former and thus lateral flow spreading.
6. The flows are assumed to evolve downstream without any influence from any external forces, and they therefore develop naturally solely due to the varying effects of the turbulence.

It is this collection of characteristics that sets these turbulent flows apart as, collectively, a special class. Many free shear flows are, in practice, affected by other complicating influences – three-dimensionality in the mean flow, significant curvature in the mean streamlines, or the influences of external turbulence, for example. We do not discuss such complications. When the simplifying conditions 1, 3, and 6 above apply, the statistical properties of the canonical free turbulent shear flows are found to maintain predictable shapes, and it is this *self-similarity* that allows us to simplify the governing equation for each flow, turning it into an ODE, which, with a simple closure assumption, can lead to a theoretical prediction of (at least) the mean flow.

6.2 The general approach

6.2.1 Introductory comments

We will initially consider two-dimensional planar flows which are homogeneous in the spanwise (z) direction and that have a dominant flow direction aligned with the x axis. The major contribution to the mean shear is thus in the y direction. Typical cases are illustrated in figure 6.1.

In this analysis, the flow is only considered far downstream from the initial source of the shear. For a jet, this initial shear is provided by a flow out of a (relatively narrow) nozzle into a zero velocity ambient. The mixing layer is effectively one side of this jet; the other side is far removed (and perhaps replaced by a continuous surface). For the wake, the initial shear is provided by a uniform flow past a body, such as a circular cylinder in the planar case or a sphere in the axisymmetric case. In

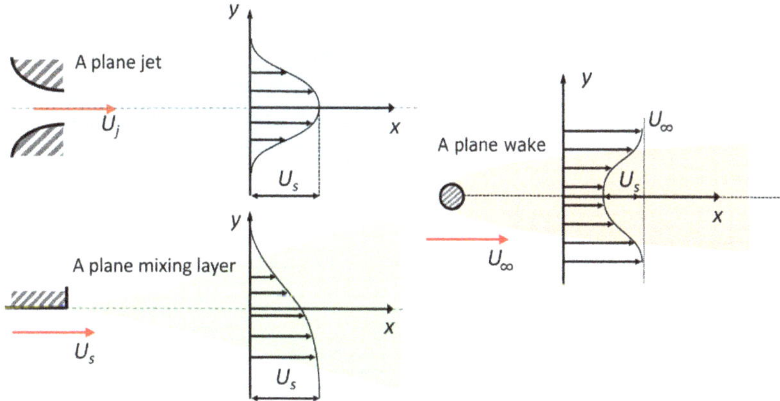

Figure 6.1. Sketches of a two-dimensional planar jet, mixing layer, and wake.

all cases, it is assumed that sufficiently far downstream, the flow has forgotten most of the details of its creation and begins to behave in a self-similar manner (as defined in section 6.2.6) particular to each flow. We first emphasise that the lack of any history effect – i.e. that the secondary details of the particular initial conditions are forgotten – is *only* an assumption and, as we will see, there is much evidence that the initial conditions may never be forgotten, or at least only at extremely large distances downstream. (Note that, in any case, the initial momentum flux in the case of a jet, or the obstacle drag in the case of a wake, can naturally never be forgotten, because it provides the fixed momentum flux requirement for the flow.)

6.2.2 Order-of-magnitude analysis

We define L as the streamwise length scale representing the (large) distance downstream from the origin of the flow. It is a matter of experimental fact that the cross-stream dimension of the flow, whose length scale we define as δ, is very much smaller than L, i.e. $\delta/L \ll 1$, so that the flows are 'slender'. There are various ways δ might be specified. In the case of a jet, for example, one might use the jet half width, meaning either the distance between the locations where the mean velocity is one half of its maximum, or the square root of the second central moment of the velocity profile. However, the precise definition is immaterial for the scaling arguments below.

Velocity scales can also be defined by U_s and $\hat{u}$, say, which characterise the scale of the cross-stream variation in mean velocity (see figure 6.1) and the scale of the fluctuating velocities, respectively. The former could refer to the maximum velocity difference across the flow, e.g. the maximum (centreline) velocity in a jet issuing into still air or, for a mixing layer, the free stream velocity on one side of the flow, with a zero velocity on the other (although if both sides have nonzero velocities, U_s could represent the velocity difference between the two external streams). The following discussion is based on these characteristic length and velocity scales, which are appropriate for a jet or mixing layer; the arguments are somewhat more complicated for a wake, for which there are *two* separate mean velocity scales – U_s and U_∞ (see figure 6.1). The ideas are essentially the same, however, and the resulting approximated equations turn out to be very similar if not identical.

We consider planar flows and make order-of-magnitude estimates of various important statistical quantities. First, note that $\partial/\partial x = \mathcal{O}(L^{-1})$ whereas $\partial/\partial y = \mathcal{O}(\delta^{-1})$. Thus, because $\delta/L \ll 1$, it is clear that

$$\underbrace{\frac{\partial U}{\partial x}}_{\mathcal{O}\left(\frac{U_s}{L}\right)} \ll \underbrace{\frac{\partial U}{\partial y}}_{\mathcal{O}\left(\frac{U_s}{\delta}\right)} . \tag{6.1}$$

The two nonzero terms in the continuity equation ($\partial U/\partial x + \partial V/\partial y = 0$) must obviously balance and using equation (6.1) this immediately implies that

$$V = \mathcal{O}\left(\frac{U_s \delta}{L}\right), \tag{6.2}$$

provided that U does not change significantly over a distance L (see [49]).

Given the scale of the velocity fluctuations defined above ($\hat{u}$), we can also say that the four individual nonzero Reynolds stresses are given by

$$\overline{u'^2} = \mathcal{O}(\hat{u}^2), \quad \overline{v'^2} = \mathcal{O}(\hat{u}^2), \quad \overline{w'^2} = \mathcal{O}(\hat{u}^2), \quad \text{and} \quad \overline{u'v'} = \mathcal{O}(\hat{u}^2) \tag{6.3}$$

(noting that $\overline{u'w'}$ and $\overline{v'w'}$ are identically zero in these two-dimensional flows). This does *not* mean that the turbulent fluctuations are equal. Often, the variance of the streamwise fluctuations is nearly double the magnitude of the normal stresses in the other two directions. However, this order-of-magnitude analysis is sufficient to gain insight into the relative importance of the various quantities appearing in the equations of motion, which are discussed next.

The exact equation of mean motion resulting from Reynolds averaging is, restating equation (2.16),

$$\frac{\partial U_i}{\partial t} + U_j \frac{\partial U_i}{\partial x_j} = -\frac{1}{\rho}\frac{\partial P}{\partial x_i} - \frac{\partial \overline{u_i' u_j'}}{\partial x_j} + \nu \frac{\partial^2 U_i}{\partial x_j \partial x_j}. \tag{6.4}$$

Consider first the steady flow **transverse momentum equation** (i.e. the expansion of equation (6.4) for $i = 2$), given here with the appropriate order of magnitude of each term written below,

$$\underbrace{U\frac{\partial V}{\partial x}}_{\frac{U_s^2}{L}\frac{\delta}{L}} + \underbrace{V\frac{\partial V}{\partial y}}_{\frac{U_s^2}{L}\frac{\delta}{L}} = \underbrace{-\frac{1}{\rho}\frac{\partial P}{\partial y}}_{?} - \underbrace{\frac{\partial \overline{u'v'}}{\partial x}}_{\frac{\hat{u}^2}{L}} - \underbrace{\frac{\partial \overline{v'^2}}{\partial y}}_{\frac{\hat{u}^2}{L}\frac{L}{\delta}}. \tag{6.5}$$

The viscous term, which is $\mathcal{O}((1/\mathrm{Re})U_s^2/\delta)$, has been omitted for an important reason. It is assumed that the Reynolds number is sufficiently high to make this term negligible (just as in the case of homogeneous shear flow discussed in section 5.2), so that none of the flows considered are influenced by viscous effects. This independence from the Reynolds number (often called *Reynolds number similarity*) is fundamental to the classical study of turbulence. It asserts that if the flow is turbulent, the structure of the energy-containing motions and the development of the mean flow are independent of the Reynolds number and are controlled solely by the initial boundary conditions (at the x origin in these shear flow cases). Note that these initial conditions *may* be Reynolds number dependent, but that is a separate issue. The notion of Reynolds number similarity in the downstream flow is quite separate from the concept of self-similarity, discussed in section 6.2.6, and is independent of the 'thin shear layer approximation' developed below. It is also more accurate than the latter, increasingly so as the Reynolds number rises.

Applying the slender flow assumption, $\delta/L \ll 1$, the orders of magnitude listed below the equation suggest that only the $\overline{v'^2}$ term remains. But this is certainly nonzero, so it must be balanced by the pressure gradient term. Thus the equation becomes

$$0 \approx -\frac{1}{\rho}\frac{\partial P}{\partial y} - \frac{\partial \overline{v'^2}}{\partial y}. \tag{6.6}$$

This implies that

$$P(x, y) \approx P_\infty - \rho\overline{v'^2} \tag{6.7}$$

assuming that the external free stream flow contains negligible velocity fluctuations. In most cases, the influence of $\rho\overline{v'^2}$ is relatively small and the pressure in the free stream is thus representative of the pressure throughout the turbulent flow. Furthermore, the streamwise pressure gradient can be estimated using

$$-\frac{1}{\rho}\frac{\partial P}{\partial x} \approx \frac{\partial \overline{v'^2}}{\partial x}, \tag{6.8}$$

noting, however, that under the slender flow assumption, the gradient of any quantity in the streamwise direction is relatively small; these two terms are therefore not too important. Clearly, variations of pressure within the turbulent region are only significant at second order and, to leading order, the axial pressure gradient $\partial P/\partial x$ is the same as that in the free stream, $\partial P_\infty/\partial x$.

6.2.3 Developing the thin shear layer equations

We now consider the streamwise momentum equation, which can be written as follows (with the magnitudes of each term again included below):

$$U\underbrace{\frac{\partial U}{\partial x}}_{\frac{U_s^2}{L}} + V\underbrace{\frac{\partial U}{\partial y}}_{\frac{U_s^2}{L}} = -\frac{1}{\rho}\frac{\mathrm{d}P_\infty}{\mathrm{d}x} - \underbrace{\frac{\partial(\overline{u'^2} - \overline{v'^2})}{\partial x}}_{\frac{\hat{u}^2}{L}} - \underbrace{\frac{\partial \overline{u'v'}}{\partial y}}_{\frac{\hat{u}^2}{L}\frac{L}{\delta}}, \tag{6.9}$$

where the usual pressure gradient term has been replaced using equation (6.7). Now, the fourth term in this equation is clearly $\mathcal{O}(\hat{u}^2/U_s^2)$ smaller than the terms on the left-hand side, and we assume that $\hat{u}/U_s$ is small so, to first order, the term can be neglected and the axial momentum equation becomes

$$U\frac{\partial U}{\partial x} + V\frac{\partial U}{\partial y} = -\frac{1}{\rho}\frac{\mathrm{d}P_\infty}{\mathrm{d}x} - \frac{\partial \overline{u'v'}}{\partial y}. \tag{6.10}$$

The continuity equation and this simplified governing equation are known as the lowest-order *thin shear layer equations*. In all specific cases considered in this chapter, it is assumed that the flow develops in the absence of any pressure gradient in the free stream (i.e. $\mathrm{d}P_\infty/\mathrm{d}x = 0$).

Incidentally, this equation differs from that governing the spread of laminar shear flows only by the appearance of the turbulence term instead of the laminar (viscous) equivalent. Both describe a momentum flux across the flow and at high Reynolds numbers, defined by $\mathrm{Re} = U_s\delta/\nu$, say, the cross-stream momentum flux in a turbulent shear flow is much greater than in the equivalent laminar flow, i.e. $-\rho\partial\overline{u'v'}/\partial y \gg \mu\partial^2 U/\partial y^2$. Therefore, $\rho\overline{u'v'}/\delta \gg \mu U_s/\delta^2$ and hence $\overline{u'v'}/U_s^2 \gg 1/\mathrm{Re}$. This will always be true for a high enough Re, however small $\hat{u}/U_s$ is. We expect the momentum flux term to be significant, so this term must be of comparable magnitude to the terms on the left-hand side, i.e. $\hat{u}^2/\delta = \mathcal{O}(U_s^2/L))$. It follows from the scalings in equation (6.9) that

$$\frac{\delta}{L} = \mathcal{O}\left(\frac{\overline{u'v'}}{U_s^2}\right) < \mathcal{O}\left(\frac{\hat{u}^2}{U_s^2}\right), \tag{6.11}$$

from which it is clear that the smallness of δ/L is an inevitable consequence of the smallness of $\hat{u}/U_s$ – the fundamental requirement for the thin shear layer approximation. Once typical turbulence fluctuations become comparable with mean velocities (i.e. $\hat{u} = \mathcal{O}(U_s)$), which can occur in certain regions of many flows, the thin shear layer equations no longer apply. Note that equations (6.2) and (6.11) imply that $V/U_s \ll \hat{u}/U_s \ll 1$, which means that the spreading rate is small compared with $\hat{u}$. This is not very surprising, as the turbulence cannot spread faster than the spreading of the mean flow profile, since it extracts its energy from the gradient of the mean flow. Equation (6.9) is the *second-order* thin shear layer equation. Often, the difference in the Reynolds normal stresses is considerably smaller than either of them separately, so that the leading-order equations ((6.10) plus continuity) are actually more accurate than might be anticipated.

For completeness, we give below the axisymmetric forms of the thin shear layer equations, when there is no underlying axial pressure gradient. These can be deduced from the full governing equations using exactly the same scaling arguments.

$$\frac{\partial U}{\partial x} + \frac{1}{r}\frac{\partial(rV)}{\partial r} = 0, \qquad \text{(Continuity)} \tag{6.12}$$

$$U\frac{\partial U}{\partial x} + V\frac{\partial U}{\partial r} = -\frac{1}{r}\frac{\partial}{\partial r}\left(r\overline{u'v'}\right). \qquad \text{(Mean momentum)} \qquad (6.13)$$

The second of these is the first-order thin-shear-layer-approximated form of the momentum equation, corresponding to equation (6.10) for planar flows, but the Reynolds normal stress terms could easily be included (as in equation (6.9) for the planar case).

6.2.4 The thin shear layer version of the TKE equation

Similar order-of-magnitude arguments can be employed to simplify the turbulent kinetic energy (TKE) equation – the two-dimensional version of equation (2.28). We note that

1. the $\partial U/\partial y$ strain is much larger than the other components in the production term (I),
2. $\partial/\partial y \gg \partial/\partial x$ in the diffusive transport terms (III), and
3. away from walls, we can ignore the viscous diffusive transfer term, as done to obtain equation (2.28).

These approximations lead to a reduced TKE, which can be written as

$$\underbrace{U\frac{\partial k}{\partial x} + V\frac{\partial k}{\partial y}}_{I} = \underbrace{-\overline{u'v'}\frac{\partial U}{\partial y}}_{II} - \underbrace{\frac{\partial}{\partial y}\overline{v'\left(k + \frac{p'}{\rho}\right)}}_{IIIa+IIIb} - \underbrace{\epsilon}_{IV}\,, \qquad (6.14)$$

with the same numbering of terms as that used for the full equation (2.28). This is the thin-shear-layer-approximated form of the TKE equation. As in section 2.6, I represents the advection of TKE by the mean flow, II is the production of TKE, III is the transport of TKE via the turbulence itself and the pressure fluctuations, and IV is the dissipation of TKE.

6.2.5 The momentum integral equation

Finally, consider the integral with respect to y of the mean momentum equation (6.10). In the absence of any free stream pressure gradient and using the continuity equation, this is just

$$\rho\frac{\mathrm{d}}{\mathrm{d}x}\int_{-\infty}^{+\infty} U^2 dy = 0, \qquad (6.15)$$

which is commonly called *the momentum integral equation*. It assumes that there is no free stream turbulence and that V is zero outside the jet (so that both $\overline{u'v'} = 0$ and $V = 0$ at $y = \pm\infty$). It also reflects the fact that (in the absence of any external forces) the momentum flux must be invariant with downstream distance. Furthermore, its x-integral is constant and, for specific flows, this is set by the initial condition – e.g. the thrust of a jet. In the case of a wake, the appropriate version of equation (6.15) (not developed here) integrates to give

$$\rho \frac{\mathrm{d}}{\mathrm{d}x} \int_{-\infty}^{+\infty} U_\infty(U_\infty - U)\,dy = 0, \qquad (6.16)$$

where the x-integrated constant represents the drag of the object producing the wake. These two equations can, of course, be directly derived by an integration (across the whole flow) of the momentum flux through a thin slice dy at a given y.

6.2.6 Self-similarity

Self-similarity can be defined as occurring in any shear flow when cross-stream profiles of any quantity (mean velocity, Reynolds stresses, etc.) at different downstream locations can be made to collapse using simple scale factors which depend on only one variable. Thus, the appropriately normalised mean velocity profile, for example, is a function only of a normalised transverse position. The implication is that the flow reaches a kind of equilibrium – an asymptotic state – in which all its dynamical influences evolve together. This can only possibly happen, if it happens at all, sufficiently far downstream from the origin.

Following Pope's argument [38], we consider a quantity $Q(x, y)$ which is generally dependent on both x and y. We define characteristic scales for the dependent and independent variables of $Q_o(x)$ and $\delta(x)$, respectively, so that scaled variables can be defined by

$$\eta = \frac{y}{\delta} \quad \text{and} \quad \hat{f} = \frac{Q(x, y)}{Q_o}.$$

If, and only if, the scaled dependent variable ($\hat{f}$) is independent of x, i.e. there is some function $f(\eta)$ such that $\hat{f}(\eta, x) = f(\eta)$, then $Q(x, y)$ is said to be **self similar**. The characteristic scales must, of course, be chosen appropriately. In sections 6.3–6.5 we use this concept of self-similarity to transform the thin-shear-layer-approximated forms of the equations of motion into simpler ordinary differential equations governing a number of different free shear flows.

6.2.7 Further remarks

Most (although not all) of the simplifications discussed in the sections above also apply to boundary layers, as we will see in chapter 8. In fact, historically, the development of scaling arguments and the notion of self-similarity ('self-preservation', to use Townsend's expression [49]) probably began with Blasius in the earliest years of the last century, specifically in the context of laminar boundary layers. These ideas were later extended to turbulent boundary layers. For that reason, the phrase 'boundary layer approximation' is therefore often used (e.g. [38]). To avoid confusion and to emphasise that many of the assumptions rely on the flows being slender, in all the discussion above, we have used, instead, the phrase 'thin shear layer approximation'; the simplifications used originally for boundary layers turn out to be applicable to free shear flows and are precisely those we have introduced above.

6.3 The axisymmetric jet

We illustrate the implications of self-similarity arguments by considering, as a first example, the axisymmetric jet – probably one of the most studied of all shear flows remote from walls, for obvious practical reasons. Such jets issue from a round nozzle (of diameter d) and, depending on the flow within the nozzle, there may first be a short *potential core region* (typically about $6d$ long) bounded by an axisymmetric mixing layer, which grows in thickness so that it eventually merges together, marking the end of the core region. However, flow development continues for some distance, and only further downstream can the jet attain self-similarity. Solutions for the mean velocity and shear stress would then necessarily have the forms

$$\frac{U}{U_s(x)} = f(\eta) \quad \text{and} \quad \frac{-\overline{u'v'}}{R_s(x)} = g(\eta), \tag{6.17}$$

where $\eta = r/\delta$ and δ and R_s are functions of x (with R_s defined as the Reynolds stress scaling). This means that profiles of the mean velocity and the Reynolds stress at all downstream locations (sufficiently far from the jet) collapse, because all the streamwise variations are encapsulated by three functions, $\delta(x)$, $U_s(x)$, and $R_s(x)$. A proper determination of these functions must allow *all* the terms in the equations of motion (and, indeed, in other dynamical equations like the TKE) to maintain the same relative balance between them at all x. Such a determination is not always possible but, when it is, the flow is in equilibrium and is a fully self-similar flow.

6.3.1 A dimensional analysis approach

The jet provides an initial source of momentum flux, ρM_o, say (the jet's thrust). The integral (with respect to r) of the momentum equation (6.16) is,

$$\frac{\mathrm{d}}{\mathrm{d}x} \int_0^{+\infty} 2\pi U^2 r\, dr = 0, \tag{6.18}$$

where we have used the continuity equation and the fact that both V and $\overline{u'v'}$ are zero external to the jet. Integrating this with respect to x yields a constant, which must equal M_o, and inserting the self-similar forms for U (equation (6.17)) allows the resulting equation to be expressed by

$$U_s^2 \delta^2 \int_0^{+\infty} 2\pi \eta f^2\, d\eta = M_o. \tag{6.19}$$

Note that the integral is just a (non-dimensional) number whose value depends only on the shape of the f profile. (This will clearly be true for all integrals that contain only functions of η and f in the integrand.) Real jets provide other initial features in addition to M_o, e.g. a source of mass. If we ignore such possibilities for the moment by assuming that the jet is simply a point source of momentum, then U_s, δ, and R_s must be dependent on x and on the momentum flux ρM_o, but *only* on these two quantities – i.e. $U_s = U_s(x, \rho M_o)$, $\delta = \delta(x, \rho M_o)$, and $R_s = R_s(x, \rho M_o)$. It follows immediately on dimensional grounds alone that

$$\delta \sim x, \tag{6.20}$$

$$U_s \sim M_o^{1/2}/x, \quad \text{and} \tag{6.21}$$

$$R_s \sim M_o/x^2 \sim U_s^2. \tag{6.22}$$

Note that δU_s is constant and that this also follows directly from equation (6.19).

It is clear that if any kind of self-similar solution is possible, then the jet must spread linearly and its centreline velocity must decay according to x^{-1}. Since there are no other parameters, the solution must, in fact, be universal – i.e. independent of all other details of the initial conditions, with no other possible solutions. It is also clear that the velocity scale used for the Reynolds stress must be the same as that used for the velocity, i.e. U_s – equation (6.22). All of this is the classical dimensionality approach used in some texts, e.g. [31].

6.3.2 The more common approach

If, instead of using dimensional analysis, one assumes *a priori* that the self-similar solution depends only on a single length scale (δ) and a single velocity scale, then the R_s appearing in equation (6.17) for the self-similar form of the shear stress must be proportional to U_s^2 (since it is U_s that scales the mean velocity). Substituting the self-similar forms for U and $\overline{u'v'}$ into the mean momentum equation (6.13) and then integrating with respect to η transforms the equation to the following ODE:

$$\left[U_s\frac{\mathrm{d}U_s}{\mathrm{d}x}\right]f^2 - \left(\left[U_s\frac{\mathrm{d}U_s}{\mathrm{d}x}\right] + 2\left[\frac{U_s^2}{\delta}\frac{\mathrm{d}\delta}{\mathrm{d}x}\right]\right)\frac{f'}{\eta}\int_o^\eta f\eta\mathrm{d}\eta = -\left[\frac{R_s}{\delta}\right]\frac{(\eta g)'}{\eta}, \tag{6.23}$$

in which the dashed quantities denote differentiation with respect to η. The continuity equation, the fact that the flow is symmetric about $\eta = 0$, and an assumption that there is no turbulence in the free stream, are all required in deriving this equation. It can be conveniently multiplied by δ/U_s^2 to give

$$\left[\frac{\delta}{U_s}\frac{\mathrm{d}U_s}{\mathrm{d}x}\right]f^2 - \left(\left[\frac{\delta}{U_s}\frac{\mathrm{d}U_s}{\mathrm{d}x}\right] + 2\left[\frac{\mathrm{d}\delta}{\mathrm{d}x}\right]\right)\frac{f'}{\eta}\int_o^\eta f\eta\mathrm{d}\eta = -\left[\frac{R_s}{U_s^2}\right]\frac{(\eta g)'}{\eta}. \tag{6.24}$$

Note that no assumptions about the source conditions have been made in generating these forms of the mean momentum equations. However, we know from the momentum integral equation (6.19) that $U_s \sim \delta^{-1}$ (so that the factor multiplying the integral is constant). Substituting this into the above equation reduces it to

$$\left[\frac{\mathrm{d}\delta}{\mathrm{d}x}\right]\left(f^2 + \frac{f'}{\eta}\int_o^\eta f\eta\mathrm{d}\eta\right) = \left[\frac{R_s}{U_s^2}\right]\frac{(\eta g_{12})'}{\eta}. \tag{6.25}$$

For a self-similar solution, the two terms in square brackets must have the same dependence on x, so that

$$\frac{U_s^2}{R_s}\frac{d\delta}{dx} = \text{constant}, \tag{6.26}$$

but we emphasise that, in principle, the two terms do not separately have to be constant. (This is not always made clear in the classical texts.) It should be obvious that if such a solution exists, it must be the asymptotic one, for if one of the bracketed terms has an x-dependence which differs from that of the other term, then it will eventually either dominate or decay away, leaving, in this case, an impossibly unbalanced equation or, in other free shear flow cases, a different equation set.

The self-similar approach outlined above is discussed in many texts, e.g. [46, 49] and readers interested in seeing more complete analyses could consult those classic texts where this is done. Note, however, that with the initial assumption mentioned above, that $R_s \sim U_s^2$, the second of the bracketed terms above must be constant and thus so is $d\delta/dx$, leading again to the linear growth of the flow width, $\delta \sim x$, and the $U_s \sim x^{-1}$ decay of the velocity scale. Incidentally, if we define a local Reynolds number, i.e. $\text{Re} = U_s\delta/\nu$, then it follows that Re remains constant at all distances downstream in the self-similar region (unlike the case for some other free shear flows). If the initial assumption that there is only one velocity scale (i.e. $R_s \sim U_s^2$) were relaxed by retaining R_s, then the second bracketed term in the above equation would remain and all we could say about a possible self-similar solution is that both bracketed terms must vary with x in the same way. However, both laboratory and numerical experiments over many decades have shown that typical axisymmetric jets do indeed have a self-similar form governed by a linear growth and a x^{-1} decay in the centreline velocity. This implies that the bracketed terms in equation (6.25) are in fact both constant (as suggested by the first approach to the problem, using dimensional analysis) and *not* functions of x. It turns out that under a certain assumption about the turbulence kinetic energy dissipation, explored more fully in section 6.6.3, consideration of the TKE with appropriate scaling arguments for each term leads to a further constraint on the similarity variables, forcing $R_s \sim U_s^2$, even if that were not imposed as an initial assumption; linear growth ($\delta \sim x$) thus really *does* represent the genuine self-similar solution for this flow.

Strictly speaking, we expect the self-similar solution to apply only some way downstream from the jet origin, so we should replace x in all the above by $x - x_o$, where x_o is a *virtual origin*. This accounts for the fact that there is always likely to be an upstream portion of the flow which is not in a self-similar equilibrium state. For this axisymmetric jet, as noted earlier, there is an initial potential core region, and self-similarity only emerges further downstream – starting at a point defined by x_o. Finally, it should also be noted that using the second-order form of the momentum equation does not affect the results of this scaling analysis.

6.3.3 An analytical solution

Obtaining a full theoretical solution for the jet flow – i.e. determining, at least, the shape of the mean velocity profile – requires solving equation (6.25), an ODE. Although, in principle, this should be much easier than solving the PDE (equation (6.16)), it naturally requires a strategy with which to overcome the inevitable closure problem, i.e., in this case, to link the shear stress (represented by $g(\eta)$) to the velocity field ($f(\eta)$). One of the simplest closure models, linking the turbulent shear stress *directly* to the mean velocity gradient, was introduced in section 2.4. This eddy viscosity model (for the round jet flow) can be written as

$$\overline{u'v'} = -\nu_T \frac{\partial U}{\partial r} \tag{6.27}$$

(to leading order). Assuming that ν_T is constant across the wake, writing it nondimensionally as $\hat{\nu}_T = \nu_T/(\delta U_s)$, and using the similarity scaling for $\overline{u'v'}$ and U – equations (6.17) – it follows that

$$g = -\hat{\nu}_T f'. \tag{6.28}$$

Note that although $1/\hat{\nu}_T$ may look like a Reynolds number, it is not a genuine Reynolds number characterising the turbulence, as it is scaled using the mean flow parameters. It is perhaps better to consider it simply as a flow constant, as Townsend did [49]; we will call it R_T. The governing equation (6.25) becomes

$$\beta R_T \left(\eta f^2 + f' \int_o^\eta f\eta d\eta \right) + \eta f'' + f' = 0, \tag{6.29}$$

with equation (6.26) reducing to the spreading rate $\beta = \mathrm{d}\delta/\mathrm{d}x$. This can be shown to have the solution given by

$$f = (1 + a\eta^2)^{-2} \quad \text{with} \quad a = \beta R_T/8. \tag{6.30}$$

If δ is defined so that $f = \frac{1}{2}$ at $\eta = 1$, it follows that $a = \sqrt{2} - 1$.

6.3.4 Some typical data

Some typical mean velocity profiles are shown in figure 6.2(a) for a jet of diameter d and a Reynolds number $\mathrm{Re} = U_{jet}d/\nu \approx 23000$. Note from figure 6.2(b) that by $x/d = 100$, the centreline velocity has dropped to about 6% of the jet's exit velocity (i.e. only about $2\,\mathrm{m\,s^{-1}}$ in this experiment), so the velocities at the edges of the jet are extremely small that far downstream. In addition, at any x/d, the turbulence intensity (u'_{rms}/U), which even on the centreline is around 25%, becomes of the order of unity towards the jet edges. The cross-stream mean velocity V also eventually becomes comparable to U. These features make measurements using ordinary hot-wire anemometry (HWA) somewhat problematic, particularly for the higher-order moments such as the Reynolds stresses. The data in figure 6.2 were obtained using a combination of specialist HWA, classical Pitot tubes, and particle image velocimetry (PIV). Similar data, often plotted (equivalently) against $\eta = r/\delta$

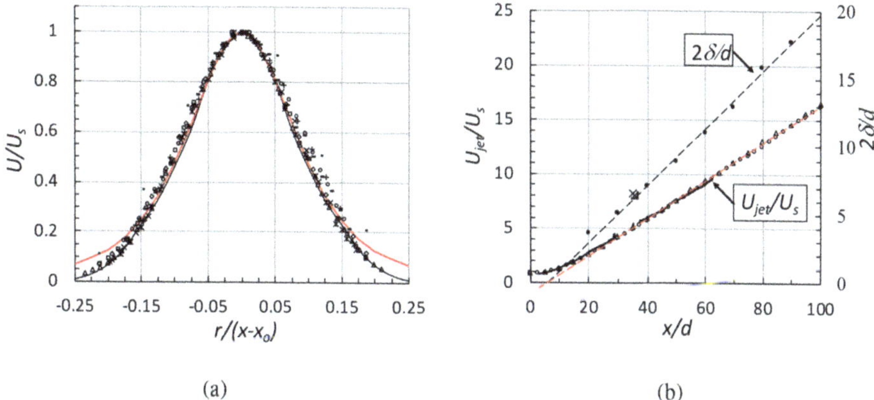

<table>
<tr><td align="center">(a)</td><td align="center">(b)</td></tr>
</table>

Figure 6.2. (a): Mean velocity profiles for a jet with nozzle a diameter of d and an exit velocity of U_{jet} over a downstream distance range of $20 \leqslant x/d \leqslant 100$, each normalised by its centreline value, U_s, and using the x_o obtained from the data in (b). The solid black line represents LDA data from Hussein *et al* [25] and the red line represents the analytic solution obtained using equation (6.30). (b): x-development of the jet width, where δ is defined by the r-location at which the velocity is $\frac{1}{2}U_s$, and U_{jet}/U_s. All symbols, and the black line in (b), are data obtained in the authors' laboratory (and see [52]) using a combination of Pitot tubes, single and crossed HWA, and PIV.

rather than $r/(x - x_o)$, have been obtained many times, e.g. by Hussein *et al* [25], whose mean velocity profile, obtained using laser Doppler anemometry (LDA), is included in figure 6.2(a). Incidentally, they showed that Reynolds stresses obtained using ordinary HWA and a flying hot-wire technique designed to minimise errors yielded significantly different Reynolds stresses, underlying the importance of using modern instrumentation for this flow. Nonetheless, even much older data were adequate to demonstrate the self-similarity of the mean velocity field, e.g. [55]. Note also from figure 6.2(b) that both the velocity decay and the growth of the jet width suggest a virtual origin of, in this case, around $7d$ and that the subsequent self-similar behaviour does not begin before at least $x/d = 30$. These numbers are typical of laboratory jets. The outliers in figure 6.2(a) are the $x/d = 20$ data, but the remaining profiles confirm the self-similar nature of the flow further downstream.

Recall the analytic solution given by equation (6.30). The particular jet with the behaviour shown in figure 6.2 yielded $\beta \approx 0.11$ and (noting that $r/(x - x_o) \approx \eta/\beta$) the equation is plotted in figure 6.2(a) for $R_T = 40$. It is evident that a constant eddy viscosity model provides a not unreasonable solution for the mean velocity, although it overestimates the velocity at the edges of the jet – a result of an overestimate of the eddy viscosity there (and thus an underestimate of the mean flow gradient).

6.3.5 Universality

We emphasise that the universality of this self-similar solution – i.e. the fact that the value of $d\delta/dx$ is the same for *all* axisymmetric jets (and, more fully, that the β represented by equation (6.26) is fixed) – relies crucially on the fundamental assumption that the initial conditions have no influence, other than setting the momentum source M_o and affecting the location of the virtual origin, x_o. However, even quite early experiments indicated that different jets can actually have very different (linear) spreading rates. For example, a factor-of-two variation in $d\delta/dx$ was documented for different round jets in the 1980s [21]. (Some jet flows have even been shown to expand significantly faster than linearly [41].) There is now plenty of evidence that the initial conditions *do* have a significant and very long-lasting effect on the exact form of self-similar jets, see [57] for a more recent example, and it has also been argued, using a fuller similarity analysis, that the Reynolds stresses, for example, should be scaled in line with equation (6.26) – i.e. using $U_s^2 d\delta/dx$ [18, 25]. Both laboratory experiments and direct numerical simulation (DNS) have shown that this yields a significantly better collapse (e.g. for the scaled $\overline{u'v'}$) than if just U_s^2 is used; see, for example, [8].

As we will see in subsequent sections, this apparent lack of universality has, in recent decades, become even more evident for other canonical free shear flows such as planar mixing layers and, in particular, axisymmetric wakes, so we will make further remarks in due course. However, it is worth saying immediately that, despite the longstanding assumption that initial features eventually die out, at the time of writing there is no conclusive evidence that they always do. In the case of axisymmetric wakes, however, there *is* some evidence that different initial conditions are forgotten, so that true universality eventually appears, but only after extraordinarily large downstream distances [39]. We consider such wakes in the next section, following this with some discussion of other free shear flows. We will also revisit the round jet case to consider its turbulence kinetic energy balance (shown in figure 6.5(a)) and how it differs from that of the round wake.

6.4 Axisymmetric wakes

6.4.1 The common approach

For wake flows, in addition to the velocity scale U_s, denoting the maximum variation in velocity across the whole flow, there is an important second velocity scale, U_∞ – the free stream velocity, see figure 6.1. Equation (6.17) must therefore be rewritten as

$$\frac{U_\infty - U}{U_s} = f(\eta) \quad \text{and} \quad \frac{-\overline{u'v'}}{R_s(x)} = g(\eta) \tag{6.31}$$

with $\eta = r/\delta$ as before. In this case, a self-similar solution is possible only in the far wake, where $U_s/U_\infty \ll 1$. The first term in the momentum equation (6.13) can then be written as $U_\infty \partial U/\partial x$ and the second term is much smaller than this and can thus be ignored. We assume again that there is no axial pressure gradient in the free stream (so $U_\infty \neq f(x)$), nor any free stream turbulence. Following the same common approach used for the axisymmetric jet (i.e. substituting the self-similar forms for U and $\overline{u'v'}$ into the simpler mean momentum equation), the governing equation becomes

$$-\left[\frac{\delta}{U_s}\frac{dU_s}{dx}\right]f^2 + \left[\frac{d\delta}{dx}\right]\eta f' = \left[\frac{R_s}{U_\infty U_s}\right]\frac{(\eta g)'}{\eta}. \tag{6.32}$$

In this far-wake situation, the momentum integral equation becomes (using the self-similar variables)

$$2\pi\rho\, U_\infty U_s \delta^2 \int_o^{+\infty} \eta f\, d\eta = D \tag{6.33}$$

where D is the drag of the body creating the wake. It is common to define a momentum thickness, θ, for the wake, as follows:

$$\theta^2 = \frac{1}{U_\infty^2} \int_o^\infty U\left(U_\infty - U\right) r\, dr,$$

which, for the far-wake situation ($U_s \ll U_\infty$), can be written using the self-similar variables as

$$\theta^2 = \frac{U_s}{U_\infty}\delta^2 \int_o^\infty \eta f\, d\eta \tag{6.34}$$

so that the drag is just $D = 2\pi\rho\, U_\infty^2\theta^2$ (and note that θ is constant). Clearly, $U_s\delta^2$ must be constant and the governing equation can be written

$$2\eta f + \eta^2 f' = \left[\frac{R_s}{U_\infty U_s}\bigg/\frac{d\delta}{dx}\right](\eta g)'. \tag{6.35}$$

As for the axisymmetric jet case, if we were to assume just one relevant velocity scale, so that $R_s \sim U_s^2$ and noting that $U_s \sim \delta^{-2}$, it follows from the constant bracketed term that $\delta^2 d\delta/dx$ is constant, so that the variations of δ and U_s with x must satisfy

$$\delta \sim (x - x_o)^{1/3} \quad \text{and} \quad U_s \sim (x - x_o)^{-2/3}. \tag{6.36}$$

However, a consideration of the scaled TKE equation, with the (high Reynolds number) assumption that the dissipation term scales according to U_s^3/δ, shows that $R_s = U_s^2$ is, in fact, a requirement for self-similarity (as in the jet case), so the above solution is exact – provided, of course, that $U_s \ll U_\infty$. Using $R_s = U_s^2$, the inverse of the bracketed term in equation (6.35), which can be called the spreading parameter β, is given by

$$\beta = \frac{U_\infty}{U_s}\frac{d\delta}{dx}, \tag{6.37}$$

and a single integration of this equation then leads to

$$\beta\eta f - g = 0. \tag{6.38}$$

Note that the local Reynolds number within the wake, $\mathrm{Re} = U_s\delta/\nu$, falls like $x^{-1/3}$, so one might expect that viscous effects will begin to appear and eventually dominate far enough downstream.

6.4.2 An analytic solution

To solve the governing equation (6.38), we can again use the simple eddy viscosity closure model. Scaling the eddy viscosity using δU_s, as before, so that $\hat{\nu}_T = \nu_T/(\delta U_s) = 1/R_T$, and applying the eddy viscosity relation (equation (6.27)) to give $g = -\hat{\nu}_T f'$, equation (6.38) becomes simply

$$\beta R_T \eta f + f' = 0, \tag{6.39}$$

which has the solution

$$f = e^{-a\eta^2}, \tag{6.40}$$

with $a = \beta R_T/2$. Defining δ so that $\eta = 1$ at the half-velocity location (i.e. where $f = \frac{1}{2}$) gives $a = \ln 2$. We now consider some of the extant data for the mean velocity, turbulence stresses, and the TKE equation balance.

6.4.3 Implications obtained from data

There have been numerous studies of axisymmetric wakes and, for sufficiently high Re, they generally satisfy the above predictions for δ and U_s and have the expected self-similar form. Figure 6.3 shows examples of a set of mean flow profiles obtained many decades ago, in the wake of a long body of revolution [36]. The analytic solution, like that for the jet, is a reasonable match for the data, but is increasingly less accurate towards the edge of the wake because of the overprediction of the eddy viscosity there (see section 6.3.4). What has become very clear, however, is that the effect of initial conditions is extremely long-lasting, so that the value of the spreading parameter β varies widely from experiment to experiment. For example, it was

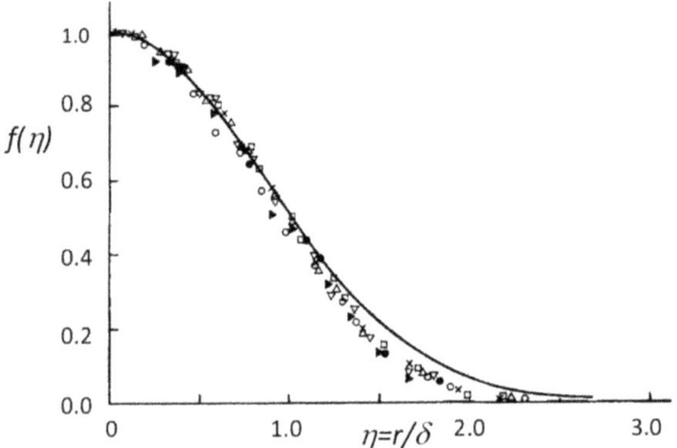

Figure 6.3. Mean velocity profiles over one half of an axisymmetric wake; the different symbols denote downstream distances in the range $8.3 < x/d < 66$ (d is the diameter of the wake-generating body)[1]. The black line is the analytic (constant eddy viscosity) solution, $f = \exp(-\eta^2 \ln 2)$.

---

[1] Reproduced with permission from [36], copyright Cambridge University Press.

discovered some decades ago [6] that even wakes from two different axisymmetric bodies that had exactly the same drag could yield very different values of β in the self-similar region. β values for a porous disc and a sphere were 0.21 and 0.27, respectively, but these are not universal. Pope's [38] collection of six published values of β from experiments with different wake generators (including a sphere) range from 0.064 to 0.8, despite the fact that all individual cases show good agreement with the self-similar behaviour (i.e. $\delta \sim x^{1/3}$, $U_s \sim x^{-2/3}$). These variations are relatively much larger than the range of spreading rates seen in self-similar round jets.

The issue of universality, introduced in section 6.3.5, thus seems even more pertinent for the axisymmetric wake. Townsend's universal self-similarity hypothesis is based on the notion that 'by the time the turbulence reaches the far field, it will have been subjected to a universal 'eddy scrambling' process that causes it to 'forget' the near-field characteristics (such as geometry-dependent vortical structures), set by the initial conditions' [39]. If this hypothesis is to be shown to be correct, it might be necessary to make measurements a very long way downstream. But this is very challenging; the velocity deficit, U_s is small compared with U_∞ and it decays quite slowly. Furthermore, the turbulence velocities, although small compared to those in the jet, are similar in magnitude to U_s. Since, to ensure a sufficiently high Re, one cannot use too small a wake-generating body in most laboratory facilities, it is often impossible to explore beyond a few hundred body diameters downstream at most. Furthermore, since the *local* Re falls with x, it could be that the viscous-dominated flow appears before the initial condition effects die out. (Note that the laminar equivalent to equation (6.32), with an appropriate scaling analysis, shows that the viscous wake parameters, δ and U_s vary similarly to $x^{1/2}$ and x^{-1}, respectively.)

Universality remains a topic of current research. It is worth noting, in the specific context of the axisymmetric wake, that DNS studies have suggested that two wakes with very different initial eddy structures are self-similar but different in ways that reflect the different initial conditions, as fuller analysis suggests [19, 27]. Far enough downstream, however, they reverted to the *same*, likely universal, state, and there was no way that the different initial eddy structures could be deduced from the characteristics of that state [39]. This regime only began once the deficit velocity was a mere 1% of its initial value. Note too that even in this far-wake region, the viscous term in the mean momentum equation remained very small compared to the Reynolds stress term – i.e. $\nu/\nu_T \ll 1$. These were time-dependent wakes (easier for DNS to address than would be the more common spatial cases) and the start of the universal region corresponded to nearly 100 m downstream from a 0.025 m-diameter generating body in an equivalent spatial case! This result implies, first, that universal self-similarity is not likely to be seen in practice (a fact attested to by the majority of the literature) and, secondly and importantly, the classical Townsend hypothesis is essentially correct.

A further feature of many experiments, which pointed to self-similar but nonuniversal behaviour, is the nature of the Reynolds stress profiles. An example of such profiles is given in figure 6.4, from [42]. In this experiment, the maximum velocity deficit at $x/d = 65$ (the first of the profile locations in the figure) was less

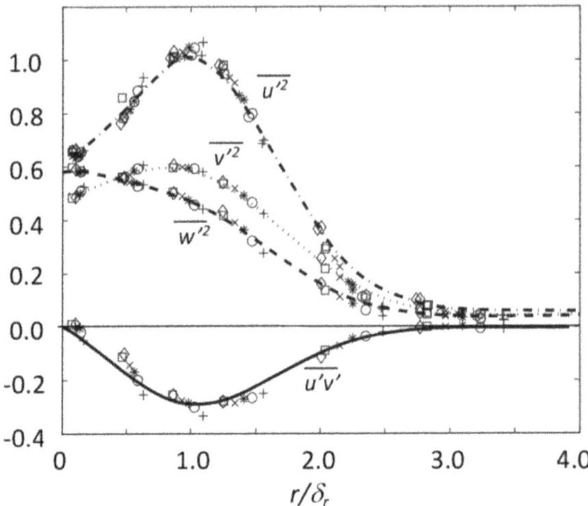

Figure 6.4. Reynolds stresses at five locations behind a solid disc of diameter d, in the downstream range $65 < x/d < 115$, all normalised using the maximum value of the axial stress, $\overline{u'^2}_m$. From the author's laboratory[2]. Here, $\overline{v'^2}$ and $\overline{w'^2}$ are the radial and circumferential components, respectively. The smooth lines through the data are for clarity and δ_r is defined as the location at which the axial stress reaches a maximum.

than 3% of U_∞, and to minimise experimental scatter, the data for this figure were normalised by the maximum value of the axial stress, $\overline{u'^2}_m$, say, with the wake width δ_r defined here as the twice the distance between the axis and the location of maximum axial stress. Note that $\overline{u'^2}_m \delta_r^2$ was found to be closely constant, as required for self-similarity, and the wake width and velocity deficit closely satisfied the expected behaviour, equation (6.36).

An immediate conclusion from the stress profiles is that the turbulence is far from being isotropic. The streamwise normal stress, for example, is, at some points, about twice the spanwise stress. Such differences between all the normal stresses are common to all free shear flows. The profile shapes are similar to those of other data available in the literature (and, incidentally, not too dissimilar to those for the axisymmetric jet), but they are not identical. For example, the ratio of the maximum to the centreline axial stress was about 1.18 for a porous disc [10], 1.32 for a solid disc [27], 1.62 and 1.46 for a porous disc and a sphere, respectively [6], and 1.62 for the data in figure 6.4 for a solid disc. These are substantial differences and they are a result not least of the different initial conditions, pointing again to clear non-universality within the downstream ranges covered by the experiments. None of these authors were able to reach anywhere near the distance that has been suggested as being necessary for all initial conditions to be forgotten [39]. The shear stress data in figure 6.4, along with the corresponding mean flow profile, gave $R_T = 14.5$ and thus $\beta = 0.095$, which are both fairly typical of many experiments although, as noted above, by no means all.

---

[2] Reprinted by permission from [42], copyright (2012) with permission of Springer.

6.4.4 The TKE balance

It is instructive to compare the turbulence kinetic energy balance for the wake with that for the jet. These are shown in figure 6.5. Note that, in both flows, the energy production near the centreline is close to zero, because both the velocity gradient and the shear stress are zero at $r = 0$. In the jet, the production is a little further from zero because of the relatively larger contribution from the normal stress terms, which are nonzero at $r = 0$ (and the turbulence intensities are much higher in the jet than in the wake). The maximum production occurs, for both flows, close to where the turbulence shear stress and the mean velocity gradient reach their maxima. Production never reaches much more than about one half of the maximum transport in the wake, but in the jet their maximum values are very similar. Energy reaches the central region (around $r = 0$) from further out by both advection (by the mean flow) and turbulent transport, where these are balanced largely by dissipation. The jet data were obtained in the 1990s by LDA, a significantly more accurate technique than standard hot-wire methods in regions of relatively high turbulence intensity, as mentioned previously. Note, however, that the TKE balance for the jet in figure 6.5(b) is not too dissimilar to an early one measured in the 1960s using HWA [55]. Finally, note that although in homogeneous shear flows there is an exact balance between production and dissipation (see section 5.2), this is not even closely true anywhere in either the wake or the jet, which emphasises the influence of the

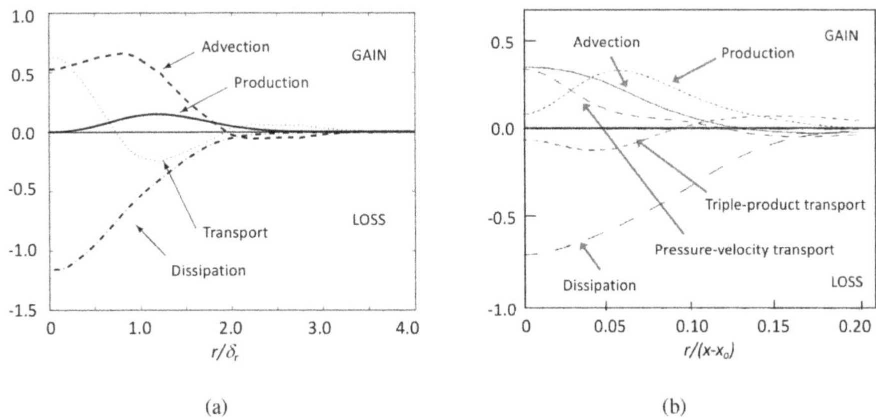

Figure 6.5. TKE balances (equation (6.12)), but with the second-order normal stress contribution included in the production term. (a) The axisymmetric wake, from the authors' laboratory[3]. The advection and production were deduced using the smoothed mean velocity and stress profiles (figure 6.4), the dissipation term was estimated using spectral data and Taylor's hypothesis, and the total of the transport terms was obtained by difference, ensuring that the sum of all terms is zero at each r. Terms are normalised using $\hat{u}^3$ and δ_r. (b) The axisymmetric jet[4], with terms normalised by U_s^3 and $(x - x_o)$. Transport terms are plotted separately, with the pressure–velocity contribution deduced by difference, ensuring that the sum of all terms is zero at each r.

---

[3] Reprinted by permission from [42], copyright (2012) with permission of Springer.
[4] Reproduced with permission from [25], copyright Cambridge University Press.

inhomogeneity in the cross-stream direction in ensuring obvious nonzero contributions from terms which are identically zero in the homogeneous (constant shear) case.

6.5 Planar flows

6.5.1 Jets

The scaling of the mean velocity and Reynolds stresses is exactly the same in planar as in axisymmetric flows, so, for the jet, equation (6.17) applies (with $\eta = y/\delta$). The (integrated) momentum equation is

$$U_s^2 \delta \int_{-\infty}^{+\infty} f^2 \, d\eta = M_o, \tag{6.41}$$

where, again, ρM_o is (loosely) the jet thrust. Strictly speaking, M_o is the first-order momentum flux in the self-similar region; its precise relation to the initial jet thrust may depend somewhat on the jet's initial configuration; see [24], for example. In any case, it follows from equation (6.41) that $U_s^2 \delta$ is constant. We could repeat the dimensional argument (used for the axisymmetric jet in section 6.3.1) to deduce that $\delta \sim x$ and $U_s \sim x^{-1/2}$. Alternatively, we can follow the standard approach (as shown in section 6.3.2), substituting the self-similar forms for U and $\overline{u'v'}$ into the momentum equation (6.10) (with zero axial pressure gradient) and using $U_s^2 \delta = $ const., to obtain

$$\frac{1}{2}\left[\frac{d\delta}{dx}\right]\left(f^2 + f'\int_{-\infty}^{+\infty} f d\eta\right) = g'. \tag{6.42}$$

For self-similarity, this clearly requires $\delta \sim x$ and thus $U_s \sim x^{-1/2}$. We again denote the spreading parameter $d\delta/dx$ by β. Note that it is implicit in this governing equation that the scaling velocity used for the shear stress is again U_s, but this can once more be substantiated by considering the TKE equation. Note also that in this flow, the local Reynolds number (which, as before, scales with $U_s\delta$) increases with downstream distance, like $x^{1/2}$, so that even very far downstream, viscous effects are not important. Like the round jet, this flow is exactly self-similar in that all the terms in the TKE equation, as well as the mean momentum equation, scale properly using just one length and one velocity scale.

Assuming a constant eddy viscosity (ν_T) across the jet, g' in equation (6.42) can again be replaced by $-(1/R_T)f''$ (with $R_T = 1/\hat{\nu}_T$) and the equation then has the solution

$$f = \mathrm{sech}^2(a\eta), \tag{6.43}$$

where $a = \frac{1}{2}\ln(1 + \sqrt{2})^2 = \sqrt{\beta R_T/4}$. (Detailed steps leading to the solution can be found in [38].) For a typical spreading rate of $\beta = 0.1$, this yields $R_T \approx 31$. The mean flow profiles are not too dissimilar to those of axisymmetric wakes and the analytic solution given above fits well except, again, towards the outer edge of the jet. Although most studies have suggested that $\beta \approx 0.1$ and that $U_s^{-2} \sim x$, there is considerable variation in the slope of U_s^{-2} versus x. This can vary by at least a factor

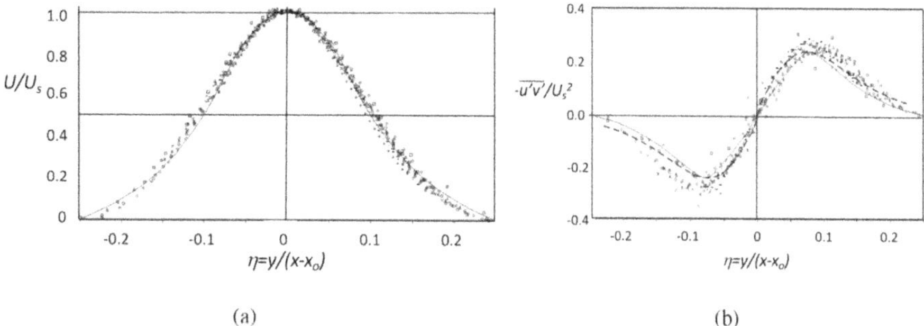

Figure 6.6. Profiles of mean velocity (a) and turbulence shear stress (b) measured using PIV in a planar jet at numerous locations in the range $15 \leqslant x/d \leqslant 105$ (from $x/d = 25$ for shear stress). Unpublished data from the authors' laboratory [51]. For this experiment, $d\delta/dx$ was 0.105. The open squares in (b) are for $x/d = 25$, just before the self-similar region began for the shear stress. The solid lines are the profile from [22] and the dashed line in (b) was calculated from the mean flow profile in (a).

of two (for example, see [4]), so that $U_s^2\delta/U_{jet}^2 d$ (with d the jet nozzle width) is not a universal constant but depends on the initial conditions.

Figure 6.6 shows a set of profiles of the mean velocity and shear stress for a planar jet with an exit Re of about 18000; these were obtained using particle image velocimetry (PIV). The degree of scatter is fairly typical for such measurements, but the data collapse satisfactorily over a wide range of x and the shear stress profile agrees reasonably well with the profile computed from the momentum equation, given the mean flow profile. (If each shear stress profile had the shear stress normalised by its maximum value, the scatter in figure 6.6(b) would be significantly reduced; a more recent example of data presented in that way can be found in [9].) The peak stress occurs, as usual, close to the point where the mean velocity gradient (in figure 6.6(a)) is at its greatest (just inboard of $y/\delta = 1$, or $\eta \approx 0.1$), so this is the region in which the production of turbulence energy is at its greatest, just as for axisymmetric jets. Symmetry ensures that the shear stress is zero on the jet centreline ($\eta = 0$). The normalised rms of the streamwise stress, u'_{rms}/U_s, is around 25% there and it rises to nearer 30% where the shear stress is at a maximum. U/U_s is about one half at the latter location, so the *local* turbulence intensity there, u'_{rms}/U, is more than 50% and increasingly higher further out, which emphasises the difficulties in obtaining accurate data with classical HWA in jet flows.

Some of early authors who studied planar jets sought to minimise these errors by using jets that emerged into a nonzero but parallel free stream (e.g. [16].) This also reduces the possible influence of external draughts in the laboratory, which are known to be a possible source of errors, especially in the far field, as are particular features of the rig geometry at the jet exit [25]. If the external velocity is small compared with the excess velocity on the centreline, then the behaviour is similar to that of a jet issuing into a zero velocity field, but if it is large, the flow becomes more like a planar wake. In both cases, the flow can only be approximately self-similar. On the other hand, if the external velocity is neither very large nor very small compared to the excess velocity, then the mean flow equation does not allow self-similarity [49].

Finally, we mention that alternative assumptions about how the dissipation term in the TKE equation should be scaled lead to rather different conclusions about, amongst other things, the growth rate. As a precursor to section 6.6.3, we mention the work of, for example, Cafiero *et al* [9], who, making such an alternative assumption, showed that it led to a spreading rate governed by $\beta \sim (x - x_o)^{2/3}$ instead of the classical $\beta \sim (x - x_o)$. Their experimental data were not inconsistent with that, although they also showed a reasonable fit (over a somewhat limited range of x) to the classical scaling result. We discuss this issue more fully in section 6.6.3.

6.5.2 Wakes

Like the axisymmetric wakes discussed in section 6.4.1, self-similarity can only occur asymptotically in the far wake, as $U_s/U_\infty \to 0$. Using the scaling given by equation (6.31), the momentum integral equation (6.16) can be expressed as

$$\rho U_\infty U_s \delta \int_{-\infty}^{+\infty} f\,\mathrm{d}\eta = D \qquad (6.44)$$

where D is the object's drag per unit spanwise length (and is equal to the momentum deficit flow rate [5]). This immediately implies that $U_s\delta$ is independent of x. We recognise again that for $U_s/U_\infty \ll 1$, the first term of the general mean momentum equation (6.10) becomes $U_\infty \partial U/\partial x$ and the second term is negligible. For a zero external pressure gradient, the momentum equation in self-similarity form becomes

$$\beta(\eta f)' = g',$$

which integrates to give

$$\beta \eta f - g = 0. \qquad (6.45)$$

Interestingly, this is exactly the governing equation for axisymmetric wakes, equation (6.38). The spreading parameter β is again given by equation (6.37), i.e. $(U_\infty/U_s)\mathrm{d}\delta/\mathrm{d}x$, which must be constant for self-similarity to apply. Because of the constancy of $U_s\delta$, this implies that δ and U_s must vary according to $x^{1/2}$ and $x^{-1/2}$, respectively.

With the same constant eddy viscosity assumption as that used in the earlier cases, the solution to equation (6.45) is again

$$f = e^{-a\eta^2} \quad \text{with} \quad a = \beta R_T/2. \qquad (6.46)$$

The flow constant is $R_T = 2 \ln 2/\beta$ using the same definition for δ (so that $f = \frac{1}{2}$ at $\eta = 1$), $a = \ln 2$. With a typical value of β found in experiments (0.11) $R_T = 12.5$. Just as for the three flows discussed earlier, this solution provides a good match for the measured mean flow profiles except near the wake's edges, where it overestimates the velocity. However, similarly to the case for axisymmetric wakes, there is a considerable variation of at least 25% in the measured values of β, even when the wake-generating bodies (e.g. a cylinder, flat plate, porous plate, or aerofoil) have equal drag and yield the same Reynolds number based on U_∞ and d, the body width (e.g. [56]). This kind of variation, although rather smaller than what has been found for the axisymmetric wake, is reminiscent of what was noted in section 6.4.3 for the

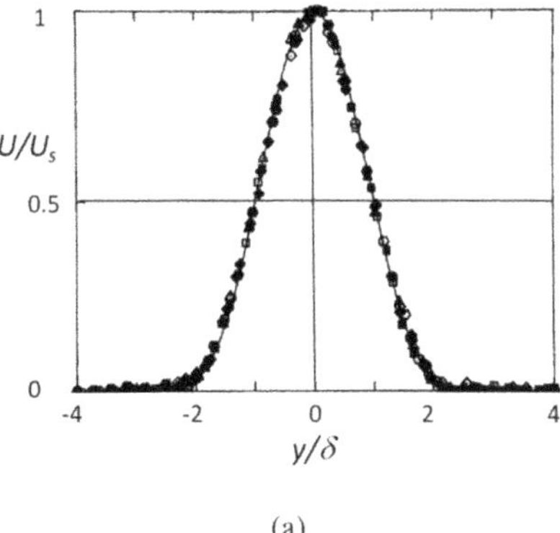
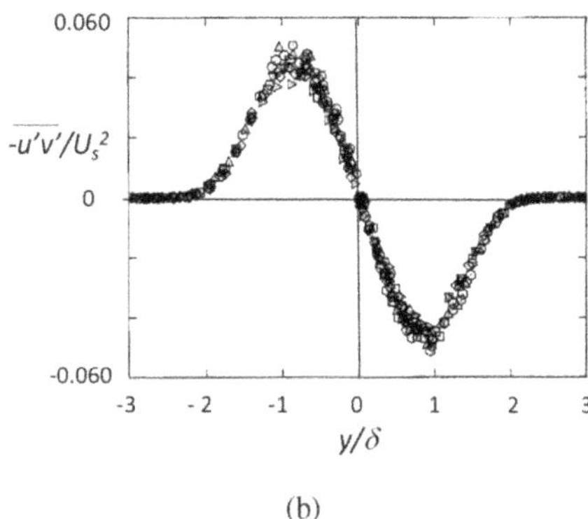

(a)   (b)

Figure 6.7. Plane wake data for a 30% solidity flat plate: (a) mean velocity, (b) shear stress. The different symbols indicate different downstream locations covering the range $400 < (x - x_0)/\theta < 14000$ with the virtual origin at $x_0/\theta = 116$. Reproduced with permission from [56], copyright Cambridge University Press.

latter and again points to a lack of universality, at least within the downstream region that has been explored thus far. It has been shown that this lack of universality reflects the dependence of the large-scale turbulence structures in the wake on the geometrical form of the wake generator [58]. Therefore, there is a link between universal self-similarity and large-scale vortical structures, which depend on the initial conditions, as argued for the case of the axisymmetric wake [6, 18].

As an example of plane wake data, figure 6.7 shows the mean velocity and shear stress profiles obtained in a wake behind a perforated plate [56]. Profiles are shown over a considerable downstream range, for which U_s/U_∞ varied from 0.15 to 0.03. Given these low velocity deficits, the degree of scatter, particularly in the shear stress data, is encouragingly small. For this particular wake, β was about 0.108 (giving $R_T = 12.8$); this implies a maximum shear stress, $g'_{\max} = (\overline{u'v'}/U_s^2)_{\max} = \beta e^{-0.5}/\sqrt{2 \ln 2} = 0.056$ (using equations (6.45) and (6.46)), which is reasonably close to the measured maximum seen in figure 6.7(b). Recall, however, the wide variation in β (from 0.1 to 0.13) for the different wake-generating bodies in these experiments. It is perhaps significant that some of the bodies (e.g. the cylinder) produced Kármán vortex shedding, which decays only slowly downstream, and others (e.g. the aerofoil) did not.

6.5.3 Mixing layers

The final free shear flow we discuss, albeit relatively briefly, is the planar mixing layer between two streams oriented in the x direction, see figure 6.1. Note that because of the boundary and geometry of an axisymmetric mixing layer, which typically exists in the near-field of an axisymmetric jet in the potential core region, there cannot be a self-similar solution [49], although some data do suggest profiles that can be collapsed for suitable choices of the thickness scale. For simplicity, we consider the simplest (single-stream) case, in which U_∞ is the constant velocity on one side and there is zero velocity on the other. The characteristic length and velocity scales are again defined as δ and U_s $(=U_\infty$ for this case), and we may specify the layer thickness δ as the distance between the locations at which $U = 0.1\,U_\infty$ and $U = 0.9\,U_\infty$. Alternative definitions (e.g. the $0.25\,U_\infty$ and $0.75\,U_\infty$ locations) are equally convenient, but do not change the essential results. Note that the characteristic velocity does not change with x, unlike the situations for jets and wakes. The scaled y-coordinate and mean velocity are then

$$\eta = (y - y_o)/\delta \quad \text{and} \quad f(\eta) = U/U_s. \tag{6.47}$$

With the y-origin at the trailing edge from which the mixing layer springs, y_o could be chosen as zero, or the location where $U = \frac{1}{2}U_\infty$, or where the turbulence shear stress is at its maximum (as used in [38]). However, in these latter two cases, y_o must change with x, so that $\mathrm{d}y_o/\mathrm{d}x \neq 0$; this complicates the analysis somewhat. It is simplest to choose axes such that $V = 0$ at $y = 0$; as a result of continuity, this means that in similarity coordinates

$$V = U_s \frac{\mathrm{d}\delta}{\mathrm{d}x} \int_0^\eta \eta f' \mathrm{d}\eta,$$

because $V(0) = 0$ by definition. It is then straightforward to show that the momentum equation in similarity form can be written as

$$-\left[\frac{\mathrm{d}\delta}{\mathrm{d}x}\right] f' \int_0^\eta f \mathrm{d}\eta = g', \tag{6.48}$$

where we have again chosen $R_s = U_s^2$ (and recall that U_s is constant for this flow). Notice that this choice of axes means that the shear stress is a maximum at $\eta = 0$. For the more general two-stream mixing layer case, with U_1 denoting the lower of the two stream velocities, there is an additional factor that multiplies the growth rate $\mathrm{d}\delta/\mathrm{d}x$, which is given by $-(U_1/U_s)\eta f'$.

It is evident from equation (6.48) that the spreading rate, $\beta = \mathrm{d}\delta/\mathrm{d}x$, must be constant. The layer therefore grows linearly with x ($\delta \sim x$ or, more strictly $\delta \sim x - x_0$), just as for a plane jet. Since U_s is constant, the local Reynolds number also therefore increases linearly with x. With the usual application of the constant eddy viscosity assumption (equation (6.28)), the governing equation (6.48) becomes

$$\beta R_T f' \int_0^\eta f \mathrm{d}\eta - f'' = 0. \tag{6.49}$$

This does not have a simple analytic solution (and nor does the equation governing a two-stream mixing layer). For a more detailed analysis, see [38, 46, 49].

As an example of typical mean flow and stress profiles, figure 6.8 shows results from one of a series of experiments on two-stream mixing layers covering a range of ratios of the two external velocities, r_v, say [29]. The mean flow profile has a shape that is quite close to that of the error function, i.e.

$$f = \frac{1}{2}\left(1 + \mathrm{erf}(\eta/\sigma\sqrt{2})\right) \quad \text{where} \quad \mathrm{erf}\left(\frac{\eta}{\sigma\sqrt{2}}\right) = \frac{1}{\sigma\sqrt{2\pi}} \int_0^\eta e^{-\eta^2/2\sigma^2} \mathrm{d}\eta, \tag{6.50}$$

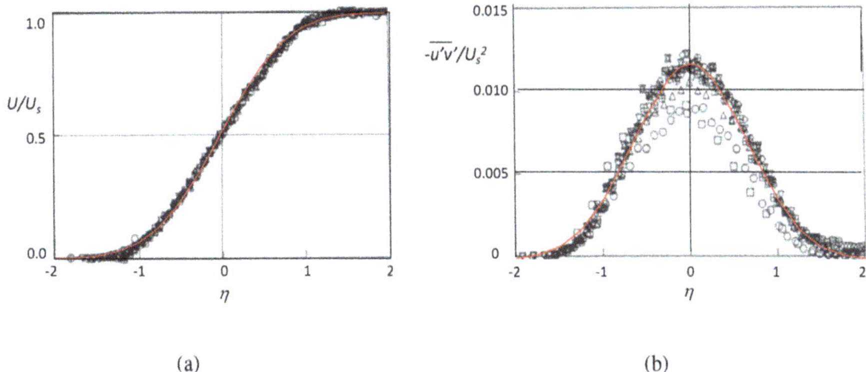

(a)   (b)

Figure 6.8. A mixing layer between two parallel streams that have a velocity ratio of $r_v = 0.5$[5]. U_s is the velocity difference between the streams and U is the local velocity minus the low-speed stream velocity. (a) Mean velocity profiles at numerous downstream locations. The red line represents equation (6.50) with $\sigma = 0.65$. (b) Shear stress data for the same mixing layer. The red line represents the shear stress profile calculated from $g = -f'/R_T$ with $R_T = 24.7$. Open squares, circles, and triangles refer to locations prior to the self-similar region.

but we emphasise that this expression for f is *not* a solution of equation (6.49). However, it is precisely the uniform eddy viscosity solution of the *temporal* mixing layer's self-similar governing equation. This is the spatially fixed but time-developing flow corresponding to the steady spatially developing case (see [38], for example). It has been studied using DNS [43] and shown to have a mean velocity profile that is close to the constant eddy viscosity solution, equation (6.50), although interestingly not as close as that shown in figure 6.7(a) for the spatially developing case. In that figure (taken for the results with $r = 0.5$), the best fit error function requires $\sigma \approx 0.65$. Figure 6.7(b) shows that the stress profiles measured at the three most upstream locations are significantly lower than the rest; self-similarity clearly occurs later for the turbulence quantities than it does for the mean velocity – a behaviour that is common across all the free shear flows.

Most authors have studied single-stream mixing layers (because they are easier to set up in the laboratory) and for these the spreading rate β is typically around 0.11. For two-stream layers, it is well established that β depends on the ratio of the two external velocities, r_v, say, (as fuller analyses suggest [1, 49]). Typically, β decreases with increasing r_v, taking values of e.g. 0.032 and 0.0073 for $r_v = 0.5$ and 0.9, respectively [29, 30]. It has also been found that high values of r_v require a much longer downstream distance before the flow reaches a self-similar state. Just as for other free shear flows, the initial conditions can have a significant effect. This is especially true for a two-stream mixing layer, for that usually requires a setup using a splitter plate initially separating the streams. The particular states of the two boundary layers at the trailing edges on each side of the plate (e.g. whether laminar,

---

[5] Reprinted by permission from [29], copyright (1991) with permission of Springer.

or turbulent, and probably with different momentum thicknesses) can have a long-lasting influence, as can effects arising from different thicknesses of the trailing edge.

The shear stress profile determined from $g = -f'/R_T$, using equation (6.50), is included in figure 6.8(b). This uses the measured value of the maximum stress (0.0115), giving $R_T = 26.5$ (for $\sigma = 0.65$). The profile shows good agreement with the data, but recall that the error function is *not* the exact solution to the governing equation of the spatially developing mixing layer. It is simply a good approximation, in that it fits experimental data reasonably well. Normal turbulence stress profiles, $\overline{u'^2}$, $\overline{v'^2}$, and $\overline{w'^2}$, have a similar shape to that of the shear stress profile (and are thus not shown here), with maxima close to $\eta = 0$ and typical values (when normalised by $\overline{u'v'}_{max}$) of about 3.3, 2.3, and 1.8, respectively [35]. There is some conflicting evidence, however, for whether or not the turbulence structure (including these ratios) depends on the velocity ratio r_v.

6.6 Concluding discussion

6.6.1 General remarks

For the reader's convenience, table 6.1 presents, for each of the flows discussed above, the self-similar behaviour of the δ and U_s scales along with the mean velocity solution of the governing equation, derived assuming a constant eddy viscosity. In the case of the mixing layer, recall that there is no straightforward analytic solution, but the error function provides a good approximation to the measured profile. And the wakes are, of course, 'small deficit' wakes, i.e. $U_s/U_\infty \ll 1$, because at larger relative values of the deficit velocity, the equations do not admit self-similar solutions. The table includes the behaviour of the local flow Reynolds number ($\sim \delta U_s/\nu$), approximate values of the flow constant R_T, and the spreading parameter β in each case. These latter two numerical values should be viewed as merely typical. As mentioned in the earlier sections, there can be a great deal of variability because of the long-lasting affects of initial conditions. Nonetheless, some general comments are appropriate.

First, it is clear that wakes, whether they are planar or axisymmetric, have significantly smaller values of R_T than those of jets and mixing layers. R_T (the inverse of the eddy viscosity normalised by δU_s) can be thought of as an entrainment parameter. As mentioned in section 6.1 (see item 5), all the flows are characterised by intermittency; that is, at each location sufficiently remote from the centreline, the flow will sometimes be turbulent and sometimes not, as a result of the large-eddy

Table 6.1. The self-similar solutions and parameters for various flows; x should be understood throughout to be $\sim(x - x_0)$.

Flow	δ	U_s	Re	f	a	R_T	β
Axi. jet	x	x^{-1}	constant	$(1 + a\eta^2)^{-2}$	$\beta R_T/8$	40	0.11
Axi. wake	$x^{1/3}$	$x^{-2/3}$	$x^{-1/3}$	$e^{-a\eta^2}$	$\beta R_T/2$	14.5	0.095
Plane jet	x	$x^{-1/2}$	$x^{1/2}$	$\mathrm{sech}^2(a\eta)$	$\sqrt{\beta R_T/4}$	31	0.10
Plane wake	$x^{1/2}$	$x^{-1/2}$	constant	$e^{-a\eta^2}$	$\beta R_T/2$	12.5	0.11
Plane mixing layer (single stream)	x	constant	x	$\frac{1}{2}(1 + \mathrm{erf}(\eta/(\sigma\sqrt{2})))$	—	27	0.10

structures in the flow engulfing the external fluid. Defining γ_i as the standard deviation of the y-location of the interface position, it can be shown (rather loosely) that $R_T \sim (\delta/\gamma_i)^2$ [17]. Measurements show that γ_i/δ is significantly larger in wakes than in jets (or mixing layers), explaining the smaller value of R_T in wakes. One of the reasons for the larger extent of the turbulent flow's bounding surface is that local turbulence kinetic energy, relative to the scaling velocity U_s, is significantly larger in wakes than in jets. In jets, k/U_s^2 is around 0.07, whereas it is about 0.11 in wakes; the eddy structures are thus rather more energetic relative to the deficit velocity in wakes, even though the absolute turbulence intensities (e.g. u'_{rms}/U) are much smaller. On the other hand, because the absolute turbulence energies in jets are much larger than those in wakes, jets approach the asymptotic self-similar state much more rapidly than do wakes. Indeed, Townsend himself suggested that 'observations of the plane wake may not be representative of the asymptotic flow even after long development' [49]. This accords with modern data and our earlier discussion in section 6.4.3.

Second, recall that in all the jet and wake cases, the analytic, constant eddy viscosity solution for the mean velocity does not decay as rapidly at the edges of the flow as is indicated by the data, regardless of whether they are laboratory or DNS data. This indicates that the eddy viscosity falls towards the flow edges and suggests that an improved closure model would be necessary before one could expect good agreement between data and prediction across the whole flow, whether it is a jet, a wake, or a mixing layer.

Third, recall that throughout our exploration of the various flows, it was noted that the influence of initial conditions is not always insignificant and may be long-lasting; see, in particular, the discussion in section 6.3.5. We emphasise, therefore, that the flow 'constants' in table 6.1 are merely typical, although they may be close to those expected in the genuine asymptotic self-similar state. Specific experiments on notionally self-similar flows may well, and often do, demonstrate closely self-similar profiles, yet lead to significantly different values of the flow constants that are not representative of the universal asymptotic state. In addition to the assumption underlying all the analyses, that the initial conditions are forgotten, we emphasise too the additional assumption that the same velocity scale can be used for the Reynolds stress as well as for the mean velocity; this is validated (for all the flows) only by a specific form for the scaling of the dissipation term in the TKE equation. Alternative forms are possible, however, and we conclude this chapter with a brief discussion of one of these, along with its implications for the self-similar behaviour of the various flows. First, however, we must discuss briefly the whole issue of entrainment, i.e. the processes that occur at the edges of free shear flows (and also boundary layers) through which external irrotational fluid is engulfed in the turbulent region of the flow and thus becomes turbulent itself.

6.6.2 Entrainment

All free shear flows and boundary layers have at least one boundary region that separates an external irrotational flow from the fully turbulent flow within, see

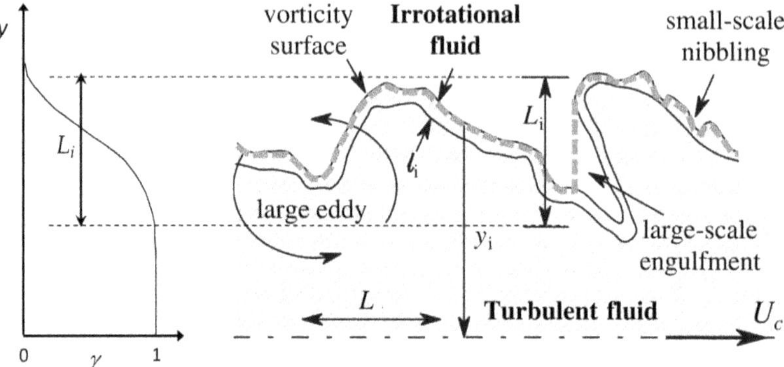

Figure 6.9. A rough sketch showing a snapshot of a vertical cut of a short streamwise fetch of the intermittent zone between fully turbulent and irrotational regions near the edge of a shear flow at a particular time. Here, y_i is the (fluctuating) location of the superlayer at a particular x. The outer edge of the superlayer is shown by the grey dashed line at the top of the superlayer. Adapted with permission from [7, 53]; copyrighted by the American Physical Society.

figure 6.9. As noted in point five of section 6.1, this zone is bounded by an *intermittency surface*, whose distance from $y = 0$ at (x, t) is defined by y_i on the figure. Typically, the probability density of y_i is roughly Gaussian and may have a standard deviation of as much as one half of the flow's width [53]. At any point (x, y) within the zone, there are periods during which the flow is turbulent, separated by periods during which the flow is nonturbulent, so the whole zone is often called the intermittent region. One can define an intermittency function as unity in the turbulent fluid and zero in the nonturbulent fluid and assume that at any point, one can somehow distinguish (at a time t) between turbulent and nonturbulent fluid. It is then possible to deduce the *conditional* time averages of any quantity q, say, defined as $\overline{q_T}$ and $\overline{q_N}$ – the averages in the turbulent and nonturbulent regions, respectively. It is common to define an *intermittency factor*, $\gamma(y)$, as the probability that some quantity that discriminates between turbulent and nonturbulent fluids exceeds its threshold value for nonturbulent flow, so that $\gamma(y)$ varies monotonically between unity in the fully turbulent region and zero outside it. The usual value of the (unconditional) time average, $\bar{q}$ can then clearly be written as

$$\bar{q} = \gamma\overline{q_T} + \left(1 - \gamma\right)\overline{q_N}. \tag{6.51}$$

This makes it possible to explore the details of the turbulence in the intermittent zone, without allowing them to be contaminated by nonturbulent contributions, as happens in the more usual time-averaging process. The crucial issues are the choice of the discriminating quantity and what its threshold value (below which the flow is deemed to be nonturbulent) should be. Since vorticity (or enstrophy) is nonzero in the turbulent region and zero in the nonturbulent region, this is often chosen, but many other choices are possible and have been used in both laboratory and DNS experiments. One common approach has been to use heat as a scalar within the turbulent flow, since it is relatively simple to discriminate (using temperature) between 'hot' turbulent fluid and the ambient, even if the latter is actually turbulent, see [23] for a typical example. A number of profiles of unconditional averaged quantities compared directly with the turbulent and nonturbulent contributions for different flows are given, for example, in Pope [38]. It is worth noting that early theoretical work [37] suggested that the nonturbulent normal stresses (induced by the large eddies in the fully turbulent flow) should decrease with distance according to $(y - \bar{y_i})^{-4}$, where $\bar{y_i}$ is the mean position of the interface; this has been largely confirmed by experiment. Figure 6.9 is a sketch of the intermittent zone, with a plot of the intermittency factor associated with it. The intermittent zone, with a thickness of L_i, which is often a significant portion of the flow's total width (depending on the particular flow), has within it a thin, continuous surface which classically is called the superlayer. This was first proposed long ago by Corrsin and Kistler [13], who hypothesised that there is a step change in velocity (and vorticity) across it and that the spreading process is similar to turbulent diffusion. However, it has since been clarified that the velocity jump occurs not as a singularity (nature abhors singularities) but over a finite, albeit very thin region – with a thickness of l_i in the figure – within which molecular diffusion occurs. After all, it is only through molecular processes that vorticity can be increased from zero. It is also now confirmed that the spreading and entrainment process can be significantly enhanced by the large-scale motions, as Townsend suggested [48, 49]. It is these large-scale motions which bring nonturbulent fluid into contact with the superlayer where it is converted to turbulent fluid by small-scale ('nibbling') eddies and molecular diffusion.

There has been much discussion about the relative importance of the large-scale, engulfing process – a result of the large contortions of the superlayer as illustrated in figure 6.9 and generated by the large eddies – and the small-scale processes across the superlayer itself. It has become clear that the superlayer has a fractal nature and thus a large surface area [44], so that even though it is very thin and increasingly so as Reynolds number rises, the 'total entrainment is independent of viscosity even though the actual conversion of nonturbulent fluid to turbulent fluid must be carried out by the small-scale action of the Kolmogorov scales' [12]. Note that the small-scale processes have an approximate streamwise length scale of the order of the Taylor microscale [7, 12, 54].

All the features described above are characteristic of all free shear flows (i.e. wakes, jets, and mixing layers) and also of the outer regions of all boundary layers. There are differences between flows, however, in terms of the consequent entrainment rate and hence the shear layer growth rate. The mean rate of formation of turbulent fluid per unit area (projected onto the xOy plane) can be defined as the entrainment velocity, which we could call U_e, and can be related to the intermittency factor and calculated for self-similar flows, as done by Townsend [48], who argued for an entrainment process controlled by the stable and unstable growth and decay of the large eddies. We noted earlier the differences in spreading rates, β, and the entrainment parameter, R_T, between the free shear flows, see section 6.6.1. The boundary layer's entrainment parameter is different again (typically $R_T \approx 55$) to those of the free flows given in table 6.1. There are concomitant differences between the different flows in the appropriately normalised U_e [48, 49]. We emphasise that although (in canonical free shear flows) the streamwise momentum is conserved, the mass flux of turbulent fluid always increases with downstream distance, as a result of the entrainment processes.

These global parameters have been the most studied, but increasing attention is being paid to the details of the interface motion itself, the statistics of its outward velocity, for example, and how such local features are related to the global entrainment velocity U_e and are affected by (large-scale) global processes. Detailed studies discussing both the local and global entrainment phenomena and their interactions include, for example, those of Chauhan et al [12] and the very recent works of Jahanbakhshi and of van Reeuwijk, et al [26, 40], which provide good introductions to the increasingly extensive literature on turbulent entrainment, driven partly by the increasing power of DNS.

6.6.3 Dissipation scaling

We have not explicitly demonstrated that a particular scaling of the dissipation term in the turbulence kinetic energy equation is needed to confirm that $R_s \sim U_s^2$, although the fact has been mentioned (in sections 6.3.2 and 6.4.1 in the context of the axisymmetric jet and wake, respectively). Such a scaling of the Reynolds stress quantities leads to a self-similarity behaviour of the TKE equation (and indeed other aspects of the flow dynamics) that is entirely consistent with the scaling of the momentum equation, with the resulting form of flow spreading rate in each case. However, given that nearly all free shear flows of industrial or environmental importance do not reach a sufficiently far field region to expect genuine universality, there has been some effort over the last two or three decades to explore this non-universal (non-equilibrium) region in new ways – partly by identifying the importance of the initial conditions theoretically – see [18] for an extensive discussion. But in addition, importantly, the implications of choosing a different scaling for the dissipation have been explored.

The (classical) scaling of the dissipation term (ϵ) in the TKE equation uses δ and U_s and can be expressed by

$$\epsilon = C_\epsilon \frac{U_s^3}{\delta}, \tag{6.52}$$

where C_ϵ is assumed to be constant and we have taken (i) $U_s^3 \sim k_0^{3/2}$ with k_0 the turbulence kinetic energy on the flow centreline and (ii) $L \sim \delta$, for which there is good evidence. This is therefore essentially the equilibrium dissipation law (equation (5.26), discussed in section 5.1.3), and assumes that the dissipation is independent of the Reynolds number. However, there is evidence that C_ϵ (on the flow centreline, for example) does not remain constant with x. This has been found to be the case in axisymmetric wakes [14, 34] and even, some decades ago, for planar jets in regions far downstream [4, 22]. The former authors found that $\epsilon\delta/U_s^3$ decreased with x even beyond $x/d = 120$. Figure 6.10 shows data obtained for a planar jet [4], but in a replotted form as shown by Cafiero et al [9], with U_s replaced by u'_{rms} on the flow centreline; the ratio of these two quantities remains constant with x, as anticipated.

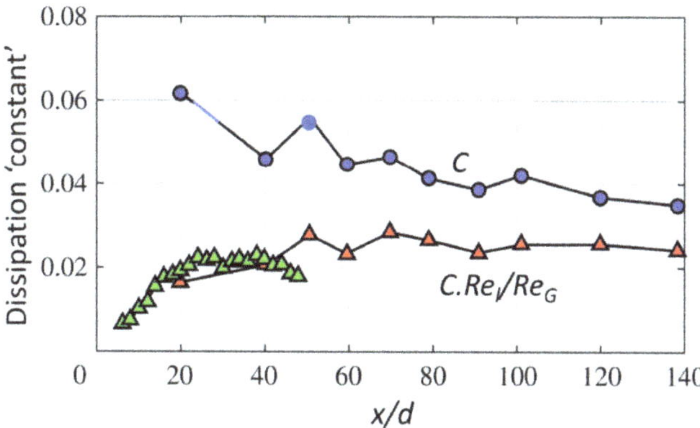

Figure 6.10. The dissipation 'constant'[6]. The blue dots show $C = \epsilon\delta/1.5u_{\text{rms}}'^{3}$, from the data of [4]. The red triangles show $C\text{Re}_\delta/\text{Re}_G$, derived from those data. The green triangles are the same, but from [9].

Nedic *et al* [33] proposed a Reynolds-number-dependent scaling which can be written as

$$\epsilon \sim \frac{\text{Re}_G^{m}}{\text{Re}_\delta^{n}} U_s^3 \Big/ \delta. \tag{6.53}$$

The C_ϵ factor is thus taken as a ratio of the global Reynolds number (based on the initial conditions) to a local Reynolds number, defined by $\text{Re}_\delta = \sqrt{k_0}\,\delta/\nu$. The former is fixed, but the latter varies with the downstream distance in a way that depends on the particular flow. Note that this particular form remains independent of viscosity whenever $m = n$. If both indices are zero, the usual scaling is recovered. If $m = 0$ and $n = 1$, the classical low-Reynolds-number scaling results, since then we are effectively assuming that $\epsilon \sim \nu U_s^2/\delta^2$, as in [46]. Note, however, that there is some evidence that, provided the eddy viscosity is constant everywhere, this scaling can emerge even at high Reynolds numbers in a self-similar region *before* the final, universal scaling emerges [39]. The interesting case is $m = n = 1$, when equation (6.53) is termed the *non-equilibrium dissipation law* – a high Reynolds number relation, as it does not depend on viscosity. This alternative to the classical scaling of ϵ leads to different self-similarity results.

For example, in the case of an axisymmetric wake, it leads to $\delta \sim x^{1/2}$ and $U_s \sim x^{-1}$, rather than the classical result in table 6.1. For a planar jet, one finds that $\delta \sim x^{2/3}$ and $U_s \sim x^{-1/3}$. Incidentally, in this latter case, one also finds that the ratio of U_s to the velocity used to scale the cross-stream mean velocity (implicitly $\sim U_s$ in the classical analysis) is not constant, as it is in the classical analysis; in fact the ratio varies like $x^{1/3}$. These behaviours appear to satisfy experimental (and DNS) data a little better [9, 33]. In particular, for example, figure 6.10 shows that the earlier values of C_ϵ [4], when factored by $\text{Re}_\delta/\text{Re}_G$, remain reasonably constant beyond about $x/d = 40$. The whole topic has been discussed and continues to be discussed

---

[6] Republished from [9] with permission from The Royal Society. Permission conveyed through the Copyright Clearance Centre.

by a number of authors, building on earlier considerations [3, 20, 50]. Given the comment by Taylor long ago that C_ϵ 'is unlikely to be universal', this is all perhaps unsurprising [45]; it also seems now that C_ϵ may not even be constant within any particular flow in which there is a self-similar but nonuniversal region. (See [11] for a further discussion, in particular, of the 'new' non-equilibrium dissipation relation, equation (6.53) in the context of axisymmetric wakes.)

6.7 Subject giants: A A Townsend

Alan Townsend (1917–2010) was a talented experimentalist and is famous for his book on *The Structure of Turbulent Shear Flow* [47, 49]. He was born in Melbourne, Australia, and completed his Bachelor's and Master of Science degrees at the

Figure 6.11. Alan Townsend (1917–2010) pictured at his office in Cambridge, 1975. Image reproduced with permission from [28], copyright Cambridge University Press.

University of Melbourne. At the time, there were no PhD programmes in Australia so in 1938 he went to the University of Cambridge, earning a scholarship to study nuclear physics. During the Second World War he paused his studies, working partially in the UK and partially back in Melbourne, which is when he met George Batchelor. Batchelor convinced Townsend to return to Cambridge after the war and to work with him on turbulence, under the supervision of G I Taylor. They worked together studying the decay of homogeneous turbulence behind a grid using Townsend's expertise in instrumentation to obtain detailed measurements of the turbulence properties using HWA.

Townsend went on to study wakes and jets; in this field of study, he was one of the first to think of turbulence in terms of what we now call 'coherent structures'. To paraphrase part of his 1948 paper, he described the turbulent motion within jets and wakes as consisting of small-scale fluctuations superimposed on a distribution of slower large-scale eddies [28]. In the context of boundary layers, *Townsend's attached eddy hypothesis* was the idea that the main energy-containing eddies in the flow control many of the turbulence properties and, as the size of these eddies tends to scale with their distance from the wall in bounded flows, they can be considered to be attached to the wall. His more general ideas about the structure of turbulent shear flows were the topic of his book [47], the success of which, in part, led to his election as a Fellow of the Royal Society in 1960. It is astonishing to consider that he obtained his numerous three-dimensional insights into the structure of turbulence based solely on point measurements made using hot wires. One of his PhD students, Ron Adrian, published more recent advances in our interpretation of coherent structures in wall turbulence using PIV [2]. In the context of the present book, it is perhaps a curious fact of history that it is, at the time of writing, almost 50 years ago that Townsend was the external examiner for the PhD thesis of the first author (IPC), who, at that time, was struggling to understand the first edition (1956) of his book.

Sample exercises

6.1. Deduce the momentum integral equation (6.15) from the mean streamwise momentum equation (6.10) by taking its integral with respect to y.

6.2. The file 'MixingLayerData.txt'[7] contains properties of a planar mixing layer, interpolated from the hot-wire measurements of Delville and Bonnet [15]. This dataset focusses on the results taken at a distance of $x = 800$ mm downstream from the splitter plate, which separates two flows with free stream velocities of $U_B = 41.54$ m s^{-1} (upper) and $U_A = 22.40$ m s^{-1} (lower) in a wind tunnel. The data include y (mm), U (m/s), $\overline{u'v'}$ (m^2 s^{-2}), k (m^2 s^{-2}), and ϵ (m^2 s^{-3}).

(a) Plot the velocity profile along with the approximate self-similar solution given by equation (6.50) on the same axes. Note that the centreline drifts away from $y = 0$ as a mixing layer develops.

---

[7] https://github.com/cvanderwel/TurbulentFlows/blob/main/data/MixingLayerData.txt

(b) One way of defining the width of the mixing layer is the distance between the points where the velocity equals $U_A + 0.9(U_B - U_A)$ and the point where it equals $U_A + 0.1(U_B - U_A)$. Determine the width of this mixing layer. Assuming self-similarity, how far downstream would you expect the width of the mixing layer to double?

(c) Plot the Reynolds shear stress and TKE profiles and explain why they are shaped as they are.

(d) Compute and plot the production and dissipation profiles and explain why they are shaped as they are.

6.3. The file 'WakeData.txt'[8] contains measurements in the planar wake behind a 2D aerofoil from the experiment of Nakayama [32]. The file contains hot-wire measurements of the mean velocity, U, and the Reynolds stress, $\overline{u'v'}$, among other data, obtained as vertical profiles behind the aerofoil at zero degree angles of attack. The data only cover the near wake behind the aerofoil (i.e. at distances ranging from 0.01 to 2.00 chord lengths behind the trailing edge of the aerofoil), so in this exercise, one can evaluate whether this flow can be described as self-similar.

(a) Plot the mean streamwise velocity profile U/U_∞ versus the vertical position y/c. Notice how the wake spreads and the velocity deficit decreases with downstream distance.

(b) Now, define a new variable equal to the velocity deficit function (i.e. $(1 - U/U_\infty)$) and plot that as a function y/c.

(c) The self-similar profile for a planar wake is given by $f = e^{-a\eta^2}$ (equation (6.46)). Fit a function of this form to the velocity deficit data (i.e. $1 - U/U_\infty = c_1 e^{c_2(x-c_3)^2}$, where c_1, c_2, and c_3 are tuneable coefficients). Hint: either use a curve fitting tool to evaluate the coefficients or guess them by trial and error. Note that the best fit of c_1 represents the velocity scale U_S, the best fit of c_2 is related to the half width (see (d)) and the best fit of c_3 accounts for any shift in the centreline of the wake.

(d) What is the half width of this wake? One way to determine the half width is from the second central moment of the profile. This should be roughly equal to $\sqrt{1 - 1/(2c_2)}$, where c_2 is the coefficient from the curve fit performed in (c). From these estimates, determine the half width at half-maximum velocity (HWHM), which is the definition of δ that we use in this chapter, by multiplying by a factor of $\sqrt{2\ln(2)}$.

(e) Replot the velocity data using the self-similar variables determined above (i.e. $f = (1 - U/U_\infty)/(U_S/U_\infty)$ and $\eta = (y - c_3)/\delta$). Do the data collapse?

(f) Check whether $U_S\delta$ and $\beta = U_\infty/U_S\, d\delta/dx$ are constant, whether the half width grows according to $\delta \sim x^{1/2}$ and whether the maximum velocity deficit decays according to $U_S \sim x^{-1/2}$, as predicted by the

[8] https://github.com/cvanderwel/TurbulentFlows/blob/main/data/WakeData.txt

similarity solutions. If these relations are violated, it indicates that the assumptions for self-similarity are, perhaps, not applicable in this near wake. Why might this be the case?

(g) Plot the Reynolds shear stress profiles, also using self-similar variables (i.e. $g = -(\overline{u'v'}/U_\infty^2)/(U_S/U_\infty)^2$). Do the data collapse? Often, while the mean properties look self-similar, it takes longer for the turbulence properties to truly collapse.

References

[1] Abramovich G N 1963 *The Theory of Turbulent Jets* (Cambridge, MA: MIT Press)

[2] Adrian R 2007 Hairpin vortex organization in wall turbulence *Phys. Fluids.* **19** 301

[3] Antonia R A and Pearson B R 2000 Effect of initial conditions on the mean energy dissipation rate and the scaling exponent *Phys. Rev.* E **62** 8086–90

[4] Antonia R A, Satyaprakash B R and Hussain A K M 1980 Measurements of dissipation rate and some other characteristics of turbulent plane and circular jets *Phys. Fluids.* **23** 695–700

[5] Batchelor G K 1967 *An Introduction to Fluid Dynamics* (Cambridge: Cambridge University Press)

[6] Bevilaqua P M and Lykoudis P S 1978 Turbulence memory in self preserving wakes *J. Fluid. Mech.* **89** 589–606

[7] Bisset D, Hunt J C R and Rogers M M 2002 The turbulent/non-turbulent interface bounding a far wake *J. Fluid. Mech.* **451** 383–410

[8] Boersma B, Brethouwer G and Nieuwstadt F T M 1998 A numerical investigation on the effect of the inflow conditions on the self-similar region of a round jet *Phys. Fluid.* **10** 899-909

[9] Cafiero G and Vassilicos C 2019 Non-equilibrium turbuence scalings and self-similarity inn turbulent planar jets *Proc. R. Soc.* A **475** 1–25

[10] Cannon S, Champagne F and Glezer A 1993 Observations of large-scale structures in wakes behind axisymmetric bodies *Expt. Fluids.* **14** 447–50

[11] Castro I P 2016 Dissipative distinctions *J. Fluid. Mech.* **788** 1–4

[12] Chauhan K, Philip J, de Silva C M, Hutchins N and Marusic I 2014 The turbulent/non-turbulent interface and entrainment in a boundary layer *J. Fluid. Mech.* **742** 119–51

[13] Corrsin S and Kistler A L 1954 *The Free-Stream Boundaries of Turbulent Flows* (Washington, DC: National Advisory Committee for Aeronautics) NACA TN No. 3133.

[14] Dairay T, Obligardo M and Vassilicos J C 2015 Non-equilibrium scaling laws in axisymmetric turbulent wakes *J. Fluid. Mech.* **781** 166–95

[15] Delville J and Bonnet J P 1997 Database of a plane turbulent mixing layer from C.E.A.T. Poitiers *A Selection of Test Cases for the Validation of Large-Eddy Simulations of Turbulent Flows* AGARD Advisory Report NO 345

[16] Everitt K W and Robins A R 1978 The development and structure of turbulent plane jets *J. Fluid. Mech.* **88** 563–83

[17] Gartshore I S 1966 An experimental examination of the large-eddy equilibrium hypothesis *J. Fluid. Mech.* **24** 89–98

[18] George W K 1989 The self-preservation of turbulent flows and its relation to initial conditions and coherent structure *Advances in Turbulence* ed R Arndt and W K George (New York: Hemisphere) pp 39–73

[19] George W and Davidson L 2004 Role of initial conditions in establishing asymptotic flow behaviour *AIAA J.* **42** 438–46

[20] Goto S and Vassilicos J C 2009 The dissipation rate coefficient of turbulence is not universal and depends on the internal stagnation point structure *Phys. Fluids.* **21** 104

[21] Gutmark E and Ho C-M 1983 Preferred modes and spreading rates of jets *Phys. Fluids.* **10** 2932–8

[22] Gutmark E and Wygnanski I 1976 The plane turbulent jet *J. Fluid. Mech.* **73** 465–95

[23] Hancock P and Bradshaw P 1989 Turbulence structure of a boundary layer beneath a turbulent free stream *J. Fluid. Mech.* **205** 45–76

[24] Hussein A K M F and Clark A R 1977 Upstream influence on the near field of a plane turbulent jet *Phys. Fluids.* **20** 1416–26

[25] Hussein H, Capp S and George W 1994 Velocity measurements in a high-Reynolds-number, momentum-conserving, axisymmetric, turbulent jet *J. Fluid. Mech.* **258** 31–75

[26] Jahanbakhshi R 2021 Mechanisms of entrainment in a turbulent boundary layer *Phys. Fluids.* **33** 1–17

[27] Johansson P, George W and Gorley M 2003 Equilibrium similarity, effects of initial conditions and local Reynolds number *Phys. Fluids.* **15** 603–17

[28] Marusic I and Nickels T B 2011 A A Townsend *A Voyage Through Turbulence* ed P A Davidson, Y Kaneda, K Moffatt and K R Sreenivasan (Cambridge: Cambridge University Press) pp 305–28

[29] Mehta R D 1991 Effect of velocity ratio on plane mixing layer development: influence of the splitter plate wake *Exp. Fluids.* **10** 194–204

[30] Mehta R D and Westphal R V 1986 Near-field turbulence properties of sinngle- and two-stream plane mixing layers *Exp. Fluids.* **4** 257–66

[31] Monin A and Yaglom A 1972 *Statistical Fluid Mechanics* vol 1 (Cambridge, MA: MIT Press)

[32] Nakayama A 1985 Characteristics of the flow around conventional and supercritical airfoils *J. Fluid. Mech.* **160** 155–79

[33] Nedic J, Vassilicos J C and Ganapathisubramani B 2013 Axisymmetric wakes with new nonequilibrium similarity scalings *Phys. Rev. Lett.* **111** 144503

[34] Obligado M, Dairay T and Vassilicos J C 2016 Nonequilibrium scaling of turbulent wakes *Phys. Rev. Fluids.* **1** 044409

[35] Oster D and Wygnanski I 1982 The forced mixing layer between parallel streams *J. Fluid. Mech.* **123** 91–130

[36] Peterson L and Hama F 1978 Instability and transition of the axisymmetric wake of a slender body of revolution *J. Fluid. Mech.* **88** 71–96

[37] Phillips O M 1955 The irrotational motion outside a free turbulent boundary layer *Proc. Camb. Phil. Soc.* **51** 220–9

[38] Pope S B 2000 *Turbulent Flows* (Cambridge: Cambridge University Press)

[39] Redford J, Castro I and Coleman G 2012 On the universality of turbulent axisymmetric wakes *J. Fluid. Mech.* **710** 419–52

[40] van Reeuwijk M, Vassilicos J C and Craske J 2021 Unified description of turbulent entrainment *J. Fluid. Mech.* **908** 12–21

[41] Reynolds W, Parekh D, Juvet P and Lee M 2003 Bifurcating and blooming jets *Annu. Rev. Fluid. Mech.* **35** 295–315

[42] Rind E and Castro I 2012 On the effects of free-stream turbulence on axisymmetric wakes *Exp. Fluids.* **53** 301–18

[43] Rogers M M and Moser R D 1994 Direct simulation of a self-similar turbulent mixing layer *Phys. Fluids.* **6** 903–23

[44] Sreenivasan K R, Ramshankar R and Meneveau C 1989 Mixing, entrainment and fractal dimensions of surfaces in turbulent flow *Proc. R. Soc. Lond.* A **421** 79–108

[45] Taylor G I 1935 Statistical theory of turbulence parts i-iv *Proc. R. Soc. Lond.* A **151** 421–78

[46] Tennekes H and Lumley J L 1972 *A First Course in Turbulence* (Cambridge, MA: MIT Press)

[47] Townsend A A 1956 *Structure of Turbulent Shear Flow* 1st edn (Cambridge: Cambridge University Press)

[48] Townsend A A 1966 The mechanism of entrainment in free turbulent flows *J. Fluid. Mech.* **26** 689–715

[49] Townsend A A 1976 *Structure of Turbulent Shear Flow* 2nd edn (Cambridge: University Press)

[50] Vassilicos J C 2015 Dissipation in turbulent flows *Annu. Rev. Fluid. Mech.* **47** 95–114

[51] Webb S 2006 Jet impingement on porous surfaces *PhD Thesis* University of Southampton, UK

[52] Webb S and Castro I 2006 Axisymmetric jets impinging on porous walls *Exp. Fluids.* **40** 951–61

[53] Westerweel J, Fukushima C, Pedersen J M and RHunt J C 2005 Mechanics of the turbulent/non-turbulent interface *Phys. Rev. Lett.* **95** 1–4

[54] Westerweel J, Fukushima C, Pedersen J and Hunt J C R 2009 Momentum and scalar transport at the turbulent/non-turbulent interface *J. Fluid. Mech.* **631** 199–230

[55] Wygnanski I and Fiedler H 1969 Some measurements in the self-preserving jet *J. Fluid. Mech.* **38** 577–612

[56] Wygnanski I, Champagne F and Marasli B 1986 On the large-scale structures in two dimensional, small deficit, turbulent wakes *J. Fluid. Mech.* **168** 31 71

[57] Xu G and Antonia R 2002 Effect of different initial conditions on a turbulent round free jet *Exp. Fluids.* **33** 677–83

[58] Zhou Y and Antonia R A 1995 Memory effects in a turbulent plane wake *Exp. Fluids.* **19** 112–20

Chapter 7

Internal wall-bounded flows

All the specific types of turbulent flow discussed in earlier chapters have been assumed to be remote from any solid walls. This chapter and the next explore the shear flows that directly result from the presence of a wall. Classic examples include turbulent pipe flow (an internal flow) and the turbulent boundary layer (an external flow). Such flows are necessarily more complex than the free shear flows discussed earlier, not least because sufficiently near to the wall, they must include significant viscous effects. The Reynolds number (however defined) thus almost always assumes greater importance than in the free shear flows discussed in chapter 6. After some initial remarks, we consider the major types of internal flow in sections 7.2 (Couette flows), 7.3 (channel flows), and 7.4 (pipe flows). External wall-bounded flows are considered in the next chapter (chapter 8). In all cases, the discussion starts with a brief consideration of the corresponding laminar flows. In the case of internal flows, these are among the very few exact solutions of the Navier–Stokes equations and set the scene for exploration of the turbulent versions which inevitably arise when the Reynolds number is high enough.

7.1 Initial remarks

Wall-bounded turbulent shear flows are exceedingly common, arguably much more common than the free shear flows discussed earlier. Internal flows through pipes and channels along with external flows around vehicles and over buildings are examples of flows in which mean flow shear is created in the boundary layer formed by the flow near the solid surface. In fact, the frictional effects near the wall provide the most common source of shear. Without these effects, there would be no turbulence. They arise as a result of *the no-slip condition* – the requirement that, because of nonzero viscosity, the velocity of the fluid at the wall must equal the wall's velocity (whether zero or not).

Consider, for example, the flow through a long pipe – an internal flow. One could reasonably claim, given the preponderance of pipes (and channels) of all sizes and in such numerous contexts, that this is *the* most common kind of fluid flow resulting from human endeavour. Indeed, the measurement of the flow rate in pipes might well be the most important measurement made daily and often continuously, throughout large sections of industry: an error of only a fraction of one percent could lead to the loss or gain of very large sums of money (for the supplier or the consumer). Recall section 2.8, which mentions Reynolds' early work on pipe flow. He showed that if the pipe Reynolds number exceeds a few thousand, the flow becomes fully turbulent. Most industrial pipework contains flows that have very much larger Reynolds numbers, so a large fraction of such flows are fully turbulent and thus, despite the extremely simple boundary conditions, not amenable to a full analytic solution. In addition, a crucial fact arising from the turbulent nature of such flows is that the power required to drive the flow through the pipe is usually orders of magnitude higher than would be needed if the flow remained laminar, as mentioned in section 1.3.

The planar equivalent of a pipe is a wide rectangular channel; figure 7.1 shows a snapshot of the fully turbulent flow in such a channel. We do not discuss the details here, but note that the flow typifies turbulence in that there are entangled eddy motions on a wide range of scales. It also typifies wall turbulence in particular, in that the eddy motions are of a much smaller scale nearest the wall than they are further away. Instantaneously, the cross-stream profile of the streamwise velocity is far from monotonic, reflecting the state of the flow at the instant the profile was obtained. This illustrates an important feature that we have not yet mentioned, perhaps because it should be self-evident (although it can be easily forgotten): the mean velocity profile in any turbulent flow *never* actually exists. Note, from the mean velocity profile shown in the figure, that the velocity gradient is at its largest at the two walls; the shear stress at the walls is much larger than it would be in laminar flow, emphasising the point made above, that much greater power is needed to drive the flow through the channel.

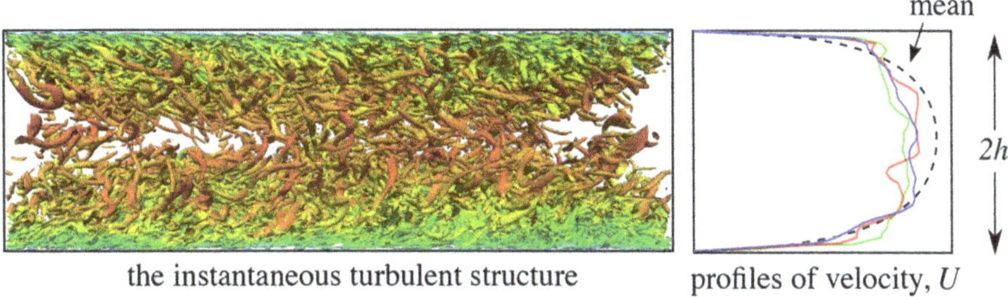

Figure 7.1. A snapshot of the eddy motions within a turbulent flow in a rectangular channel, visualised on the spanwise centreline, from a direct numerical simulation (DNS) at $Re_\tau = 590$[1]. The direction of the flow is from left to right and the eddies are visualised using the Q criterion (see section 8.5) and are coloured according to the velocity magnitude – from green (lowest) to red (highest). On the right, there are three sample instantaneous profiles of the streamwise velocity, with the time-averaged (mean) profile shown as a dashed line.

---

[1] Reprinted from [19], copyright (2017) with permission from Elsevier.

7.2 Couette flows

7.2.1 The laminar case

Many of the laminar flows discussed in this section and the following sections will be familiar to readers who have studied basic fluid dynamics. The flows represent classical ones, for which the Navier–Stokes equations yield straightforward analytic solutions when the Reynolds number is low enough to prevent turbulence from arising. We include the fundamental results here, without needing to include much detail in the analysis, or extending the discussion to involve related alternative flows (laminar flow in branching channels or curved pipes, for example). Our purpose is merely to provide the background for an exploration of the corresponding turbulent flow. We begin by discussing Couette flow, a planar flow between two walls.

Figure 7.2 sketches the ideal cases reviewed here. They are contained in a rectangular channel of internal height $2h$ and width L_z. We assume its length, L_x, is very large compared with its other dimensions, so that the entry and exit regions of the duct, where the flow is still developing, can be ignored. Attention is thus focused on the 'fully developed' region in which the axial velocity does not (and cannot) depend on x and the flow is a simple one-dimensional one in which the velocity, $U(y)$, depends only on y. This assumes, of course, that $L_z/2h \ggg 1$, so that the influence of the side walls of the duct is negligible.

In the simplest version of this situation, one wall of the duct moves with a constant velocity, $U(y = 2h) = U_w$, say, for then no pressure gradient is needed to drive the flow – frictional forces at this upper wall suffice. Under these circumstances, the flow becomes a simple shear flow; the velocity between the upper and lower walls is given by $U/U_w = y/2h$. For the laminar case, the viscous shear stress, $\mu dU/dy$, is constant across the flow and obviously given by $\mu U_w/2h$. This is a Couette flow and is common in lubricating systems, often as the axisymmetric equivalent. U_w

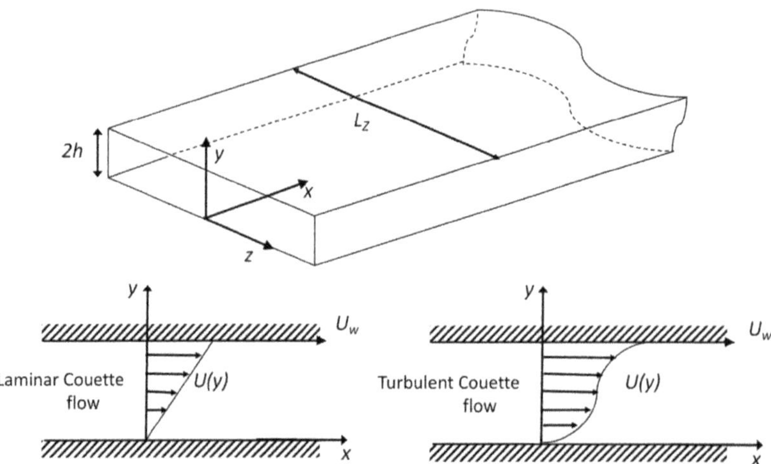

Figure 7.2. A long, straight, smooth-walled, rectangular channel with sketches of the characteristic velocity profiles for the cases of laminar and turbulent Couette flows (i.e. with a moving wall).

and h are the only pertinent velocity and length scales, respectively; therefore, as an appropriate Reynolds number for the flow, we can use either $\mathrm{Re}_w = U_w h/\nu$ or $\mathrm{Re}_c = U_c h/\nu$, (with the centreline velocity $U_c = 0.5 U_w$). Once Re_w exceeds about 1200, the flow begins the transition to turbulence; see, for example, Aydin *et al* [3]. The transition process is not discussed here, and we now turn to consider the fully turbulent Couette flow.

7.2.2 The turbulent case

As in the laminar case, in the fully developed region there is no dependence on x and, provided $L_z/h \ggg 1$, there is no dependence on z either – the flow is essentially one-dimensional. The general mean momentum equation (2.16) is repeated here:

$$\frac{\partial U_i}{\partial t} + U_j \frac{\partial U_i}{\partial x_j} = -\frac{1}{\rho}\frac{\partial P}{\partial x_i} + \nu \frac{\partial^2 U_i}{\partial x_j \partial x_j} - \frac{\partial \overline{u_i' u_j'}}{\partial x_j}. \tag{7.1}$$

Note that the continuity equation directly implies that V must be zero all across the flow, because $\partial U/\partial x$ is zero and V is zero at the walls. (This is obviously true whether or not the flow is turbulent.) With the assumptions of no dependence on either x or z, the x-direction ($i = 1$) equation simplifies greatly and can then be integrated once to yield

$$\mu \frac{dU}{dy} - \rho \overline{u'v'} = \text{constant} = \tau_w, \tag{7.2}$$

where τ_w is the wall shear stress. Equation (7.2) is simply a statement that the total mean shear, the sum of the viscous and turbulence stresses, is constant across the flow. Since $\overline{u'v'}$ is zero at the walls, it is immediately clear that the mean velocity gradient cannot be constant across the whole flow, unlike in the laminar case summarised above. The velocity gradient at the walls is larger than it is elsewhere, and the mean velocity profile takes on a characteristic S-shape, as sketched in figure 7.2. In each near-wall region, viscous stresses are significant. At sufficiently high Reynolds numbers ($\mathrm{Re}_c = U_c h/\nu \ggg 1$), however, these viscous layers are thin and one might expect the central portion of the flow to approximate a steady version of the simple shear flow discussed in section 5.2 (but see below), for which the viscous stress in equation (7.2) is negligible and the turbulent shear stress is thus constant.

The viscous region is common to all wall-bounded flows and we explore it in more detail in the following section; it is sufficient to note here that it becomes thinner and thinner as the Reynolds number rises. Note immediately, however, that we can define a characteristic velocity scale of the near-wall flow by using the wall stress; thus, the *frictional velocity*, u_τ, is defined by $\rho u_\tau^2 = \tau_w$; U and y can be normalised as

$$U^+ = U/u_\tau \quad \text{and} \quad y^+ = y u_\tau/\nu, \tag{7.3}$$

respectively, so that equation (7.2) can be written as

$$\frac{dU^+}{dy^+} - \overline{u'v'}^+ = 1. \qquad (7.4)$$

Now consider the origin of the turbulence. Recall from the earlier discussion of decaying homogeneous shear flow in section 5.2 that the only source of turbulence is the presence of the mean shear, so that there is a positive production term in the energy equation (5.29). For a fully developed, steady Couette flow, in which $dk/dt = 0$, the source of the shear stress in the central region ($-\overline{u'v'}$) must be a result of what happens at and near the two walls. This is intuitively obvious; in the absence of viscosity, the wall shear stress would be zero and whatever the velocity of the moving wall, no flow could be generated, as there would be no driving force.

This geometrically simple flow is perhaps the purest among the canonical wall-affected turbulent shear flows, not least because of its homogeneity in two space directions (x and z), the lack of any pressure gradient, and the absence of a gradient in shear stress. The flow is defined by only one parameter, the Reynolds number (in contrast to pipe flows, section 7.4, which have two controlling parameters). It has therefore been the study of numerous papers since it was first discussed by Couette in the late 1800s. But it is not, in fact, at all simple. The flow is fully turbulent above a Reynolds number (based on $U_w = 2U_c$ and the channel half-height, h) of a few thousand. Laboratory realisations of this pure, planar, flow are quite difficult to establish; they often involve a moving belt as the upper wall, which can be difficult to keep flat (at the speeds needed to establish a reasonably high Reynolds number) and to make long enough to ensure a fully developed region between the two ends. It is also usually necessary to initiate the flow in some way in order to ensure the correct mass flux. Reichardt [38] and Tillmark and Alfredsson [47, 48] are examples of early and more recent studies, respectively. To obviate the technical problems involved in setting up a truly planar flow, some laboratory studies have used the approximation provided by an annular flow in a narrow channel between two long concentric cylinders, with either the inner or outer cylinder rotating at a fixed speed. If the gap between the cylinders, g, say, is sufficiently small compared to their circumference, L_c, say, the annular flow is closely planar. There are hundreds of papers in this category, starting, perhaps, with the classical works of G I Taylor [45, 46], but nearly all of them are not concerned so much with the nature of a pure turbulent Couette flow as with the details of the transitional processes that occur once the Reynolds number is high enough. Actually, even for very large L_c/g, the influence of the well-studied secondary flows that occur prior to transition and continue beyond it (including the classic Taylor vortices) can never really allow this situation to be a genuine surrogate for a true planar Couette flow.

One might think that it would be easier these days to study the flow using DNS and, in some ways, this is true. However, this approach is not without difficulty because of the very long and wide structures that are present in the flow [17]; this makes the problem significantly more expensive than obtaining the DNS solutions for a Poiseuille (pipe) flow at comparable Reynolds numbers. Nonetheless, a number of DNS results are now available; figure 7.3 shows some results from the study of Avsarkisov *et al* [2]. We use these as being sufficiently typical to make a number of basic points about turbulent Couette flow without delving into great detail. Note,

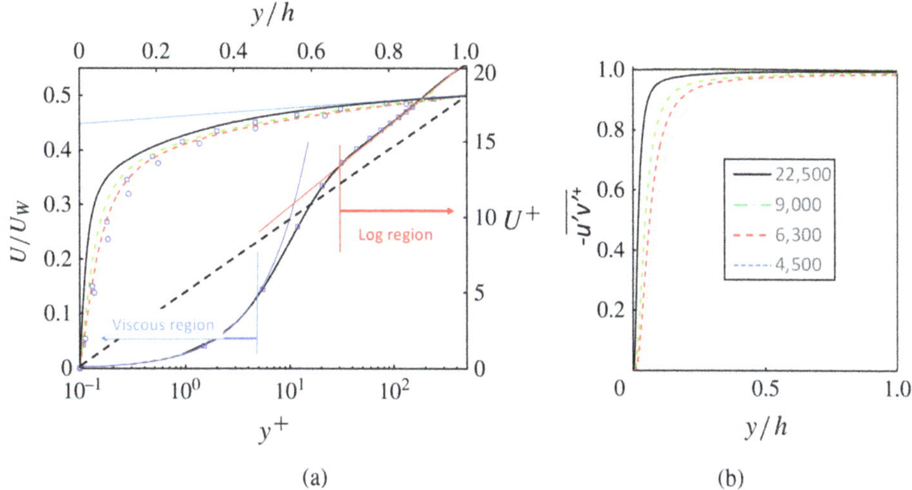

Figure 7.3. (a) Mean velocity profiles of plane Couette flows at various Reynolds numbers, over one half of the channel and normalised using the outer (top, left: h, U_w) and inner (bottom, right: ν/u_τ, u_τ) scales. (b) The turbulent shear stress. $\mathrm{Re}_w = hU_w/\nu$ values are shown in the legend. In (a) the straight, dashed black line represents the laminar profile (U/U_w versus y/h), the lower black line illustrates the same data as those shown by the upper black line but plotted using inner scaling, the solid blue line depicts the viscous law, $U^+ = y^+$, and the solid red line shows $U^+ = 1/\kappa \ln(y^+) + B$ with $\kappa = 0.41$, $B = 5.1$. The straight blue line at the top indicates the gradient of U/U_w at $y/h = 1$. The symbols are for other sets of computations at comparable Re_c, as detailed in [2]. Only one half of the channel, with a width of $2h$, is shown; the other half of the mean flow consists of an inverted mirror image, with U/U_w rising to unity at the moving wall $(y/2h = 1)^2$.

incidentally, that since the sum of the normalised viscous and Reynolds shear stresses is necessarily unity for all Reynolds numbers, as required by equation (7.4), figure 7.3 does not include the former; the frictional stress, dU^+/dy^+, is simply the difference between unity and the Reynolds shear stress data.

First, as mentioned above, the frictional layer near the wall ($y/h = 0$) becomes thinner as the Reynolds number rises. This is obvious in figure 7.3(a) from the increasingly full mean velocity profile in outer scaling (i.e. U/U_w versus y/h) and, in figure 7.3(b), from the reducing region in which the frictional shear stress is significant. Second, notice how much larger the velocity gradient near the wall is in the turbulent case, compared with the laminar case, emphasising the much larger wall stress very close to the wall and thus the much greater drag when turbulent flow occurs. It is straightforward to show that the turbulent to laminar wall stress ratio is given by $2\mathrm{Re}_\tau/(U^+|_{y=h})$, where $\mathrm{Re}_\tau = u_\tau h/\nu$; for the data in the figure at $\mathrm{Re}_c = 22500$ ($\mathrm{Re}_\tau = 550$), the ratio is about 55. Third, recall that we might have anticipated a central region of uniform velocity gradient (shear) in which the flow is similar to homogeneous shear flow (a steady version of the homogeneous shear flow turbulence (HSFT) discussed in section 5.2). Figure 7.3(a) suggests that if there is a region of constant shear, its strength weakens with increasing Reynolds number and at $\mathrm{Re}_c = 22500$ it is given by $d(U/U_w)/d(y/h)|_{y/h = 1} \approx 0.1$, shown as the thin blue line at

---

the top of the figure. It appears to be an open question whether this shear tends towards zero at an infinite Reynolds number [22] or to some small finite value [8]. What is clear, however, is that the reduction in shear is quite slow as Re_w rises.

7.2.3 The viscous sublayer

Consider now the mean velocity profile of figure 7.3(a) plotted as U^+ versus y^+. It was Prandtl [37] who first postulated that at high Reynolds numbers, there is an inner layer close to the wall, in which the mean velocity profile must be determined only by viscous scales – independently of the characteristic length and velocity scales far away from the wall – in this case, h and U_w. The natural velocity scale in the viscosity-dominated region is u_τ (as mentioned earlier) and the viscous length scale, which we call η_v, must then be $\eta_v = \nu/u_\tau$. Note that this is quite distinct from the Kolmogorov scale, η, discussed in section 3.3. Hence, $y^+ = y/\eta_v = yu_\tau/\nu$ is a wall distance appropriately normalised by this viscous scale, as previously expressed in equation (7.3). It can be thought of as a local Reynolds number, so its magnitude at any point is a measure of the relative importance of the viscous and turbulent processes. U^+ ($=U/u_\tau$) and y^+ are normally thought of as the velocity and the wall distance in *wall units* (i.e. *inner scaling*). It is clear that in the region very close to the wall, where the turbulence shear stress $\overline{u'v'}$ can be ignored when compared to the viscous stress, equation (7.2) leads directly to

$$U^+ = y^+. \tag{7.5}$$

This viscosity-dominated region is usually called the *viscous sublayer*, although sometimes it is referred to as the laminar sublayer – rather a misnomer, as that could suggest there are no velocity fluctuations within it, which is not true. Figure 7.3(a) includes the sublayer $U^+ = y^+$ relation and it can be seen to describe the profile up to $y^+ \approx 5$. Note that equation (7.5) is not exact; a fuller analysis shows that $U^+ = y^+$ is correct up to $\mathcal{O}(y^{+4})$ (i.e. a Taylor series expansion shows, on applying the boundary conditions, that near the wall terms in y^{+2} and y^{+3} are identically zero, see Pope [36], for example). Notice, too, how thin the viscous sublayer is; at $y^+ = 5$, $y/h < 1\%$, emphasising how taxing it is to undertake a laboratory experiment to measure velocities within this viscous region, which becomes increasingly thin as the Reynolds number rises.

7.2.4 Beyond the viscous sublayer

Further from the wall, the turbulence shear stress begins to be significant and eventually, after a region usually called the *buffer layer*, the profile appears to follow a log-linear relation, above around $y^+ = 30$, which can be written

$$U^+ = A \ln(y^+) + B. \tag{7.6}$$

The reasons for which one might expect such a profile are explained in section 7.3.2.

One of the practically important measures related to the energy needed to drive the flow in any wall-bounded situation (or, equivalently, the drag the flow imposes on the surface) is what is called the *skin friction coefficient*, which can be defined (for this Couette flow) in terms of the wall velocity U_w by

$$C_f = \frac{\tau_w}{\rho U_w^2/2} = 2\left(\frac{u_\tau}{U_w}\right)^2. \tag{7.7}$$

If equation (7.6) holds all the way to the centre of the channel (where $U^+ = U_w^+/2$), it follows that

$$C_f = 0.5/(A \ln(\mathrm{Re}_\tau) + B)^2, \tag{7.8}$$

showing that C_f falls with increasing Re_τ, but much more slowly than it would in the laminar Couette flow, for which $C_f = 1/\mathrm{Re}_w = 1/(2\mathrm{Re}_\tau^2)$. So, for example, taking $A = 2.5$ and $B = 5$ as typical for equation (7.5), the ratio of turbulent to laminar friction coefficients (if the flow were somehow to remain laminar) is about 11 at $\mathrm{Re}_\tau = 50$ but about 1330 at $\mathrm{Re}_\tau = 500$.

7.2.5 The turbulence

Finally, we make a few comments about the nature of the turbulence away from the wall region. We show first, in figure 7.4(a), profiles across the channel of the three normal Reynolds stresses (as rms values in viscous units, i.e. normalised by u_τ). Note first that there is a maximum in u'^+_{rms} of around 2.8 close to the wall, which turns out to be fairly typical of wall-bounded flows. Second, because the turbulence kinetic energy production term is nonzero across the whole flow, there is significant axial stress at the centreline – $u'^+_{\mathrm{rms}} \approx 2.1$. Third, it is clear that the turbulence is anisotropic throughout the flow, including around the centreline – much more so than in pure channel flow, see section 7.3.

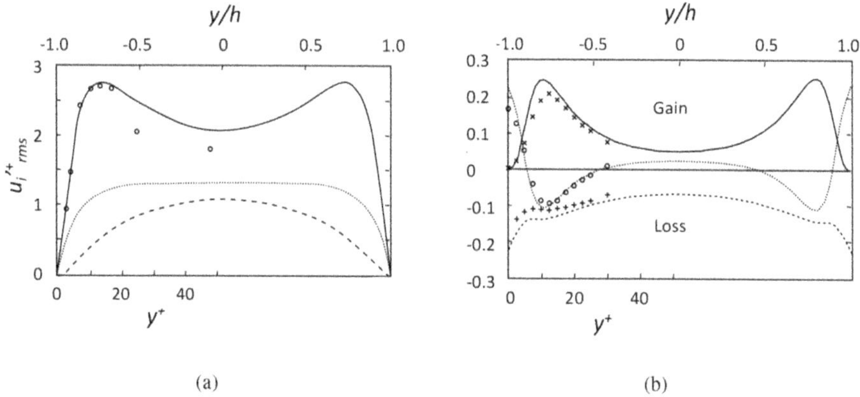

(a)   (b)

Figure 7.4. DNS data for a Couette flow at $\mathrm{Re}_\tau = 52$ from [18]. (a) normalised root-mean-square velocity fluctuations; solid line, u'^+_{rms}; dashed line, v'^+_{rms}; dotted line, w'^+_{rms}. Symbols are laboratory data at $\mathrm{Re}_\tau = 82$ from [5]. (b) Turbulent kinetic energy budget. Solid line, production; dashed line, dissipation; dotted line, transport. Symbols are from a DNS of channel flow at $\mathrm{Re}_\tau = 180$ [23][3].

---

[3] Reproduced with permission from [18], copyright Cambridge University Press.

The turbulent kinetic energy equation and the Reynolds stress equations are not exactly those presented and discussed in section 5.2, for in that case of homogeneous isotropic shear flow we assumed initial homogeneity in the turbulence and, because the velocity gradient dU/dy is invariable with y, the homogeneity remains for all time (refer to the argument just prior to equation (5.29)). However, in the case of Couette flow, although there may be a region of closely constant velocity gradient in the centre, it is certainly not constant across the whole flow. This leads to the appearance of an additional term in the Reynolds stress equation, which, for steady flow, from equation (2.27), becomes (recalling $V = W = 0$, $\partial/\partial x = \partial/\partial z = 0$)

$$0 = -\overline{u_i'u_j'}\frac{dU}{dy} - \overline{\frac{p'}{\rho}\left(\frac{\partial u_i'}{\partial x_j} + \frac{\partial u_j'}{\partial x_i}\right)} - \frac{\partial}{\partial y}\left(\overline{u_i'u_j'v'} + \frac{\overline{p'u_j'}}{\rho}\delta_{i2} + \frac{\overline{p'u_i'}}{\rho}\delta_{j2}\right)$$
$$- 2\nu\overline{\frac{\partial u_i'}{\partial x_k}\frac{\partial u_j'}{\partial x_k}}. \tag{7.9}$$

So, for example, with $i = j = 1$ (in equation (2.27), for the $\overline{u'^2}$ stress), the equation describing the transport of the streamwise component of the turbulence energy, $\frac{1}{2}\overline{u'^2}$, is

$$0 = -\overline{u'v'}\frac{dU}{dy} - \frac{1}{2}\frac{\partial}{\partial y}\overline{u'^2v'} + \frac{\overline{p'}}{\rho}\frac{\overline{\partial u'}}{\partial x} - \frac{1}{3}\epsilon, \tag{7.10}$$

which, compared with equation (5.36) for homogeneous shear flow, includes the triple-velocity-product term which would be zero if the turbulence were homogeneous. Nonetheless, DNS experiments have shown that this term is, in fact, very small except close to the wall. This equation for the axial Reynolds stress is the only one that contains the mean velocity gradient (i.e. a nonzero energy production term), so the mean flow drives the production of $\overline{u'^2}$. Energy in the other normal stress components arises as a result of the pressure strain terms (as discussed in section 5.2.2) but also, in this case, the nonzero triple-velocity-product turbulence transport. The kinetic energy equation (obtained by summing the three normal stress equations) is

$$0 = -\overline{u'v'}\frac{dU}{dy} + \frac{d}{dy}\overline{kv} + \frac{1}{\rho}\frac{d}{dy}\overline{p'v'} - \epsilon. \tag{7.11}$$

This shows directly that the imbalance between the production and dissipation of turbulence kinetic energy (k) at any point in the flow – the first and last terms – is a direct consequence of the transport of energy in the y direction. Even when the triple product term is close to zero, the pressure–velocity transport of k is not. Figure 7.3 (b) shows the turbulence kinetic energy balance for the same Couette flow (in which the fluctuation velocities are those shown in figure 7.3(a)). It is clear that the transport terms (the central two terms in the above equation, plus the viscous diffusion) is significant close to the walls, where there is thus no balance between production (the first term) and dissipation (the final term). In fact, as $y^+ \to 0$, the balance is essentially one between the inward transport of k and dissipation. Even in the central part, however, the transport term remains significantly nonzero.

We emphasise that equations (7.9)–(7.11) are written without the viscous diffusion terms, on the assumption that they are negligible. However, although at high enough Reynolds numbers this is not unreasonable in the central part of the flow, very near the wall, viscous diffusion unsurprisingly plays a significant role at any Reynolds number. In fact, it has been shown through detailed DNS studies [2] that for y^+ below about five (i.e. within the viscous sublayer), the missing viscous term in equation (7.9) balances dissipation, and all other terms are negligible. The same is true for the transport equation for $\overline{w'^2}^+$. This is reflected in the transport term shown in figure 7.4(b), which clearly becomes comparable (but of opposite sign) to dissipation below about $y^+ = 5$; this is because of the relatively large viscous diffusion, which is much larger than the other transport terms, although it is not separated from these in the figure. On the other hand, for $\overline{v'^2}^+$ the balance in that region is between pressure strain and turbulent (triple product) transport, and for $\overline{u'v'}^+$ the balance is between pressure strain and pressure diffusion. Clearly, fully developed turbulent Couette flow is complex, despite the fact that it is the flow with the simplest possible boundary conditions and, unlike all other wall flows, dependent on only one free parameter – the Reynolds number (Re_c, say).

Incidentally, since equation (7.4) shows that, at the centreline, the Reynolds shear stress, $-\overline{u'v'}^+$, is $(1 - dU^+/dy^+|_{y=h})$, the turbulence kinetic energy production there can be written as

$$P_{y=h} = \left.\frac{dU^+}{dy^+}\right|_{y=h}\left(1 - \left.\frac{dU^+}{dy^+}\right|_{y=h}\right). \tag{7.12}$$

This rather remarkable result is exact for all Couette flows and is independent of the Reynolds number. For the case shown in figure 7.3, for which the normalised velocity gradient at the centreline is about 0.05, the turbulence production on the centreline is thus about 0.048, which corresponds to the DNS result.

Overall, whilst the various turbulence statistics discussed above do depend on the Reynolds number to a greater or lesser extent, the general behaviour is similar in many respects to that for the more common case of pure channel flow, which we now discuss. Readers wishing to pursue the topic of Couette flows in more detail could refer, for example, to the joint experimental and numerical studies of Bech [5] and the DNS studies of Komminaho *et al* [18] and Tsukahara *et al* [50] and the references therein.

7.3 Channel flows

7.3.1 Governing equations

Channel flows, like Couette flows, occur between two flat plates that form a duct, see figure 7.5. The simplest case is that in which neither duct wall moves, so that the flow is driven solely by an applied streamwise pressure gradient. There are more

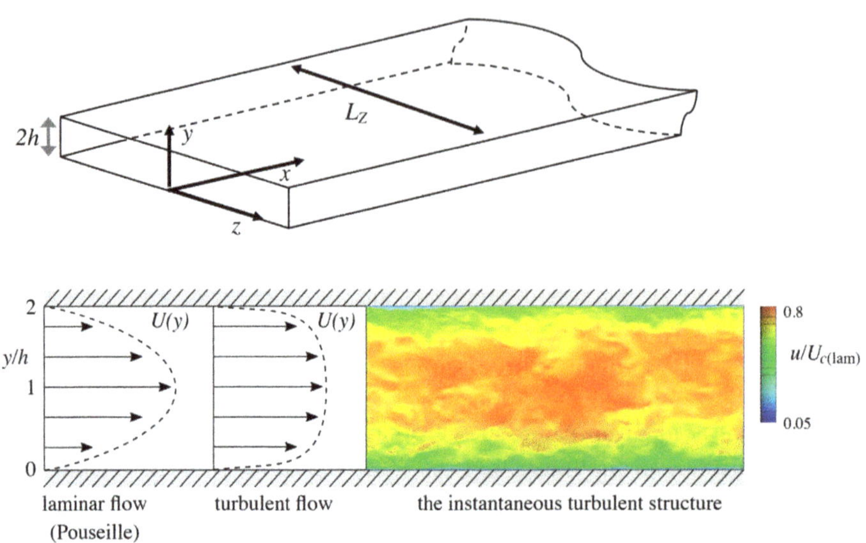

Figure 7.5. Sketch of the channel geometry, the laminar and turbulent velocity profiles in plane channel (left) and a snapshot from a numerical computation of the flow at $Re_\tau = 930$, from a video by J Lee[4]. Dark green denotes the lower velocities, shading towards red for the highest. The disorganised nature typical of turbulent flows is apparent.

complicated cases in which a pressure gradient is applied and, simultaneously, one wall moves relative to the other, but we do not consider such cases. For a laminar flow and fixed walls, consideration of the streamwise momentum equation, or a simple force balance argument, lead to a parabolic velocity profile given by

$$\frac{U}{U_m} = 2\frac{y}{h}\left(1 - \frac{y}{2h}\right), \tag{7.13}$$

where the maximum velocity on the channel centreline, U_m is $(h^2/2\mu)(dP/dx)$.

At high enough Reynolds numbers, the flow becomes transitional, and we now move on to a discussion of the fully developed turbulent channel flow. Figure 7.5 shows the geometry, a sketch of the mean velocity profile under laminar and turbulent conditions, and a snapshot from a DNS computation of the turbulent flow.

As in Couette flows, away from the entry and exit regions of a fixed-wall channel flow, i.e. in the fully developed region, the velocity statistics do not depend on x, nor, for wide enough channels and away from the side walls, on z. The flow is therefore statistically stationary and one-dimensional, just as in Couette flows. Since the flow is symmetric about the channel centreline, the statistics at y are identical to those at $2h - y$ and, in discussing the flow, we can simply consider the lower half, i.e. between $y = 0$ and $y = h$ – see figure 7.1. Defining the bulk velocity, U_B, as

---

[4] https://www.youtube.com/watch?v=t_5tEqa8rYs&ab_channel=jinLEE, courtesy of Hyung Jin Sung.

$$U_B = \frac{1}{h} \int_0^h U \, dy, \tag{7.14}$$

an appropriate Reynolds number is given by $\mathrm{Re}_B = U_B h / \nu$. It is known that the flow becomes fully turbulent once $\mathrm{Re}_B > 900$. As in Couette flow, we can consider the mean axial momentum equation or a simple force balance over the whole flow, in order to deduce how the shear stress varies. The three mean momentum equations (2.16) simplify to

$$0 = -\frac{1}{\rho}\frac{\partial P}{\partial x} - \frac{\partial \overline{u'v'}}{\partial y} + \nu \frac{\partial^2 U}{\partial y^2}, \tag{7.15}$$

$$0 = -\frac{1}{\rho}\frac{\partial P}{\partial y} - \frac{\partial \overline{v'^2}}{\partial y} \tag{7.16}$$

and

$$0 = -\frac{1}{\rho}\frac{\partial P}{\partial z}. \tag{7.17}$$

Note that continuity is automatically satisfied. These equations differ from their laminar equivalents by the appearance of the usual turbulent stress terms in equations (7.15) and (7.16), and they differ from the turbulent Couette flow because of the nonzero streamwise pressure gradient. P is clearly independent of z, and based on equation (7.16), its variation across the channel is a second-order effect:

$$\rho \overline{v'^2} + P = P_w(x), \tag{7.18}$$

where P_w is the mean pressure at the walls and depends only on x. Substituting this result into equation (7.15) yields

$$\left(\frac{\mathrm{d}P_w}{\mathrm{d}x} = \right)\frac{\mathrm{d}P}{\mathrm{d}x} = \frac{\mathrm{d}\tau}{\mathrm{d}y}, \tag{7.19}$$

where the total shear stress, τ, is given by

$$\tau = \mu \frac{\mathrm{d}U}{\mathrm{d}y} - \rho \overline{u'v'}. \tag{7.20}$$

The axial pressure gradient is, of course, what drives the flow; it acts to balance the wall stresses. Unlike the case of Couette flow, τ is clearly not constant with y. However, both sides of equation (7.19) must be constant (since one side is a function of y only and the other is a function of x only) so, using the boundary conditions, it follows that

$$\tau(y) = \tau_w\left(1 - \frac{y}{h}\right) \tag{7.21}$$

and also that

$$-\frac{\mathrm{d}P_w}{\mathrm{d}x} = \frac{\tau_w}{h}, \tag{7.22}$$

where τ_w is the wall stress (at $y = 0$). It is therefore the shear stress *gradient*, rather than the shear stress itself, that is constant across the whole channel. This is true whether or not the flow is turbulent and, in the laminar case, it is equations (7.21) and (7.22) that lead to the parabolic velocity profile quoted in equation (7.13). This laminar case is usually known as Poiseuille flow (as is the corresponding pipe flow case discussed in section 7.4), although some authors also refer to the turbulent channel case as a Poiseuille flow, despite the fact that it does not have a parabolic velocity profile.

7.3.2 The mean velocity profile

On dimensional grounds, the mean velocity gradient can depend on only two independent groups, y/h and Re. We can therefore write

$$\frac{\mathrm{d}U}{\mathrm{d}y} = \frac{u_\tau}{y}\Phi\left(\frac{y}{h}, \frac{y}{\eta_v}\right), \tag{7.23}$$

where η_v is the viscous length scale defined in section 7.2.3 as ν/u_τ. Near enough to the walls (i.e. when $y/h \ll 1$), recall that Prandtl postulated that only the viscous scales are important and the wall units are defined by equation (7.3), so that equation (7.23) must become

$$\frac{\mathrm{d}U}{\mathrm{d}y} = \frac{u_\tau}{y}\Phi_i\left(\frac{y}{\eta_v}\right) \quad \text{for} \quad \frac{y}{h} \ll 1. \tag{7.24}$$

Using wall scaling (i.e. equation (7.3), $y^+ = y/\eta_v = yu_\tau/\nu$ and $U^+ = U/u_\tau$), this becomes

$$\frac{\mathrm{d}U^+}{\mathrm{d}y^+} = \frac{1}{y^+}\Phi_i(y^+) \tag{7.25}$$

which integrates to

$$U^+ = f(y^+). \tag{7.26}$$

(As in Couette flow, in the viscosity-dominated region, this is just $U^+ = y^+ + \mathcal{O}(y^{+4})$.) In this near-wall region, we can write

$$y^+\frac{\mathrm{d}U^+}{\mathrm{d}y^+} = \Phi_i(y^+) = y^+f'(y^+). \tag{7.27}$$

Now suppose that the Reynolds number is sufficiently large that there is a region well outside the viscous sublayer in which viscosity is irrelevant but y is still much less than h. Then Φ_i in equation (7.24) must be constant; we will call this constant $1/\kappa$ and call κ the *von Kármán constant*, so that integration yields

$$U^+ = \frac{1}{\kappa}\ln(y^+) + A. \tag{7.28}$$

This is the celebrated logarithmic *law of the wall* (commonly called the *log law*) which von Kármán first deduced in 1930; it can be thought of as one of the few exact results in turbulence, although we emphasise that it is only true asymptotically as Re→∞. It requires the existence of a region in which y is large compared to η_v but small compared to h; this region is often called the *inertial layer* and sometimes the 'overlap' region. There are various alternative arguments that lead to this result but, in some ways, the above approach is the most satisfying, since it depends on minimal assumptions.

In the region beyond the viscous wall layer, one also expects that the departure of the mean velocity from its centreline value, i.e. $U_c - U(y)$, will be independent of viscosity and that the appropriate length scale will thus be h rather than η_v. Classically, the velocity scale appropriate to this velocity deficit in the outer region is still u_τ, for to an observer on the centreline and moving at the centreline velocity, the only effect of the wall is to transmit a shear stress, τ_w. (Arguments for an alternative velocity scale have been proposed but are not discussed further here.) This all implies that

$$U_c^+ - U^+(y) = g(y/h), \tag{7.29}$$

which is usually called the *velocity defect law*, and we can write the Φ in equation (7.23) as $\Phi_o(y/h)$. It follows that

$$y^+\frac{\mathrm{d}U^+}{\mathrm{d}y^+} = \Phi_o\!\left(\frac{y}{h}\right) = -\frac{y}{h}g'\!\left(\frac{y}{h}\right). \tag{7.30}$$

Since y^+ and y/h are independent variables, this equation and equation (7.27) together require that $\Phi_o = \Phi_i$. After the integration of equation (7.30) between the limits y and h and (for velocity) U^+ and U_c^+, we can express the defect velocity as

$$U_c^+ - U^+ = -\frac{1}{\kappa}\ln\!\left(\frac{y}{h}\right) + B. \tag{7.31}$$

It is worth noting that in recent decades there has been some (often heated!) discussion about whether the log law, equation (7.28), is, in fact, the most appropriate relationship for the velocity profile outside the viscous wall region. The alternative is to take it as a power law (extending well above the inertial layer), i.e. $U^+ = ay^{+n}$, where a and n are functions of the Reynolds number. This often results from an alternative assumption about the appropriate velocity scale to use for the outer flow. Interested readers might like to consult, for example, the discussion presented in [7]. Nonetheless, the weight of experimental evidence, both from laboratories and computer simulations, is, in our view, strong enough to provide powerful evidence for the efficacy of the log law in an inertial layer, although one must always bear in mind that theoretically it is only an asymptotic result requiring a sufficiently large Reynolds number.

7.3.3 Some data and their implications

We now present some experimental confirmation of the log law. First, figure 7.6 shows profiles from the laboratory channel flow experiment of Wei and Willmarth [52], undertaken in the 1980s, in which (almost for the first time) laser Doppler anemometry (LDA) was used to obtain the data, which avoids the difficulties associated with using hot-wire probes within the channel. (In retrospect, these difficulties were evident in a number of earlier studies, not least the early work of Laufer [20], and especially in the near-wall region.) The experiments also covered a wider range of Reynolds number ($\mathrm{Re}_c = hU_c/\nu$) than most previous studies – a factor in excess of 13. In figure 7.6(a), the various regions of the flow are indicated and it is clear that the data yield a reasonable collapse to a log law above the buffer layer. (Although, in principle, the authors could have used the known channel pressure gradient to deduce u_τ, they actually deduced it by fitting the near-wall data to the required $U^+ = y^+$ viscous sublayer profile; the resulting u_τ differed from that calculated from the pressure gradient by less than 6%.) The log-linear layer extends quite close to the channel centreline (which, at the different Reynolds numbers, is at different values of y^+). Since the flow is symmetric, there must be a zero velocity gradient at $y = h$, so the log-law region clearly cannot extend quite that far.

In turbulent boundary layers, explored in sections 8.3–8.7, there is usually a much more substantial region above the inertial layer, normally termed the outer layer 'wake' but, as is evident in figure 7.6(a), this region is quite narrow in channels, the excess velocity above the log law is very small and, in those particular experiments, it is only really noticeable at the highest Reynolds number. As far as U^+ is concerned, figure 7.6(a) suggests that Reynolds number effects are weak. Not surprisingly, however, they are not weak when the turbulence statistics are considered. This is most obvious in the shear stress profiles, shown in figure 7.6(b), in which the thickness of the region where viscous stresses are important seems to extend to at

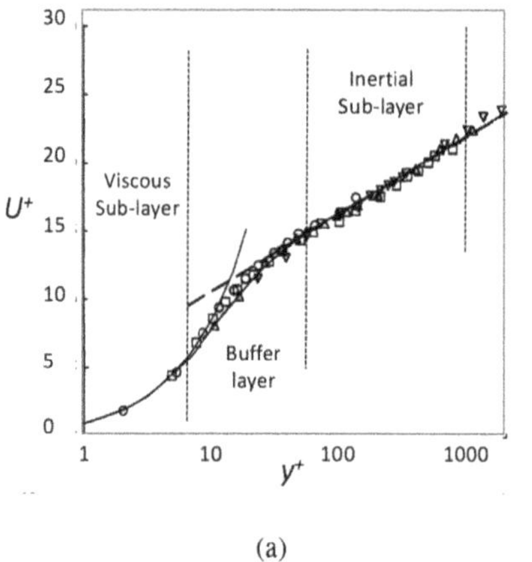

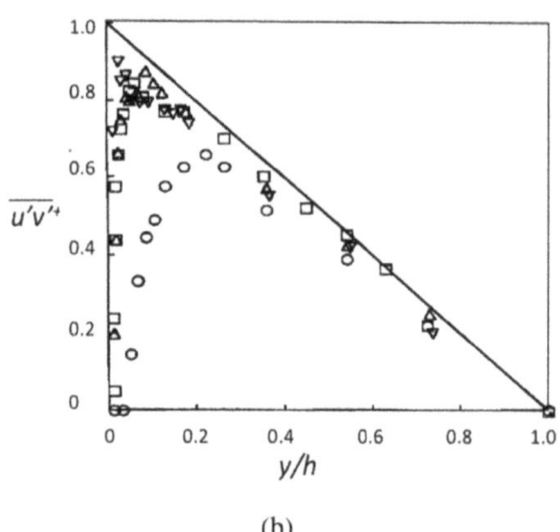

Figure 7.6. Mean velocity (a) and turbulence shear stress (b) profiles in wall units, from a turbulent channel flow, at four Reynolds numbers. $\mathrm{Re}_c(\mathrm{Re}_\tau)\approx$: O, 2970 (170); □, 14914 (710); △, 22776 (1012); ▽, 39582 (1650). Reproduced with permission from [52], copyright Cambridge University Press.

least $y/h = 0.25$ at the lowest Reynolds number. In the course of this experiment, it was shown that there is a significant interaction between the turbulence structures within one half of the channel and those in the other half; this is not surprising, for there is no 'hard' boundary at $y = h$ and eddies from below that point are able, instantaneously, to migrate to the upper half.

High-quality DNS experiments remove the inevitable uncertainties that arise in laboratory work; figure 7.7(a) shows mean velocity profiles obtained during a comprehensive set of computations by Jiménez and his co-workers [14, 15, 21]. At the higher Reynolds numbers in particular, these were expensive computations. For the reader's interest, at $\mathrm{Re}_\tau = 2003$ about 1.7×10^9 grid points were used and 6×10^6 processor hours of a supercomputer system with 2048 processors were required; 25TB of raw data were produced [14]. This kind of computation would have been inconceivable much before the beginning of this century. Except for the lowest Re_τ case, all the velocity profiles seem to collapse until near the channel centreline, just as the laboratory data shown in figure 7.6(a) suggest. However, on this kind of log-linear scale it is always difficult to be certain about the collapse in the inertial layer, and a more revealing test is to plot what is sometimes called the *Kármán measure* or the *diagnostic function*, defined by

$$\Pi = y^+ \frac{\mathrm{d}U^+}{\mathrm{d}y^+}, \tag{7.32}$$

which should equal $1/\kappa$ wherever the mean profile is logarithmic. The same data are plotted in that form in figure 7.7(b), in which it is clear that data at the highest Re_τ (4200) provide a substantial region of constant Π – extending over the approximate range $200 \leqslant y^+ \leqslant 1200$. In that region, its value is 2.58, so that $\kappa = 1/\Pi = 0.387$. The theoretical log-law line plotted in figure 7.7(a) thus uses $\kappa = 0.387$ and $A = 4.5$. This

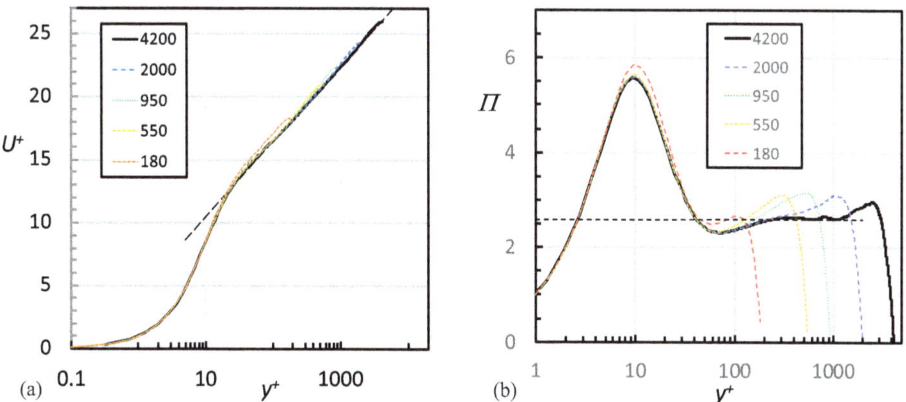

Figure 7.7. Mean velocity (a) and Kármán measure (KM) (b) profiles in wall units from a turbulent channel flow at five Reynolds numbers. Figures based on the data of Jiménez's group[5] [14, 15, 21]. Re_τ values are given in the legends. The dashed black line in (a) is equation (7.27) with $\kappa = 0.387$, $A = 4.5$.

---

[5] http://torroja.dmt.upm.es/channels

value of κ is some 3%–6% lower than the 'classical' value of 0.4–0.41, which, for most of the 20th century, was the accepted value. The data for $\mathrm{Re}_\tau = 4200$ plotted in the form of figure 7.7(a) also fit very closely to a log-law line for which $\kappa = 0.41$ and $A = 5.2$ (not shown), which illustrates how careful one should be in using such plots to deduce κ. And, we emphasise, this is in the context of knowing the wall friction velocity exactly, as it is obtained from the applied pressure gradient. In the case of external flows, such as boundary layers, the wall stress (and thus u_τ) is rarely known exactly, as we shall see in section 8.3, and this makes it even more difficult to be certain of the value of κ.

Not surprisingly, therefore, there has been some dispute over the years about the precise value of Kármán's constant, κ. Some authors believe it can be flow-dependent. This view has been well argued by Nagib and Chauhan [31], who preferred to call κ a 'coefficient' rather than a constant, but the issue remains somewhat contentious, especially among purists who believe that, in the spirit of von Kármán, it should be a universal constant. See also the discussions by Marusic [24, 25]. We will say a little more about this in our exploration of turbulent boundary layers in section 8.3.

7.3.4 The surface skin friction

Many authors have sought to measure the important skin friction coefficient. Perhaps the first comprehensive set of experiments, which also covered some of the laminar flow regime, was published by Dean [10]; he deduced a correlation for the fully turbulent channel flow which fitted the data reasonably well and is given by

$$C_f = 0.073\mathrm{Re}_b^{-0.25}, \tag{7.33}$$

where Re_b is the Reynolds number based on the bulk velocity and full channel width, $(2hU_b/\nu)$ and $C_f = 2\tau_w/(\rho U_c^2)$, respectively. Figure 7.8 shows this result and an

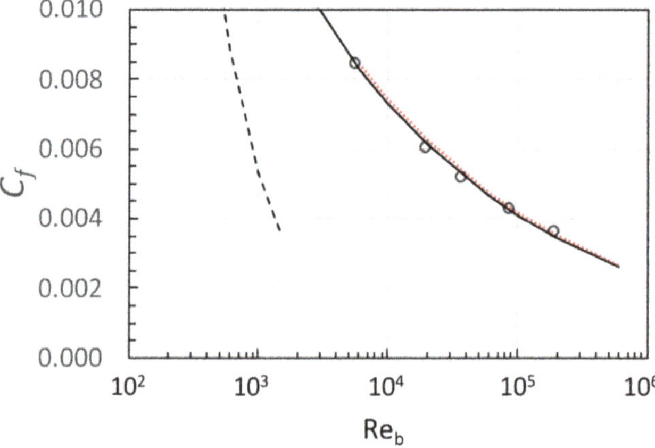

Figure 7.8. Skin friction coefficient in a turbulent channel flow. Symbols are deduced from the data of Jiménez's group[6]; the dashed line is the laminar case ($C_f = 16/(3\mathrm{Re}_b)$); the solid black and dotted red lines are the Dean and Zanoun correlations [10, 56], respectively.

---

[6] http://torroja.dmt.upm.es/channels

alternative, much later, correlation [56] based partly on more modern, and hence arguably more accurate, experimental techniques. Nonetheless, the difference is very small (0.073 is replaced by 0.0743 in the above equation). The DNS data from which figure 7.7 was derived can also be used to deduce Re_b and C_f, and the results are included in figure 7.8. Recall that transition is normally reckoned to begin at $Re_b \approx 1500$, so C_f then begins to rise from its laminar value towards the fully turbulent line, and transition is complete by about $Re_b \approx 3000$. It is also possible to deduce the C_f variation from the defect law for the velocity profile. This requires some straightforward assumptions, but the result is close to the data shown in figure 7.8; the details are given by Pope [36].

7.3.5 The turbulence

DNS computations, unlike laboratory experiments, allow deep exploration of any of the turbulence statistics that may be of interest. We conclude this section with comments about some of these, as deduced from the computations mentioned above by the Jiménez group. Figure 7.9 shows the rms axial velocity fluctuation profiles corresponding to the velocity profiles in figure 7.7(a), plotted against both y/h and y^+. A number of points should be noted. First, recall our comment in the previous section that Couette flow turbulence is noticeably more anisotropic than it is in channel flows. This is evident by comparing figure 7.9(a) with figure 7.4(a). For example, on the centreline ($y/h = 1$), the ratio of the axial to the vertical (or spanwise) rms fluctuations is 1.5, whereas in Couette flow it is between about 1.6 (for $u'^+_{\text{rms}}/w'^+_{\text{rms}}$) and 2.0 (for $u'^+_{\text{rms}}/v'^+_{\text{rms}}$). The ratio increases somewhat as y/h decreases.

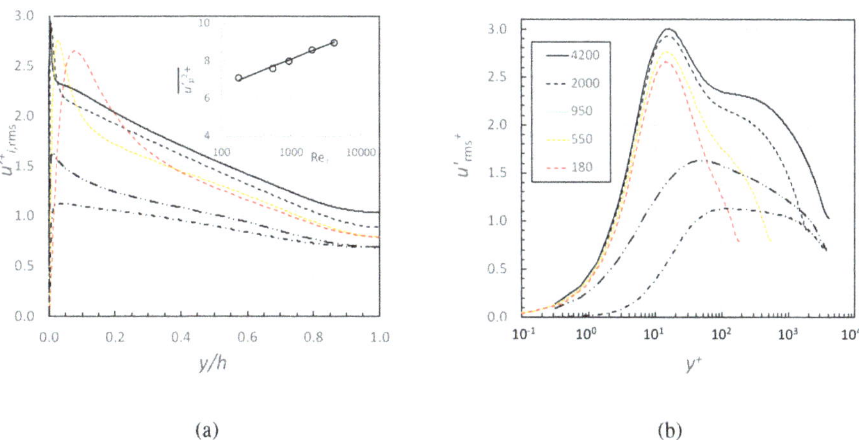

Figure 7.9. (a) Profiles of the axial velocity intensity profiles (rms values) for a turbulent channel flow at five Reynolds numbers. Figure based on the data of Jiménez's group[7] (as in figures 7.6 and 7.7); Re_τ values are given in the legend of (b). The inset in (a) shows the variation of the peak mean square values as a function of Re_τ, fitted by $\overline{u'^2}^+_p = 0.65\ln(Re_\tau) + 3.65$. The lower two (black) lines are profiles of v'^+_{rms} (the lowest line) and w'^+_{rms}. (b) The same as in (a), but plotted against y^+.

---

[7] https://torroja.dmt.upm.es/channels/data/statistics

Second, despite a reasonable collapse in the mean velocity profiles over the Reynolds number range studied (see figure 7.7(a)), there is clearly no collapse in the $u'^+_{i,\,\mathrm{rms}}$ profiles however they are plotted. The fact that fluctuation profiles are not independent of Re_τ, even within the log-law region, suggests that the usual wall scaling is inadequate for the turbulence statistics in the way that is classically expected, no doubt because the influence of the outer layer turbulence extends into the viscous region. There have been a number of attempts to introduce alternative scaling to take account of this, starting with [13], and subsequently, with varying degrees of success. We do not pursue the issue here, although rather more is said about it in section 7.4 because data (for the corresponding pipe flow case) are now available at much higher Reynolds numbers.

Third, it is evident that, as in Couette flow, there are peak values of u'^+_{rms} near the wall at a roughly constant $y^+ \approx 15$ – see figure 7.9(b). This peak value increases with the Reynolds number and, as shown in the inset of figure 7.9(a), its mean square value varies logarithmically, at least over this range of Re. Such a variation was first noted in the laboratory in the context of boundary layers by De Graaff and Eaton [13] but, since the peak is so near the wall, its accurate measurement is not without difficulty. Quality DNS avoids this difficulty, and the solid line shown in the inset is the fit proposed by the Jiménez group [42], according to the general proposal for the peak variance:

$$\overline{u'^2}^+_p = A \ln(\mathrm{Re}_\tau) + B, \qquad (7.34)$$

which has been much discussed (e.g. [26]) and initially proposed by Townsend [49]. (Note that A and B here are not related in any way to those in the standard log laws, equations (7.28) and (7.31).) It must be emphasised that this behaviour cannot continue as $\mathrm{Re} \to \infty$, for that would seem to imply asymptotically infinite dissipation in the wall region, as argued by Chen $et\ al$ [9].

We conclude by considering the turbulence kinetic energy (TKE) balance, which we call B_k here. This is exactly the same as in Couette flow, equation (7.11), and was discussed in that context there. We can summarise the full TKE (for a steady flow) as

$$B_k = P_k + T_k + Pr_t + V_k + \epsilon = 0 \qquad (7.35)$$

where P_k is the production term, T_k is the turbulent transport, Pr_t is the pressure transport, V_k is the viscous diffusion, and ϵ is the dissipation. These correspond to the four terms in equation (7.11), plus V_k. Recall that we previously ignored this latter term, viscous diffusion, which we include here and which becomes, in this case, $V_k = \nu((\mathrm{d}^2/\mathrm{d}y^2)(k + \overline{v'^2}))$; if the Reynolds number is high, we expect this to be negligible everywhere except close to the wall. In an ideal logarithmic layer in which we expect the Reynolds shear stress to be u_τ^2 (see section 8.3.2) and the mean velocity gradient to be simply $1/\kappa y$, the production term P_k would be $u_\tau^3/\kappa y$, so all the terms in the TKE might be expected to decrease roughly according to $1/y$ away from the wall, which means that an appropriate way to normalise all the terms is to multiply them by y/u_τ^3, as is done for the TKE balance shown in figure 7.10. With a logarithmic scale for y, the areas under each line are then proportional to the total (integrated) energy.

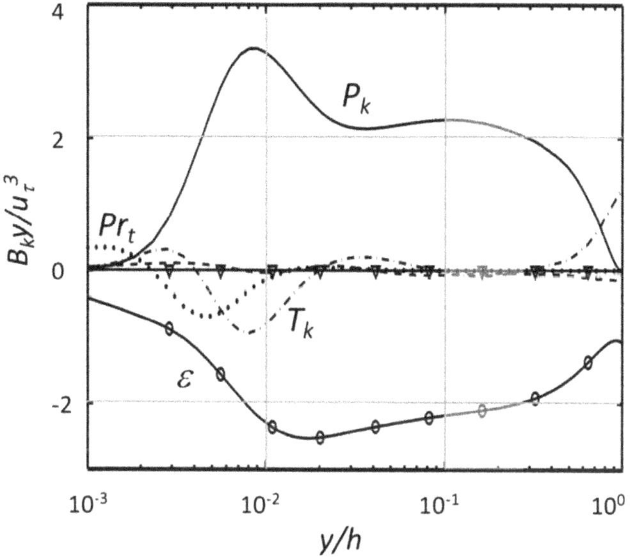

Figure 7.10. Turbulent kinetic energy balance for a turbulent channel flow, with each term normalised using y and u_τ^3: $B_k y/u_\tau^3$. The yellow region marks the approximate extent of the mean velocity log law. The data correspond to the results shown in figures 7.6 and 7.8 for the $Re_\tau = 2000$ case[8].

According to the data in figure 7.7(b), the buffer layer extends up to about $y^+ = 50$, i.e. $y/h \approx 0.025$ – beyond that, the Kármán measure Π is constant at $1/\kappa$. Note first from figure 7.10 that energy production roughly balances its dissipation in the region above this buffer layer, defined approximately by $0.025 < y/h < 0.4$. It also shows that the $1/y$ behaviour holds quite well. Above $y/h = 0.4$, the production decreases faster than that, because the mean velocity gradient reduces (and is zero at $y/h = 1.0$, of course). The dissipation then becomes increasingly balanced by turbulent transport (the triple velocity product) as the centreline is approached. Second, below $y/h = 0.025$ (i.e. in the buffer layer and below), viscous diffusion becomes important, as does turbulent transport. In this region, an adequate energy balance thus requires the inclusion of V_k, as seen for Couette flow in the previous section. Just as in the latter flow, energy enters the flow by the action of the mean flow that drives the production of the axial velocity fluctuations (recall equation (7.10) and the surrounding discussion). It is the pressure strain terms that redistribute the energy to the other components (with dissipation then removing energy from all the components).

The various features of the mean velocity and turbulence fields discussed in this and the previous section are similar to those in very many wall-bounded flows, including external flows. There are differences, nonetheless, because external wall flows have additional complicating influences. In particular, in turbulent boundary layers there is always, at the very least, some dependence on x – the flow develops in

---

[8] Figure reprinted from [15] with the permission of AIP Publishing.

the streamwise direction, so homogeneity in that direction does not usually exist. This is the topic of chapter 8.

7.4 Pipe flows

7.4.1 Introductory matters

Perhaps the most common type of internal turbulent flow worldwide is that in pipes. A straight, smooth-walled pipe (figure 7.11) is the axisymmetric equivalent of the planar channel flow we considered in the previous section. A brief reminder of the laminar case is given first, to set the scene. Consider a straight, smooth-walled pipe of internal diameter d, where d is much smaller than the pipe length. The flow is driven through the pipe by a pressure difference between its ends which, apart from relatively short regions near these ends, leads to a constant pressure gradient equal to $-dP/dx$ along the pipe's length (the pressure falls along the pipe). A simple force balance over the entire pipe cross-section leads to an equivalence between this pressure gradient and the frictional stress on the inside wall of the pipe, τ_w, i.e.

$$-\frac{dP}{dx} = \frac{4\tau_w}{d}. \tag{7.36}$$

A similar balance can be expressed by equating the force that results from the pressure difference acting on a circular cross-section between the centre of the pipe

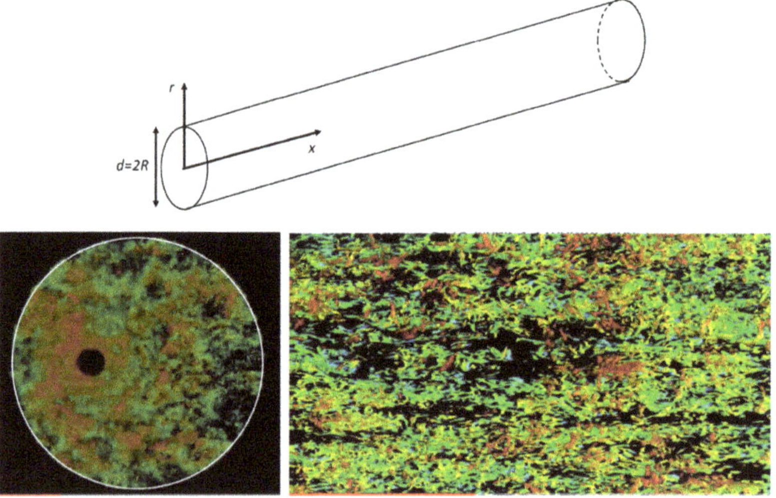

Figure 7.11. Sketch of pipe geometry and snapshots of the vortex structures, from a DNS at Re = 133000, $Re_\tau = Ru_\tau/\nu = 3008$, from the visualisations by Ahn[9] [1]. The left-hand view is down the length of the pipe along an axis slightly inclined to the pipe's axis (so that the end of the pipe is off-centre). The right-hand view is a streamwise slice through a plane normal to the pipe's axis.

---

[9] https://gfm.aps.org/meetings/dfd-2015/55f566dc69702d060d510300, copyright (2015) by the American Physical Society.

and an arbitrary radius, r, and that provided by the internal shear stress, τ, acting on the circumference of that section over the pipe length. In the absence of turbulence, only the viscous stress given by $-\mu dU/dr$ (U falls with increasing r), contributes to this internal stress, and integration of the resulting balance leads to the well-known result that the axial velocity profile is parabolic (just as in a planar channel, see figure 7.2) and given by

$$\frac{U}{U_c} = 1 - \left(\frac{r}{R}\right)^2, \tag{7.37}$$

where U_c is the centreline velocity and $R = d/2$ is the pipe radius. The bulk velocity in the pipe, U_B, is easily shown to be one half of U_c. As for the planar channel, the flow Reynolds number can be defined in terms of the bulk velocity: $\mathrm{Re} = U_B d/\nu$. It has been common practice since the Blasius era to define a *friction factor* f as

$$f \equiv -\frac{dP}{dx}\frac{d}{\frac{1}{2}\rho U_B^2}, \tag{7.38}$$

which is equal to $4C_f$, with the friction coefficient defined in the usual way ($C_f = 2\tau_w/\rho U_B^2$). Note that equations (7.36) and (7.38) imply that $f = 8(u_\tau/U_B)^2$. Using the velocity profile given by equation (7.37), which is only valid for laminar flow, it is straightforward to show that, in that case, $f = 64/\mathrm{Re}$. Reynolds' early experiments in the 1880s (see section 2.8) suggested that transition to turbulence begins at an Re of around 2000, but even after 140 years there remains much discussion both about the precise values of what are usually called the lower critical Reynolds number (below which turbulence cannot be maintained) and the upper critical Reynolds number (above which laminar flow cannot be maintained). Indeed, Reynolds himself recognised that no unique value delineates the laminar and turbulent states. It is now known that the parabolic velocity profile of equation (7.37) remains linearly stable to infinitesimally small perturbations up to infinite Re, so the value of the upper critical Re can, in fact, be made very high if the natural perturbations in the experiment are sufficiently small.

We do not discuss the fascinating topic of turbulent transition anywhere in this book, but in the context of pipe flows, the interested reader would profit from perusing [30] and Eckhardt's selection of papers [12] published as a thematic volume marking the 125th anniversary of the publication of Reynolds' historic paper, which together cite a small fraction of the thousands of papers which have addressed the topic. It is sufficient to point out, first, that uncertainties remain in trying to assign precise values to both the lower and the upper critical Reynolds numbers but, second, in practically all industrial circumstances, the 'natural' disturbances are sufficiently large and Re sufficiently high to ensure that the flow is fully turbulent. In practice, $\mathrm{Re} \approx 2300$ remains a reasonable value at which the flows in a typical (undergraduate student demonstration) experiment or an industrial gas line, for example, become turbulent. We thus turn to a consideration of fully turbulent pipe flow, recognising that this topic has also attracted huge numbers of researchers and continues to do so. Figure 7.11 shows snapshots of pipe turbulence, from a video which the reader may like to explore.

7.4.2 The friction factor

Not surprisingly, perhaps, there is an enormous literature on turbulent pipe flows. As in the laminar case, these are in some ways simply the axisymmetric equivalent of turbulent planar channel flows, depending, as they both do, only on the applied axial pressure gradient and a characteristic Reynolds number. However, there are significant differences that arise from the nonzero transverse curvature of the cross-stream coordinate in pipe flows and, consequently, the fact that there are 'side walls' whatever the value of y^+. From a practical perspective, the most important requirement is to know what the friction factor f is for a given Reynolds number (the only parameter on which it depends), for that will determine the power that is necessary to drive the fluid through the pipe. Those familiar with undergraduate texts or, indeed, industrial design codes for pipework, will be aware that for values of Re up to about 10^5, the very early relation of Blasius [6] provides a good fit, even to the subsequent large body of data not then available. Defining f by equation (7.38), his result was

$$f = 0.316/\text{Re}^{1/4}. \tag{7.39}$$

Integrating $2\pi Ur\,\mathrm{d}r$ across the pipe to obtain the bulk velocity U_B in the usual way (replacing r by R and y with y measured from the pipe wall), and using the fact that f can also be written as $8u_\tau^2/U_B^2$ (compare equations (7.36) and (7.38)), leads immediately to

$$\left(\frac{8}{f}\right)^{1/2} = \frac{U_B}{u_\tau} = 2\int_0^1 U^+\left(1 - \frac{y^+}{R^+}\right)\mathrm{d}\left(\frac{y^+}{R^+}\right). \tag{7.40}$$

This shows that the variation of f with Re depends only on the form of the velocity profile. (Note that $\text{Re} = 2(U_B/u_\tau)\text{Re}_\tau$, with $\text{Re}_\tau = Ru_\tau/\nu$.) Blasius deduced equation (7.39) using an assumption of a one-seventh power law for the velocity profile and on the basis of many extant measurements, not least the extraordinarily meticulous data produced by Saph and Schoder [39] in the first decade of the 20th century. (Readers might find a recent historical review of interest in this regard [44].)

For Re > 10^5, the Blasius relation fails to describe the data and another well-known relationship was derived by Prandtl, who assumed a log-law velocity profile. The result is known as *Prandtl's friction law for smooth pipes*; he adjusted the constants slightly so that the relation agreed well with the extant friction factor data obtained by Nikuradse [33]. The result is

$$\frac{1}{\sqrt{f}} = 2.0\log(\text{Re}\sqrt{f}) - 0.8, \tag{7.41}$$

to be compared with Blasius' law and the laminar result, $f = 64/\mathrm{Re}$. Schlichting [41] believed this result to be valid to arbitrarily high Re (and therefore thought that measurements at higher Re were not required!). Figure 7.12 shows the two relations and the laminar flow result, connected by a transitional region (marked in red). The latter should be taken as approximate; whilst there are some extant measurements covering that regime for particular experimental setups, e.g. [40], the significant uncertainties about transition that were mentioned earlier suggest that it would be unwise to attempt precision in this range of Re. Figure 7.11 includes some much more recent data obtained from the *Princeton superpipe* experiments. This facility was specifically designed to achieve very much higher Re than ever before by using highly pressurised air so as to increase its density (and hence Re) by more than an order of magnitude. Although, unlike the Blasius relation, the Prandtl one works well up to nearly 2×10^6, it deviates noticeably from the data beyond that point. Superpipe data for $\mathrm{Re} \geqslant 10^4$ and further careful study [27, 55], led to a modification of the two constants in equation (7.41) from 2.0 and 0.8 to (eventually [28]) 1.93 and 0.537, respectively, providing a good fit over more than a decade of the highest Re as well as for lower Re, as evident in the figure. At the time of writing, therefore, the best attested friction law for turbulent flow in smooth pipes at Re values in the range of $3 \times 10^5 \leqslant \mathrm{Re} \leqslant 3.5 \times 10^7$ is

$$\frac{1}{\sqrt{f}} = 1.930 \log(\mathrm{Re}\sqrt{f}) - 0.537. \tag{7.42}$$

At lower Reynolds numbers (but above the fully developed turbulence state), Prandtl's friction law applies, and the Blasius relation is also adequate at Re values up to $\mathrm{Re} = 10^5$.

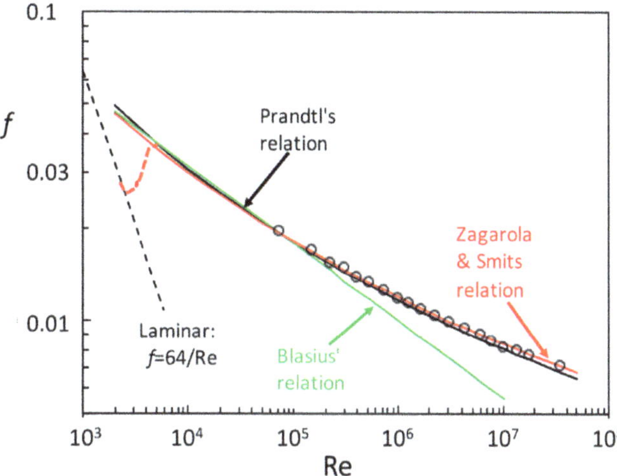

Figure 7.12. Friction factor for pipe flow. The approximate transitional region is shown as a dashed red line, rising up from the laminar result (dashed black line). The other lines are the relations of Blasius, Prandtl, and Zagarola and Smits, equations (7.39), (7.41) and (7.42), respectively. The symbols are data values given by McKeon *et al* [27], obtained in the Princeton superpipe.

The differences between these various relationships essentially result from differences in the assumed velocity profiles. At asymptotically large Reynolds numbers, for which there is sufficient separation in scales between the near-wall viscous-dominated region (where $U^+ = y^+$ with $y = R - r$) and the outer flow, one can deduce the well-known log law in essentially the same way as discussed for channels in section 7.3.2 (and used by Prandtl to obtain his friction law, (equation (7.41)). One of the early findings, however, was that the log-law regime, in terms, say, of its range of y^+ at a given Re, is not as extensive as it is in a channel flow at the same Re, even when this is well above 3000. (Recall from section 7.3.2 that the log law fits the extant channel data very well nearly up to the centre of the channel.) This difference between pipe flows and channel flows was perhaps first noted by Patel and Head [34].

7.4.3 The velocity profile

Figure 7.13 shows a selection of the mean velocity profiles obtained in the Princeton superpipe. These are similar to channel flow profiles (figure 7.7(a)), although in each case there is a more significant ('core') region between the upper end of the log-law range and the pipe centreline. The profiles collapse where they overlap in the log-law y^+ range. Note that the measured profiles extend beyond the centreline, so that the value of U_c^+ for each case can conveniently be deduced as the maximum in the profile. This clearly increases as Re increases, as does the extent of the log-law region.

One expects the existence of the log law to be most likely for higher Reynolds numbers, but a careful inspection of the profiles (not really possible using this figure)

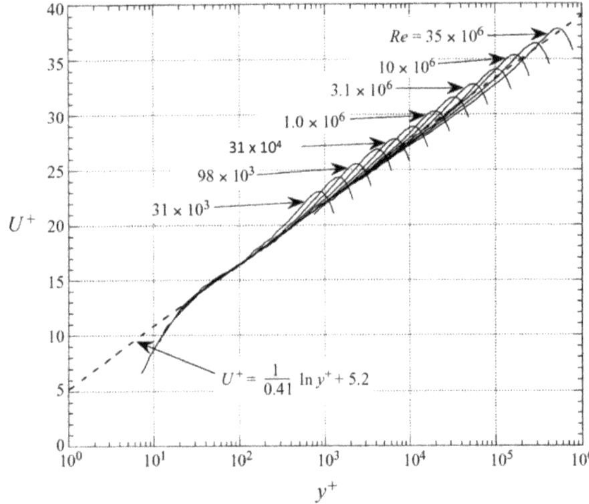

Figure 7.13. Mean velocity profiles obtained in the Princeton superpipe[10]. Reynolds numbers (Re) are shown for alternate profiles and the dashed line is the classical log law with $\kappa = 0.41$ and $A = 5.2$.

---

[10] Figure reprinted with permission from [55], copyright (1997) by the American Physical Society.

contradicts this, unless the range of the log law is taken to be, say, $600 < y^+ < 0.07\mathrm{Re}_\tau$, which is significantly more restricted than the commonly accepted limits ($50 < y^+ < 0.15\mathrm{Re}_\tau$) [55]. With these more restricted limits, the implication is that the log law cannot exist for $\mathrm{Re}_\tau < 600/0.07 = 8571$, which corresponds roughly to $\mathrm{Re} < 4 \times 10^5$. In that range, provided that Re is large enough to ensure an overlap between inner and outer regions, it turns out that a power law gives a better fit to the velocity profile, as found in the Re $=$ 44000 DNS of Wu and Moin [53], for example. (This is consistent with the Blasius frictional law, since he assumed a power law.) It has therefore been argued that pipe flow velocity profiles actually have a three-layer structure, in which the classic log law emerges at high enough values of Re_τ, but above a power-law region which is always present just above the near-wall viscous layer [11]. This latter work provides a useful entry to the literature on the nature of pipe flow velocity profiles.

Incidentally, the value of the Kármán coefficient κ is about 0.41 for good log-law fits to the profiles in figure 7.13, and this value was also found to be consistent with the implications of the centreline velocity data [28]. It is noticeably different from the value for channel flows (0.387, see section 7.3) and this issue will be discussed further in section 8.3.2. It is also worth noting that direct numerical simulations have begun to reach at least moderately high values of Re. Computations are now available at Re = 44000 [53], which is higher than the lower end of the superpipe range of cases but still significantly lower than the highest Reynolds numbers reached experimentally at the Princeton superpipe facility. These produce data that agree with the superpipe results at the same Reynolds number and yield a power law for the velocity profile, in agreement with the Princeton findings. One can expect computations at significantly higher Re values in the future, to match the largest ones obtainable in the specialist laboratory facilities typified by the Princeton superpipe. Such computations avoid, among other things, the uncertainty about the development length that is needed in a laboratory facility for fully developed conditions to be reached. This has been a contentious issue, but there is some evidence that at least $80d$ is necessary before the turbulence statistics become invariant with downstream distance, significantly less than seems to be required for planar channels; see the discussion by Marusic *et al* [24] and the references therein.

7.4.4 The turbulence

It remains to make a few comments about the fully developed turbulence statistics. Recall first that the mean velocity profiles in pipes and channels are somewhat different, not least because there is a more noticeable outer layer wake component in the pipe flow case (see figures 7.7(a) and 7.13), which was actually noted as early as the 1960s [34]. Nonetheless, more recent studies (e.g. [32]) making explicit comparisons of, for example, the Reynolds stress profiles at the same Re in pipes and channels, have indicated that despite the differences in mean velocity profiles, the stresses are very similar all across the flow – not just in the inertial layer. In the inner region, the peak axial turbulence stress rises slowly with Re in both pipes and channels (recall the channel case shown in figure 7.9(a)) and note that, anticipating our boundary layer discussion in section 8.3, the same happens in smooth wall boundary layers. However, it is perhaps currently unwise to be too dogmatic about this rise, and particularly whether or not it occurs up to the highest Re currently achieved in laboratories, given the extreme difficulty in making accurate measurements so near the wall (around $y^+ = 15$). There seems to be some evidence that it does not [16, 51].

At a sufficiently high Reynolds number, an outer peak in $\overline{u'^2}^+$ also appears, as illustrated in figure 7.14, which shows the $\overline{u'^2}^+$ profiles for a range of Re from about 2000 up to 98000, corresponding to a Re range of around 8×10^4 up to 6×10^6. A feature of these profiles is that, at least for the higher Re, the stress seems to fall logarithmically in the outer region. In fact, it was suggested long ago [49], on the basis of arguments related to those that lead to the log law in the mean velocity, that streamwise and spanwise velocity fluctuations would display a similar log law so that, for example,

$$\overline{u'^2}^+ = B_1 - A_1 \ln \frac{y}{R}. \tag{7.43}$$

It was later demonstrated that such behaviour could be derived by building on Townsend's *attached eddy hypothesis* (see section 8.5) and considering the spectral content of the fluctuation field [35]. Considerable attention has been paid to this suggestion; some data sets are in apparent conformity with the logarithmic behaviour, but others are not (in the context of boundary layers as well as channels

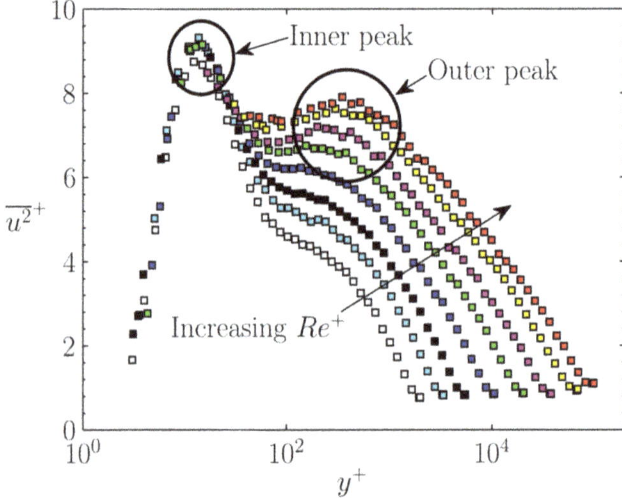

Figure 7.14. Axial Reynolds stress profiles for Re_τ values from 1985 to 98187. Princeton superpipe data[11].

---

[11] Figure reprinted with permission from [16], copyright (2012) by the American Physical Society.

and pipes). More recently, it has become apparent that only for Re_τ in excess of about 20000 can one really expect to see an extended region of such behaviour, which occurs over the same inertial range (in y^+) in which the mean flow log law applies, e.g. [51]. We discuss all this further in the context of boundary layers in section 8.3. Note finally that the similarity in stress profiles for pipes and channels also holds for the various spectra of the fluctuating velocity (and pressure); see [29], for example. Given the similarities in the stresses, similarities in the spectra are arguably not too surprising, although boundary layer spectral data are nonetheless rather different; see section 8.4.

We have thus far made no comments about the *dynamical* features of these various internal flows, but have merely presented time-averaged quantities. However, it should be obvious that these time averages, as in any turbulent flow, must depend on the nature of the structural features of the turbulence, whether those are near the wall, within the inertial layer, or in the outer region. This whole issue is discussed in the context of boundary layers in section 8.5, where we comment on the extent to which the general features differ between internal and external flows. It is worth saying here, however, that most of the important features do not differ very markedly.

7.5 Subject giant: G I Taylor

Sir Geoffrey Ingram Taylor (7 March 1886–27 June 1975) was a true giant of the field, having had a distinguished career as an English physicist and mathematician at the University of Cambridge for over 60 years. He was described in his obituary by Sir Brian Pippard as follows (quoted in [43]):

'Sir Geoffrey Ingram Taylor, who died at the age of 89, was one of the great scientists of our time and perhaps the last representative of that school of thought that includes Kelvin, Maxwell and Rayleigh, who were physicists, applied mathematicians and engineers – the distinction is irrelevant because their skill knew no such boundaries. Between 1909 and 1973 he published voluminously, and in a lifetime devoted to research left his mark on every subject he touched and on every one of his colleagues ... his outgoing manner and complete lack of pomposity conveyed, as no formal exposition could have done, the enthusiasm and intuitive understanding that informed all his work.'

Taylor was active in the field of turbulence, having multiple phenomena and concepts named after him, including the Taylor microscale, Taylor's frozen-flow hypothesis, and Taylor dispersion (in pipe flow and elsewhere), to name a few, all of which we touch upon in this text. His research spanned both fluid and solid mechanics and he published widely over his more than 60 years of a dedicated research career; his first paper appeared during his undergraduate years, studying the interference fringes formed by light waves (1909), and the last (at the age of 83), was on thunderstorms. During the First World War he contributed to the war effort, working for the Royal Flying Corps at Farnborough and helping to design and

Figure 7.15. Portrait of Sir Geoffrey Ingram Taylor (7 March 1886–27 June 1975). Reproduced with permission from [4], copyright Cambridge University Press.

improve the operation of aircraft. He also made significant contributions during the Second World War, being called upon to be part of the British delegation to the Manhattan Project. It was during this war, in 1944, that he received his knighthood.

Taylor corresponded with and influenced many of the other notable researchers in turbulence of the time including Burgers, Richardson, Prandtl, and von Kármán, notably reflected in the fact that Taylor published a bibliographical memoire of von Kármán following his death. He also left his legacy as the supervisor of both Batchelor and Townsend, who both went on to make substantial contributions of their own to the turbulence field. It was Batchelor who compiled Taylor's entire collected works into four volumes and subsequently wrote a book on *The Life and Legacy of G.I. Taylor* [4]. It is perhaps amusing for modern academics that for almost all of his life at Cambridge, he held research posts and was thus free of routine teaching, administrative, departmental, or institutional tasks; Rutherford once described Taylor as being 'paid to do no work'. There is an interesting video of G.I. (as he was affectionately known), filmed in 1967, in which he discusses low Reynolds number flow; it is part of the National Committee for Fluid Mechanics Films (NCFMF)[12].

Sample exercises

7.1. Derive the total stress equation for Couette flow given by equation (7.2) from the x-direction mean momentum equation. Are there any other flows for which this would also be valid?

7.2. Assuming that turbulent Couette flow has an S-shaped velocity profile as sketched in figure 7.2, sketch the expected shapes of the viscous stress (i.e. $\mu dU/dy$), the turbulent stress (i.e. $\rho\overline{u'v'}$), and the total stress (i.e. τ) profiles over $0 < y/h < 2$. Compare your sketches with the results presented in figure 7.3.

7.3. For a turbulent channel flow, sketch the expected shapes of the viscous stress (i.e. $\mu dU/dy$), the turbulent stress (i.e. $\rho\overline{u'v'}$), and the total stress (i.e. τ) profiles over $0 < y/h < 2$. Contrast these with those for Couette flow from the previous exercise.

7.4. The file 'ChannelData.txt'[13] contains the velocity profile within channel flow at $Re_\tau = 2000$ by Hoyas and Jiménez [14]. This exercise explores how the velocity profile compares with the expected linear relationship in the viscous sublayer and the logarithmic relationship in the inertial sublayer.

    (a) Plot the velocity profile U^+ versus y^+ using linear axes. Zoom in to the near-wall region and compare the profile with $U^+ = y^+$, which is expected in the viscous sublayer. Over what range of y^+ is this valid?

    (b) Zoom out and adjust the axes to display this plot of U^+ versus y^+ using log-linear axes. Fit the log law, equation (7.28), using $\kappa = 0.387$ and $A = 4.5$.

    (c) Plot the Kármán measure Π (given by equation (7.32)) versus y^+. Confirm the approximate extent of the inertial sublayer and the value of $1/\kappa$ indicated by the plateau.

7.5. The file 'PipeData.txt'[14] contains velocity data for a pipe flow at $Re = 10^6$ from the Princeton Superpipe by Zagarola and Smits [54]. This exercise explores how the velocity profile compares with the expected logarithmic and power law profiles.

    (a) Plot the velocity profile U^+ versus y^+ using linear axes. Zoom in to the near-wall region and compare the profile with $U^+ = y^+$, which we expect to see in the viscous sublayer. Based on the location of the closest measurement to the wall, estimate the extent (in physical units) of the viscous sublayer in this experiment.

---

[12] http://web.mit.edu/hml/ncfmf.html

[13] https://github.com/cvanderwel/TurbulentFlows/blob/main/data/ChannelData.txt

[14] https://github.com/cvanderwel/TurbulentFlows/blob/main/data/PipeData.txt

(b) Zoom out and adjust the axes to display this plot of U^+ versus y^+ using log-linear axes. Fit the log law, equation (7.28), using $\kappa = 0.41$ and $A = 5.2$.

(c) Adjust the axes to log-log scaling and consider whether a power law of the form $U^+ = C(y^+)^\gamma$ provides a better fit to the data in the near-wall overlap layer.

References

[1] Ahn J, Lee J H, Lee J, Kang J and Sung H J 2015 Direct numerical simulation of a 30R long turbulent pipe flow at $Re_\tau = 3008$ *Phys. Fluids* **27** 1–14

[2] Avsarkisov V, Hoyas S, Overlack M and Garcia-Galache J 2014 Turbulent plane Couette flow at moderately high Reynolds number *J. Fluid Mech.* **751** R1–1-R1-10

[3] Aydin E M and Leutheusser H J 1987 Experimental investigation of turbulent plane-Couette flow *Proceedings (A88-14124 03-34) Forum on Turbulent Flows (Cincinnati, OH, June 14-17, 1987)* (New York: American Society of Mechanical Engineers) 51–4

[4] Batchelor G and Taylor G I 1996 *The Life and Legacy of GI Taylor* (Cambridge University Press: Cambridge)

[5] Bech K, Tillmark N, Alfredsson P and Andersson H 1995 An investigation of plane turbulent Couette flow at low Reynolds numbers *J. Fluid Mech.* **286** 291–325

[6] Blasius H 1913 Das Aehnlichkeitsgesetz bei Reibungsvorgängen in Flüssigkeiten *Mitteilungen über Forschungsarbeiten auf dem Gebiete des Ingenieurwesens* (Berlin, Heidelburg: Springer) 1–41978-3-662-02239-9

[7] Buschmann M and el Hak M G 2003 Generalized logarithmic law and its consequences *AIAA J* **41** 565–72

[8] Busse F 1970 Bounds for turbulent shear flow *J. Fluid Mech.* **41** 219–40

[9] Chen X, Katepalli R and Sreenivasan R 2021 Reynolds number scaling of the peak turbulence intensity in wall flows *J. Fluid Mech.* **908** 1–11

[10] Dean R 1978 Reynolds number dependence on skin friction and other bulk flow variables in two-dimensional rectangular duct flow *J. Fluids Eng.* **100** 215–3

[11] Diwan S S and Morrison J M 2021 Intermediate scaling and logarithmic invariance in turbulent pipe flow *J. Fluid Mech.* **913** R1–12

[12] Eckhardt B 2009 Turbulence transition in pipe flow: 125th anniversary of the publication of Reynolds' paper *Phil. Trans. R. Soc.* A **367** Theme Issue

[13] Graaff D D and Eaton J 2000 Reynolds-number scaling of the flat plate turbulent boundary layer *J. Fluid Mech.* **422** 319–46

[14] Hoyas S and Jiménez J 2006 Scaling of the velocity fluctuations in turbulent channels up to $Re_\tau = 2003$ *Phys. Fluids* **18** 1–4

[15] Hoyas S and Jiménez J 2008 Reynolds number effects on the Reynolds-stress budgets in turbulent channels *Phys. Fluids* **20** 1–8

[16] Hultmark M, Vallikivi M, Bailey S and Smits A 2012 Turbulent pipe flow at extreme Reynolds numbers *Phys. Rev. Lett.* **108** 1–5

[17] Kitoh O, Nakabyashi K and Nishimura F 2005 Experimental study on mean velocity and turbulence characteristics of plane Couette flow: low-Reynolds-number effects and large longitudinal vortical structure *J. Fluid Mech.* **539** 199–227

[18] Komminaho J, Lundblach A and Johansson A 1996 Very large structures in plane Coutte flow *J. Fluid Mech.* **320** 259–85

[19] Krank B, Fehn N, Wall W A and Kronbichler M 2017 A high-order semi-explicit discontinuous Galerkin solver for 3D incompressible flow with application to DNS and LES of turbulent channel flow *J. Comput. Phys.* **348** 634–59

[20] Laufer J 1950 *Investigations of turbulent flow in a two-dimensional channel* California Institute of Technology 10.7907/6ZYC-HJ88

[21] Lozano-Duran A and Jiménez J 2014 Effect of the computational domain on direct simulations of turbulent channels up to $Re_\tau = 4200$ *Phys. Fluids* **26** 1–7

[22] Lund K and Bush W 1980 Asymptotic analysis of plane turbulent couette-poiseuille flows *J. Fluid Mech.* **96** 81–104

[23] Mansour N, Kim J and Moin P 1988 Reynolds stress and dissipation-rate budgets in a turbulent channel flow *J. Fluid Mech.* **194** 15–44

[24] Marusic I, McKeon B J, Monkewitz P A, Nagib H M, Smits A J and Sreenivasan K R 2010 Wall-bounded turbulent flows at high Reynolds numbers: Recent advances and key issues *Phys. Fluids* **22** 103

[25] Marusic I, JPMonty, Hultmark M and Smits A 2013 On the logarithmic region in wall turbulence *J. Fluid Mech.* **716** 1–11

[26] Marusic I, Baars W and Hutchins N 2017 Scaling of the streamwise turbulence intensity in the context of inner-outer interactions in wall turbulence *Phys. Rev. Fluids* **2** 1–22

[27] McKeon B, Li J, Jiang W, Morrison J and Smits A 2004 Further observations on the mean velocity distribution in fully developed pipe flow *J. Fluid Mech.* **501** 135–47

[28] McKeon B, Zagarola M and Smits A 2005 A new friction factor relationship for fully developed pipe flow *J. Fluid Mech.* **538** 429–43

[29] Monty J, Hutchins N, Ng H, Marusic I and Chong M 2009 A comparison of turbulent pipe, channel and boundary layers *J. Fluid Mech.* **632** 432–42

[30] Mullin T 2011 Experimental studies of transition to turbulence in a pipe *Annu Rev Fluid Mech* **43** 1–24

[31] Nagib H and Chauhan K 2008 Variations of von Kármán coefficient in canonical flows *Phys. Fluids* **20** 101518

[32] Ng H, Monty J, Hutchins N, Chong M and Marusic I 2011 Comparison of turbulent channel and pipe flows with varying Reynolds number *Exp. Fluids* **51** 1261–81

[33] Nikuradse J 1932 Strömungsgesetze in rauhen Rohren *Forschung auf dem Gebiete des Ingenieurwesens* Supplement **4**(B) (English Transl. NACA TT F-10, 359) VDI research booklet 361. Supplement to Research on the Areas of Engineering Issue B Volume 4

[34] Patel V and Head M 1969 Some observations on skin friction and velocity profiles in fully developed pipe and channel flows *J. Fluid Mech.* **38** 181–201

[35] Perry A, Henbest S and Chong M 1986 A theoretical and experimental study of wall turbulence *J. Fluid Mech.* **165** 163–99

[36] Pope S B 2000 *Turbulent Flows* (Cambridge: Cambridge University Press)

[37] Prandtl L 1925 Bericht über die Entstehung der Turbulenz *Z. Angew Math. Mech.* **5** 136–9

[38] Reichardt H 1956 Über die Geschwindigkeitsverteilung in einer geradlinigen turbulenten Couetteströmung *Z. Agnew Math. Mech.* **36** 26–9

[39] Saph A and Scoder E 1903 An experimental study of the resistances to the flow of water in pipes *Trans. Am. Soc. Civ. Eng.* **51** 253–312

[40] Schlichting H 1979 *Boundary Layer Theory* 7th edn (New York: McGraw-Hill)

[41] Schlichting H 1987 *Boundary Layer Theory* (New York: McGraw-Hill)

[42] Sillero J, Jiménez J and Moser R 2013 One-point statistics for turbulent wall-bounded flows at Reynolds numbers up to $\delta^+ \approx 2000$ *Phys. Fluids* **25** 1–15

[43] Sreenivasan K 2011 G.I. Taylor: The inspiration behind the Cambridge school *A Voyage Through Turbulence* ed P A Davidson, Y Kaneda, K Moffatt and K R Sreenivasan (Cambridge: Cambridge University Press) pp 127–86

[44] Steen P and Brutsaert W 2017 Saph and Schoder and friction law of Blasius *Annu. Rev. Fluid Mech.* **49** 575–82

[45] Taylor G 1923 Stability of a vicous liquid contained between two rotating cylinders *Phil. Trans. R. Soc.* A **223** 289–343

[46] Taylor G 1936 Fluid friction between rotating cylinders. II. Distribution of velocity between rotating cylinders when outer one is rotating and the inner one is at rest *Proc. R. Soc. Lond.* A **157** 565–78

[47] Tillmark N and Alfredsson P 1992 Experiments in transition in plane Couette flow *J. Fluid Mech.* **235** 89–102

[48] Tillmark N and Alfredsson P 1998 Large scale structures in turbulent plane Couette flow *Advances in Turbulence VII* vol 46 ed U Frisch (Dordrecht: Kluwer) pp 59–62

[49] Townsend A A 1976 *Structure of Turbulent Shear Flow* 2nd edn (Cambridge: Cambridge University Press)

[50] Tsukahara T, Kawamura H and Shingai K 2006 DNS of turbulent Couette flow with emphasis on the large-scale structure in the core region *J. Turbul.* **7** 1–16

[51] Vallikivi M, Hultmark M and Smits A 2015 Turbulent boundary layer statistics at very high Reynolds number *J. Fluid Mech.* **779** 371–89

[52] Wei T and Willmarth W 1989 Reynolds-number effects on the structure of turbulent channel flow *J. Fluid Mech.* **204** 57–95

[53] Wu X and Moin P 2008 A direct numerical simulation study on the mean velocity characteristics in turbulent pipe flow *J. Fluid Mech.* **608** 81–112

[54] Zagarola M and Smits A 1997 Scaling of the mean velocity profile for turbulent pipe flow *Phys. Rev. Lett.* **78** 239

[55] Zagarola M and Smits A 1998 Mean-flow scaling of turbulent pipe flow *J. Fluid Mech.* **373** 33–79

[56] Zanoun E-S, Nagib H and Durst F 2009 Refined c_f relation for turbulent channels and consequence for high-Re experiments *Fluid Dyn. Res.* **41** 1–12

IOP Publishing

Turbulent Flows: an Introduction

Ian P Castro and Christina Vanderwel

Chapter 8

External wall-bounded flows

This chapter considers boundary layers. After some initial remarks in section 8.1, it starts with a brief discussion of the laminar boundary laminar in section 8.2, before moving to the fully turbulent case in the remaining sections 8.3–8.7. These are first explored in their simplest form – i.e. developing on a smooth flat surface under a constant-velocity (i.e. zero-pressure-gradient), irrotational free stream. Of course, in practice, there can be many complicating influences. For flows over stationary bodies submerged in a fluid, for example, there will usually be a pressure gradient in the flow just outside the boundary layer, which could often be calculated using potential flow theory, and this provides an outer boundary condition for flow developing along the surface. Pressure gradient effects are briefly considered in section 8.6. It is also often the case that the surface is not smooth. This is most obviously true for the boundary layer on the Earth's surface, arising from geostrophic winds aloft. Often, the ratio of roughness height to boundary layer depth is sufficient to remove the dominant effects of viscosity very near the wall. Rough-wall boundary layers are considered in section 8.7. There are a number of other common complicating influences – e.g. the presence of turbulence in the external free stream, or variable density, usually generated by a wall temperature which differs from that of the free stream. It is probable that the literature on boundary layers of all kinds, which is vast and has been generated over more than a century, is comparable or perhaps even greater in size to that on pipe flows, undoubtedly because of their extreme importance in both industrial and environmental contexts. In this introductory text, we can only scratch the surface and thus we do no more than allude briefly to the effects of some of the possible complicating influences as the discussion proceeds. The chapter includes a brief consideration of the turbulence structures, section 8.5.

8.1 Introductory remarks

Recall that for internal flows, much greater power is required to drive a turbulent than a laminar flow. A similar power implication arises for external, wall-bounded turbulent flows. Flows around vehicles and over buildings are common examples of external flows. Consider flow over an aircraft wing. There is always a thin region next to the wing surface in which the fluid velocity decreases from the wing's velocity to a minimum a little way away. (In the situation of a fixed wing in a steady fluid stream in a wind or water tunnel, which is equivalent from a dynamics perspective, the fluid velocity increases from zero at the wing surface to the nonzero free-stream velocity.) This thin region was first termed 'the boundary layer' by Ludwig Prandtl in 1904 – see sections 8.2 and 8.8. For most aircraft, the appropriate Reynolds number is sufficiently high to ensure fully turbulent flow fairly soon after the leading edge of the wing, so that the aircraft's drag is much greater than it would be if laminar flow could be maintained (recall again section 1.3). Flows over and behind other kinds of vehicles, such as cars, vans, and trucks, for example, are almost always turbulent, although here, it is often the turbulence in the wake which has the most significant impact on vehicle drag.

Boundary layers differ from the internal flows considered in chapter 8 mainly because there is a 'free' boundary at the outer edge. In the simplest case, the flow develops because of an irrotational free stream moving past a solid wall, in which viscosity acts to generate vorticity that is then transported outward, creating a momentum flux, which is naturally much more rapid in turbulent than in laminar conditions. The outer region of the boundary layer (in the turbulent case) is similar in many ways to the outer edges of the free shear flows discussed in chapter 6. There is a highly convoluted boundary between the turbulent region, where there is nonzero vorticity, and the external irrotational (i.e. zero vorticity) flow. This is evident in the snapshots shown in figure 8.1 and the sketch of a typical situation in figure 8.2, which both show a boundary layer developing from the leading edge of a flat plate. One can think of the boundary layer as the region within which the vorticity generated at the surface has spread. The convoluted outer surface of this region is just the result of the advection of lumps of vorticity by large-scale eddies.

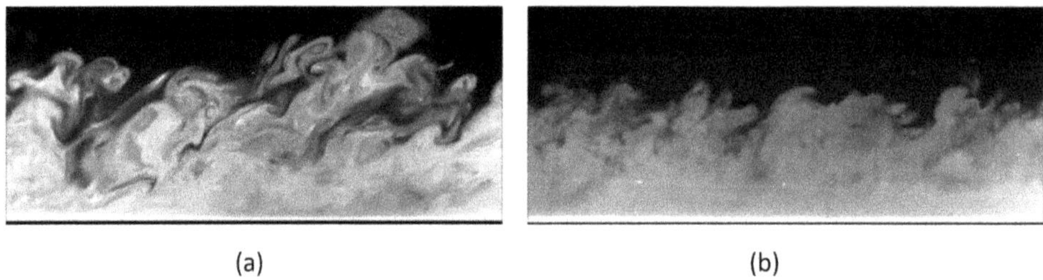

(a)   (b)

Figure 8.1. Snapshots of a boundary layer developing on a flat plate; the external flow is from left to right. The Reynolds numbers, defined as $\mathrm{Re}_\tau = \delta u_\tau/\nu$, where δ is the boundary layer thickness and u_τ is the wall friction velocity, are about 1030 on the left and 2080 on the right. The shots are taken from a movie by Lee *et al*[1] [53, 54]. (The experiment was in fact of a flow generated by a flat plate being towed through a stationary fluid.)

---

[1] https://gfm.aps.org/meetings/dfd-2014/5416804e69702d585c7b0100, reprinted with permission.

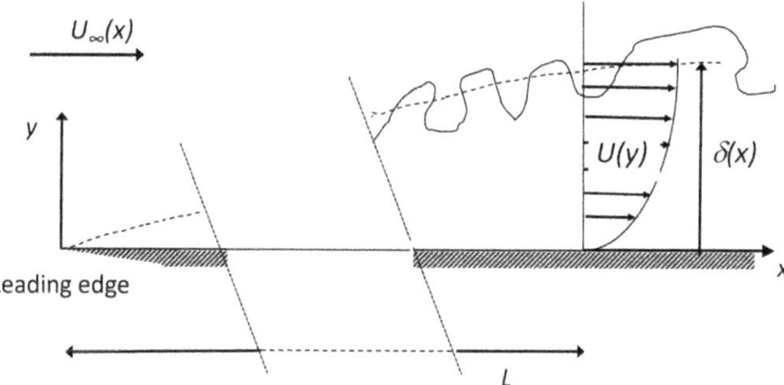

Figure 8.2. Sketch of a boundary layer growing from the leading edge of a flat plate placed along the x-axis, under a free stream of velocity U_∞. The dashed line marks the edge of the boundary layer at $y = \delta(x)$; the solid line represents an instantaneous boundary between the turbulent fluid beneath it and the irrotational fluid above it.

Another crucial difference between internal and external flows is that the wall stress is not known *a priori*. Recall that in channels and pipes, the streamwise pressure gradient is directly linked to the wall stress, so knowledge of the former provides the latter. There is (usually) no such link in boundary layers, and this makes determination of the velocity profile more difficult. On the other hand, the near-wall structure of a turbulent boundary layer is very similar in many respects to those of channels and pipes, as we will see. As in the free shear flow case, boundary layers are generally thin, in that their depth is small compared to typical axial distances within the flow, so scaling arguments, in some respects, mirror those we have already considered for free shear flows.

Incidentally, it is helpful to consider the physical processes within a general boundary layer; for instance, one developing on the surface of a stationary body – a wing with scale L, say, in a fluid moving with an upstream velocity of U_∞. (Such a boundary layer usually develops with a nonzero streamwise pressure gradient at its outer edge, so the laminar Blasius solution (see section 8.2.2) will not necessarily apply.) The vorticity within the layer originates at the surface, from which it diffuses outwards by the action of viscosity and is advected downstream by the mean flow, providing a balance between these two processes. The velocity within the boundary layer is $\mathcal{O}(U_\infty)$, so the time taken for a fluid particle to be advected over a distance L, say, is $t = \mathcal{O}(L/U_\infty)$. In that time, the vorticity will be diffused away from the surface by a distance $\mathcal{O}((\nu t)^{1/2})$. It follows that the boundary layer will have a thickness given by $\delta = \mathcal{O}((\nu L/U_\infty)^{1/2})$ and the ratio of the boundary layer thickness to the body scale is $\delta/L = \mathcal{O}(\mathrm{Re}^{-1/2})$, where Re is the body-scale Reynolds number, LU_∞/ν. The boundary layer thickness therefore decreases with increasing Re. A Reynolds number based on boundary layer thickness, i.e. $\mathrm{Re}_\delta = \delta U_\infty/\nu$, clearly increases according to $\mathcal{O}(\mathrm{Re}^{1/2})$.

Before considering the turbulent boundary layer, it is worth reviewing our knowledge of laminar boundary layers and the development of the boundary layer equations. As in the earlier chapters, we do not consider here the transitional processes through which a laminar boundary layer, even the simplest one developing on a flat wall under an external free stream without any streamwise pressure gradient, eventually becomes a fully turbulent flow. There is a very substantial literature on this topic alone, much of which is focussed on hydrodynamic stability and questions of modal dynamics. For suitable entries to it, see [44] or [100]. Despite this fascinating but complex intervening transitional region between genuinely laminar and fully turbulent flows, the process of developing the boundary layer equations for the laminar case naturally leads to a subsequent discussion of the fully turbulent case, so we start with a very brief survey of the former.

8.2 Laminar boundary layers

8.2.1 Governing equations

Prandtl realised in the early years of the last century that the layer of fluid next to the wall, in which the velocity increases monotonically from zero at the wall (because of viscosity) to the free-stream velocity further out, must be 'thin'. That is, the layer's thickness (δ, see figure 8.2) will normally be small compared with the typical dimensions of the surface on which it is developing or compared with the axial distance over which it has developed (L, say, see figure 8.2). He called this region the boundary layer. The fact that it is thin leads to significant simplifications of the Navier–Stokes equations, and these simpler equations are generally called *the boundary layer equations*; they govern the flow within the boundary layer. Their derivation parallels that leading to the thin shear layer equations discussed in the context of free shear flows, in sections 6.2.2 and 6.2.3.

We consider a two-dimensional flow in the $x - y$ plane (i.e. neither the mean flow nor the turbulence statistics depend on the spanwise, z, coordinate) with the coordinate system shown in figure 8.2 and start by recalling both the continuity equation and the Navier–Stokes equations for laminar flow, equation (2.3) in section 2.2. The basic equations for a steady, incompressible, two-dimensional laminar flow (in the absence of external forces other than the viscous force provided by the wall) reduce to

$$\frac{\partial U}{\partial x} + \frac{\partial V}{\partial y} = 0, \tag{8.1}$$

$$U\frac{\partial U}{\partial x} + V\frac{\partial U}{\partial y} = -\frac{1}{\rho}\frac{\partial P}{\partial x} + \nu\left(\frac{\partial^2 U}{\partial x^2} + \frac{\partial^2 U}{\partial y^2}\right), \tag{8.2}$$

$$U\frac{\partial V}{\partial x} + V\frac{\partial V}{\partial y} = -\frac{1}{\rho}\frac{\partial P}{\partial y} + \nu\left(\frac{\partial^2 V}{\partial x^2} + \nu\frac{\partial^2 V}{\partial y^2}\right). \tag{8.3}$$

Prandtl's thin boundary layer concept argues that the following general assumptions can be made about boundary layers (provided that they are two-dimensional and any free-stream streamwise pressure gradient is not too large):

$$V \ll U \quad \text{and} \quad \frac{\partial}{\partial x} \ll \frac{\partial}{\partial y}, \tag{8.4}$$

namely (i) that the transverse velocity is always much smaller than the streamwise velocity and (ii) that everything changes much more slowly in the streamwise direction than in the transverse direction because of the thinness of the boundary layer. These statements mirror those in section 6.2.2, which led to equations (6.1) and (6.2). Equations (8.2) and (8.3) along with scaling arguments just like those used in section 6.2 are reduced to:

$$U\frac{\partial U}{\partial x} + V\frac{\partial U}{\partial y} = -\frac{1}{\rho}\frac{\partial P}{\partial x} + \nu\frac{\partial^2 U}{\partial y^2}, \tag{8.5}$$

and

$$0 = -\frac{1}{\rho}\frac{\partial P}{\partial y}, \tag{8.6}$$

respectively. Along with the continuity equation, these are the *boundary layer equations* – clearly greatly simplified compared to equation (2.3). The streamwise momentum equation retains only one second-derivative term in only one direction. The transverse momentum equation shows that the pressure does not vary in the transverse direction so that, importantly, the local pressure outside the boundary layer (P_∞) is exactly the same as the pressure at the wall. Usefully, therefore, the pressure at the wall is set by the flow external to the boundary layer, which can often be determined using an inviscid analysis.

This set of partial differential equations is thus now mathematically parabolic and can be solved once the boundary and initial conditions are provided. Unlike the Navier–Stokes equations, which are mathematically elliptic and must be solved simultaneously over the entire flow field, the boundary layer equations can be solved numerically from, for example, a point near (but not at) the leading edge of the plate, 'marching' downstream as far as needed and only stopping near any region in which the boundary layer assumptions cease to apply – near a separation point, for example. This is exactly the tactic used by momentum integral analyses (refer to standard texts, e.g. [10, 119], or the most recent edition of Schlichting's classic text [99], for more details).

8.2.2 The Blasius flow

In the simplest case of a constant free-stream velocity (i.e. zero streamwise pressure gradient, $\mathrm{d}P_\infty/\mathrm{d}x = 0$) it was Prandtl's student Blasius who first solved the equations (in 1908) by writing them in stream-function similarity form (i.e. putting $U(x, y) = \partial\psi/\partial y = U_\infty f'(\eta)$ with the scaled variable given by $\eta = y/\delta(x) = y\sqrt{U_\infty/\nu x}$) and then using a series expansion method to solve analytically the resulting third-order ordinary differential equation ($2f''' + f''f = 0$). Later authors used numerical methods to obtain the same velocity profile. Note that the similarity approach recognises (from dimensional analysis) that the characteristic scale – the boundary layer depth δ – must be proportional to $\sqrt{U_\infty/\nu x}$. It turns out that at the point where $\eta = 5.0$, $f' = U/U_\infty \approx 0.99$ – i.e. δ is the y-value at which the velocity is quite closely equal to 99% of its free-stream value. The shear stress within the flow, τ, can be normalised using U_∞, so that $\tau_n \equiv \tau/(\rho U_\infty^2) = f'' \mathrm{Re}_x^{-1/2}$.

The boundary layer growth rate is given by

$$\frac{\delta}{x} = \frac{5.0}{\sqrt{Re_x}}, \tag{8.7}$$

where Re_x is the Reynolds number based on U_∞ and the distance from the origin of the boundary layer. As usual, the surface stress is

$$\tau_w = \mu\frac{\partial U}{\partial y}\bigg|_0 = \frac{\mu U_\infty}{\delta}f''(0). \tag{8.8}$$

The Blasius solution gives $f''(0) = 0.332\sqrt{\rho\mu U_\infty^3/x}$. This can be expressed as the usual, but here *local*, skin friction coefficient,

$$c_f = \frac{\tau_w}{\frac{1}{2}\rho U_\infty^2} = 0.664\mathrm{Re}_x^{-1/2}, \tag{8.9}$$

in which c_f is in lower case to distinguish it from the total skin friction, see below.

8.2.3 Some general relationships

Before beginning to consider turbulent boundary layers, we introduce some important integral quantities, which will be familiar to students of basic aerodynamics. Independent of whether the boundary layer is laminar or turbulent, its displacement thickness, δ^*, and momentum thickness, θ, are defined by

$$\delta^* = \int_0^\infty \left(1 - \frac{U}{U_\infty}\right)dy \tag{8.10}$$

and

$$\theta = \int_0^\infty \frac{U}{U_\infty}\left(1 - \frac{U}{U_\infty}\right)dy. \tag{8.11}$$

δ^* can be thought of as the amount by which the surface has to be displaced upwards so that the mass flux in a layer of thickness $(\delta - \delta^*)$, in which the velocity is constant at U_∞, is the same as in the actual boundary layer. Here, θ is a measure of the total drag on the surface (just as in jets and wakes, the total momentum flux deficit must equal the thrust or drag of the object generating the flow, see section 2.5). So θ must be directly related to the streamwise integral of the surface stress. This becomes evident by a direct integration of the mean momentum equation, which was first performed by von Kármán (in 1921). Integrating equation (8.5) with respect to y leads to

$$\frac{d}{dx}\left(U_\infty^2 \theta\right) + \delta^* U_\infty \frac{dU_\infty}{dx} = \frac{\tau_w}{\rho}, \tag{8.12}$$

which is true whether or not the boundary layer is turbulent (and here we have retained the axial pressure gradient). The local skin friction can therefore be expressed by

$$c_f \equiv \frac{\tau_w}{\frac{1}{2}\rho U_\infty^2} = 2\frac{d\theta}{dx} + \frac{(4\theta + 2\delta^*)}{U_\infty}\frac{dU_\infty}{dx}, \tag{8.13}$$

which is commonly called the *Kármán momentum integral equation*. In the simplest case of a zero streamwise pressure gradient, it follows immediately that $\theta = \int_0^x \frac{1}{2}c_f dx$. Note that since the total surface drag on the plate, F, say, between the leading edge and $x = L$ is just $F = \int_0^L \tau_w b\,dx$, where b is the plate width, the corresponding (total) drag coefficient, i.e. $C_d(L) = F/\left(\frac{1}{2}\rho U_\infty^2 bL\right)$, is just $\frac{1}{L}\int_0^L c_f dx$, which is the average of the local coefficient over the distance L. This is true whatever the state of the boundary layer.

If the boundary layer remains laminar, substitution of the Blasius result for c_f, equation (8.9), leads to the fact that the total drag on the surface (normalised by ρU_∞^2) up to $x = L$, say, is exactly twice the local skin friction coefficient at $x = L$. The Blasius solution also leads to the following relationships for δ^* and θ:

$$\frac{\delta^*}{x} = 1.721\mathrm{Re}_x^{-1/2} \quad \text{and} \quad \frac{\theta}{x} = 0.664\mathrm{Re}_x^{-1/2}, \tag{8.14}$$

so that both thicknesses increase like $x^{1/2}$. The *shape factor, H,* is defined as the ratio of displacement to momentum thicknesses and, for the Blasius boundary layer, is given by $H \equiv \delta^*/\theta = 2.59$. Re_x is defined by $\mathrm{Re}_x = U_\infty x/\nu$ but alternative Reynolds numbers are frequently used, not least $\mathrm{Re}_\theta = \theta U_\infty/\nu$ and $\mathrm{Re}_{\delta^*} = \delta^* U_\infty/\nu$.

We finish by emphasising that the momentum integral equation (8.13) does not depend on the boundary layer being laminar. It remains true for turbulent smooth flat-plate boundary layers, to which our attention now turns.

8.3 Turbulent boundary layers

8.3.1 Governing equations

The natural transition to turbulence in a smooth-wall flat-plate boundary layer usually begins by at least $\mathrm{Re}_x \approx 3 \times 10^6$, although it can be artificially hastened by the insertion of a suitable 'trip' wire just downstream from the leading edge. In practical situations, there are often other disturbances (such as low levels of turbulence in the free stream) which can cause early transition. We consider only the fully turbulent state when (usually) the specific route that the flow takes through the transition process (there are many) has been forgotten. Figure 8.3 shows the Blasius velocity profile for a laminar boundary layer ($U/U_\infty = f'$), along with the normalised shear stress profile, and includes a typical velocity profile for a turbulent

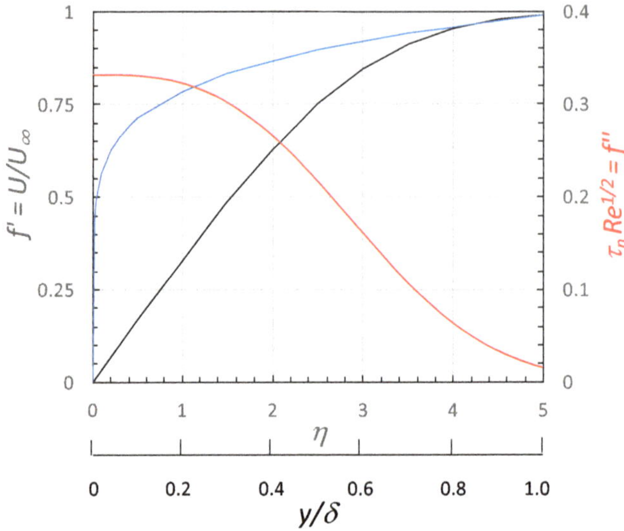

Figure 8.3. Mean velocity (black line) and shear stress (red line) for a laminar boundary layer in a zero pressure gradient. The blue line is a typical profile for a fully turbulent boundary layer at moderate values of Re.

boundary layer. It can be seen immediately that, as expected, the velocity gradient at the wall (and thus the wall stress) is very much larger than it is for a laminar boundary layer. Prandtl noticed, on the basis of the very limited (and low Reynolds number) experimental data available at the time, that the mean velocity profile was not too far from a simple power law – $U/U_\infty = (y/\delta)^{1/7}$ – and this is what is plotted in the figure. It is far from a precise representation, however, which must, in any case, depend on the Reynolds number because of the viscous region near the wall. (Note also that this profile has an infinite gradient at $y = 0$, implying an infinite wall stress, which is unphysical and clearly cannot be correct.)

Before considering the form of a more accurate velocity profile, we restate the boundary layer equations, but with the leading-order turbulence terms included:

$$U\frac{\partial U}{\partial x} + V\frac{\partial U}{\partial y} = -\frac{1}{\rho}\frac{\partial P}{\partial x} - \frac{\partial \overline{u'v'}}{\partial y} + \nu\frac{\partial^2 U}{\partial y^2}, \tag{8.15}$$

$$0 = -\frac{1}{\rho}\frac{\partial P}{\partial y} - \frac{\partial \overline{v'^2}}{\partial y}. \tag{8.16}$$

These mirror the laminar boundary layer equations (8.5) and (8.6). They are exactly the thin shear layer equations presented in the context of free shear flows (equations (6.10) and (6.6)), which result from the same kind of scaling analysis but with the important inclusion of the frictional term in the streamwise momentum equation, since this cannot be ignored in the near-wall region.

Using exactly the same arguments that follow equation (6.6), a rearrangement of the transverse momentum equation (8.16) reveals that the pressure within the boundary layer is very close to that in the free stream and that $\rho\overline{v'^2}$ is a second-order quantity. The streamwise pressure gradient in equation (8.15) can therefore be replaced by $\partial P_\infty/\partial x$ (or $-U_\infty(\mathrm{d}U_\infty/\mathrm{d}x)$). Noting that the total shear stress, τ, is given by

$$\tau = \mu\frac{\partial U}{\partial y} - \rho\overline{u'v'}, \tag{8.17}$$

equation (8.15) can thus be rewritten as follows:

$$U\frac{\partial U}{\partial x} + V\frac{\partial U}{\partial y} = U_\infty\frac{\partial U_\infty}{\partial x} + \frac{1}{\rho}\frac{\partial \tau}{\partial y}. \tag{8.18}$$

In the absence of a streamwise pressure gradient, the third term clearly drops out.

One might expect that at high Reynolds numbers, for which the thickness of the viscous region near the wall is extremely small compared with δ, it could be possible to introduce scaling parameters such as η and U_s, so that $\eta = y/\delta$ and $(U_\infty - U) = U_s f(\eta)$, and make a self-similarity argument similar to those made for free shear flows. This is indeed possible and it turns out that self-similar solutions for all of the flow above the viscous sublayer can only arise (asymptotically as Re$\to\infty$) in rather restricted circumstances – in particular, if the parameter $(\Delta/u_\tau)(\mathrm{d}U_\infty/\mathrm{d}x)$ is constant, where Δ (used instead of δ) is a normalised boundary layer thickness defined by $\Delta u_\tau = \int_0^\infty (U_\infty - U)\mathrm{d}y$, see equation (8.24) below. In general, this requires specific forms of nonzero streamwise pressure gradient, so the results are not really as useful as those obtained for free shear flows in chapter 6. Furthermore, the resulting velocity profile has to be matched to that in the viscous sublayer region, which causes additional complications. The exception is the case in which the wall is sufficiently rough, as a particular kind of roughness variation with x can lead to self-similar boundary layers. Interested readers should consult classic texts for the details [109, 114]. Rather than pursue it here, we consider first the form of the velocity profile and its dependence on the Reynolds number which arises through a combination of heuristic arguments and experimental data.

8.3.2 The velocity profile

It was the early pioneers, Prandtl (1925, 1933) and his student von Kármán (1930), who first postulated that a turbulent boundary layer should be thought of as being divided into separate and perhaps overlapping sublayers. This has been the standard approach ever since, as it fits very well with all the experimental evidence, whether from the laboratory or (more recently) numerical computations, in particular, direct numerical simulation (DNS). Perhaps unsurprisingly, however, precise details of the extent and form of each sublayer have evolved over the intervening decades. Our aim here is to first present the classical view, without going into too much detail of many of the possible nuances. The arguments essentially rest on the idea that there is an inner region that scales with u_τ and ν/u_τ, an outer region that scales with u_τ and δ, and an overlap of these two regions, in which both viscosity and δ are irrelevant, leading (at high enough Reynolds numbers) to the famous log law in that overlap region.

The discussion focusses in turn on each of the regions illustrated by the typical profile shown in figure 8.4, but we will be concerned mostly with the inertial layer and the outer region.

The inner region
The inner part of the boundary layer comprises three subregions, all of which we discussed earlier in the context of internal flows. On the basis that when $y \ll \delta$ only the viscous length scale (ν/u_τ) is important, Prandtl [82] proposed that this inner layer could be described by the similarity relation:

$$\frac{U}{u_\tau} = f_i\left(\frac{yu_\tau}{\nu}\right). \tag{8.19}$$

In the region nearest the wall, the arguments were given in the context of Couette flows (section 7.2.3) and led to the fact that, to a high degree of accuracy,

$$U^+ = y^+. \tag{8.20}$$

We are again using the 'wall' units as described in section 7.2.2 leading to equation (7.3). This is the viscous sublayer, and we do not repeat the discussion here; it is relevant to practically all wall-bounded flows and is largely impervious, at least as far as the profile shape is concerned, to major changes in the regions above. However, strong pressure gradients, in particular, can have an influence, as discussed in section 8.6. Classically, it extends to around $y^+ = 5$, although this upper boundary is not precise and can depend somewhat on what is happening in the outer flow. We repeat the earlier comment that it is often called the laminar sublayer, even though there are generally significant velocity fluctuations within it. However,

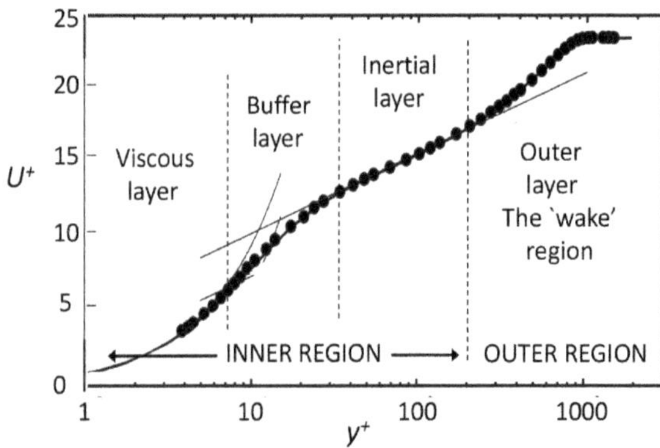

Figure 8.4. A typical mean velocity profile. The symbols represent laboratory data and the solid line through them shows the DNS profile, both from [96], for $\mathrm{Re}_x \approx 1.3 \times 10^6$, $\mathrm{Re}_\theta = 2541$. The other two solid lines show the log law (see below) and the viscous law, $U^+ = y^+$.

the turbulent stress, $\rho\overline{u'v'}$, is negligible compared with the viscous stress which, on that basis, is clearly constant within this sublayer, i.e.

$$\tau = \mu\frac{\partial U}{\partial y} = \rho u_\tau^2 = \tau_w, \tag{8.21}$$

because $\partial U^+/\partial y^+ = 1$ and $U^+ = U/u_\tau$, $y^+ = yu_\tau/\nu$, where the friction velocity, u_τ, is *defined* by $\rho u_\tau^2 = \tau_w$.

Above the viscous sublayer, turbulence fluctuations become increasingly important but only further out does the viscous stress become negligible compared with $\rho\overline{u'v'}$. The region where both play a role is often called the buffer layer, which typically extends, according to the classical literature, to about $y^+ = 30$. However, in recent decades, extensive studies at high Reynolds numbers have suggested that even this may not be conservative enough, as we will see.

Above the buffer region lies the inertial layer, throughout which viscosity is assumed to be irrelevant (because $y \gg \nu/u_\tau$), but y is still very small compared with the boundary layer depth, δ, so wall scaling remains appropriate. This can only happen at high enough Reynolds numbers, but if it does, arguments typified by that in section 7.3.2 lead to the celebrated log-law profile, equation (7.28), which we repeat here:

$$U^+ = \frac{1}{\kappa}\ln y^+ + A. \tag{8.22}$$

Note that in this region, where viscosity is unimportant, $\partial U/\partial y = u_\tau/\kappa y$ (which actually also follows directly from dimensional arguments), so the ratio of the turbulent shear stress to the total shear stress can be shown to be $1 - \nu/(\kappa u_\tau y)$, and thus the classical limit of $yu_\tau/\nu > 30$ for the bottom of the logarithmic region is not too unreasonable. We delay further discussion of this inertial region, including consideration of its upper and lower bounds (in y^+) and the values of κ and A, until after an introduction to the outer layer. However, it is important at this point to emphasise that the entire inner region only occupies, at most, some 20% of the boundary layer thickness, depending on the Reynolds number. Figure 8.4 suggests that at the relatively low Reynolds number of the data shown there, the upper limit of the log law is around $y^+ = 200$, compared with the boundary layer thickness of about $y^+ = 1000$. The figure also emphasises the thinness of the region below the log layer: for the data shown, the boundary layer thickness is about 25 mm, so $y^+ = 30$ is less than 1 mm above the surface. Special hot-wire anemometry was needed to allow the experimental data to reach the lowest measurement point, about 0.1 mm above the surface [96]. Note, incidentally, the excellent agreement between the experimental data and those obtained by DNS. At the time (2009), $\mathrm{Re}_\theta \approx 2500$ was about the highest that could satisfactorily be reached by DNS, and obtaining accurate values experimentally is not easy below that Reynolds number.

The outer region
It was through an extensive examination of the then available data that Coles [24] (interestingly, in the very first issue of the Journal of Fluid Mechanics) first showed

that the velocity profile over the entire boundary layer could be represented by a sum of two similarity functions – the *law of the wall*, f_i in equation (8.19), and the *law of the wake* (or defect law):

$$\frac{U_\infty - U}{u_\tau} = f_o\left(\frac{y}{\delta}, \frac{u_\tau}{U_\infty}\right).$$

(8.23)

This has two independent variables; the first describes the position scaled on δ and the other, $u_\tau/U_\infty = 1/U_\infty^+$, describes a weak Reynolds-number dependence. Typically, U_∞^+ varies between about 19 and 35 in the range $5000 < \mathrm{Re}_\theta < 200000$ [27]. Because of this dependence and because the 'physical' boundary layer thickness δ is rather ill-defined, it is common to use the Rotta–Clauser integral thickness [22, 89], defined by

$$\Delta = \int_0^\infty \frac{U_\infty - U}{u_\tau}\,\mathrm{d}y = \frac{U_\infty}{u_\tau}\delta^*,$$

(8.24)

where the second equality follows from the definition of the displacement thickness, equation (8.10). The defect law can thus be written as

$$\frac{U_\infty - U}{u_\tau} = f_o\left(\frac{y}{\Delta}\right).$$

(8.25)

Note that there is an integral constraint on f_o, given by $\int_0^\infty f_o\,\mathrm{d}(y/\Delta) = 1$, which must be satisfied for all boundary layers at all Reynolds numbers. Note also that the ratio of the inner to the outer length scales is given by $(\nu/u_\tau)/\Delta = 1/\mathrm{Re}_{\delta^*} \ll 1$. The same argument outlined in section 7.3.2 for channel flow, yielding equation (7.31), gives the necessary expression for f_o, and the deficit law can be written as

$$U_\infty^+ - U^+ = -\frac{1}{\kappa}\ln\left(\frac{y}{\Delta}\right) + B.$$

(8.26)

The use of the Rotta–Clauser length scale, Δ, often leads to a better collapse of the data in the outer region over a wide range of Reynolds numbers because of the implicit inclusion of changes in u_τ/U_∞ with Re_θ. The degree of collapse is illustrated in figure 8.5, in which the data cover $32000 < \mathrm{Re}_\theta < 211000$. Data in the range $2500 < \mathrm{Re}_\theta < 58000$ collapse equally well onto these data, except at the lowest Reynolds numbers, because of the reducing extent of the inner log law, equation (8.22). The overall collapse extends over a range of Re_θ of well over an order of magnitude, within which the velocity profiles display similarity in outer coordinates (defined by u_τ and Δ). These results come from a comprehensive compendium of laboratory and (the then just emerging) DNS data, in the mid-1990s [27]. The degree of collapse in the outer region (i.e. for y/Δ greater than about 0.1) is particularly impressive.

For the moment, we will continue by using the more obvious y/δ rather than y/Δ in equation (8.25), although determining the value of δ, unlike Δ, from any mean flow profile has an inherent degree of arbitrariness, as noted above. Ignoring the viscous sublayer and thus using the log law for f_i, we can write the velocity profile as

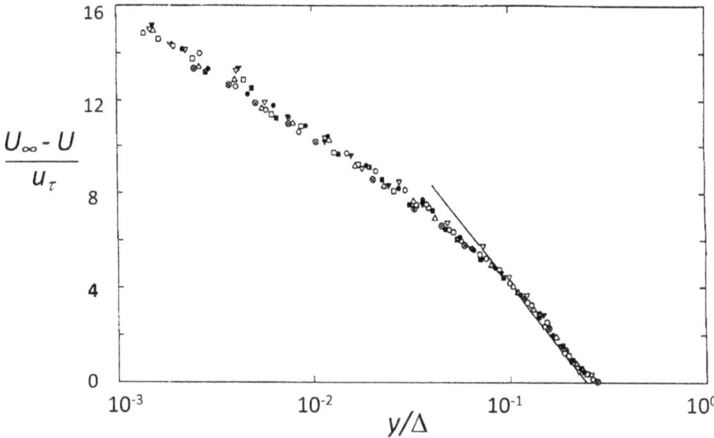

Figure 8.5. Mean velocity profiles in outer layer coordinates. The different symbols refer to experiments at different Re_θ. Reprinted from [27], copyright (1996) with permission from Elsevier.

$$U^+ = \frac{1}{\kappa}\ln(y^+) + A + \frac{\Pi}{\kappa}w\left(\frac{y}{\delta}\right). \tag{8.27}$$

Note that this Π, usually called the *wake strength parameter*, is not the same as that used to define the Kármán measure, introduced by equation (7.32). w is the (assumed universal) wake function, which must have boundary conditions specified by $w(0) = 0$ and $w(1) = 2$. Coles tabulated this to fit the available data but it has often been conveniently approximated by

$$w\left(\frac{y}{\delta}\right) = 2\sin^2\left(\frac{\pi}{2}\frac{y}{\delta}\right), \tag{8.28}$$

a shape not dissimilar to (one half of) a planar self-similar wake profile. There have been numerous alternative analytical models for the wake function. Whatever is used, equation (8.27) can be written in deficit form as :

$$\frac{U_\infty - U}{u_\tau} = -\frac{1}{\kappa}\ln\left(\frac{y}{\delta}\right) + \frac{\Pi}{\kappa}\left[2 - w\left(\frac{y}{\delta}\right)\right], \tag{8.29}$$

where the right-hand side of this relation is the outer function f_o in equation (8.25), but using δ as the outer length scale. We emphasise that the complete velocity profile requires (at least) two similarity variables for its description: $y^+ = yu_\tau/\nu$ and y/δ (or y/Δ). Since, at high Reynolds numbers, the outer layer dominates the boundary layer, it is reasonable to define the 'equilibrium' state of a turbulent boundary layer as one in which the velocity deficit in the outer part exhibits self-similarity.

Figure 8.6 shows data from a modern DNS of a smooth-wall, zero-pressure-gradient boundary layer at a momentum thickness Reynolds number of $Re_\theta = 2540$, compared with the full Coles' expressions, plotted as U^+, equation (8.27) in figure 8.6(a), and the

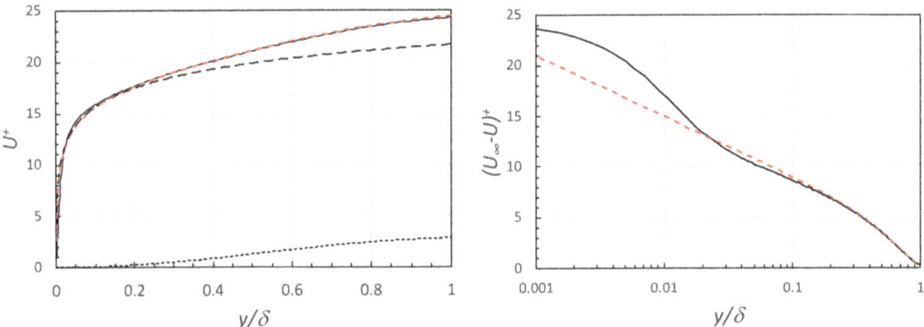

Figure 8.6. Velocity profile for $\mathrm{Re}_\theta = 2540$ (prepared from tabulations of the data available from KTH[2], Stockholm [96] and using $\kappa = 0.384$, $A = 4.173$ and $\Pi = 0.55$). (a) U^+; the black line represents the DNS data, the dashed black line depicts the log law (equation (8.22)), the black dotted line shows the wake function (equation (8.27) $\times\Pi/\kappa$), and the dashed red line illustrates their sum (equation (8.27)). (b) deficit profile; the black line represents the data; the dashed red line depicts the deficit relation, equation (8.29). See sample exercise 8.1.

deficit, equation (8.29) in figure 8.6(b). The values chosen for κ and A are the same as those used for the log-law plot in figure 8.4, and the wake strength parameter Π takes the original Coles' value of 0.55. Note first that the wake contribution, calculated using the approximation of equation (8.28), is only significant once y/δ exceeds about 0.2. Secondly, figure 8.6(b) emphasises that the inertial and wake regions together constitute about 97% of the boundary layer; the buffer region, at least for this Re_θ, starts to peel away from the log law at $y/\delta \approx 0.03$.

A well-resolved DNS allows accurate deduction of the wall friction (via the velocity gradient at the wall) but, compared with experiments, is limited in the Re_θ that can be achieved. On the other hand, within the last quarter-century, experimental techniques have been devised (principally oil-film interferometry) that allow direct measurement of the wall stress, up to high Re_θ. Figure 8.7(a) shows a set of velocity profiles over a wide range of Re_θ, plotted using inner scaling but, crucially, with u_τ obtained from independent c_f measurements. The collapse over the whole inner region – up to around $y^+ = 200$ – is impressive. An equally impressive collapse is achieved in the outer region, figure 8.7(b). (A similar collapse is achieved using δ as the outer layer length scale [72].) It has therefore become possible to assess the adequacy of classical scaling more convincingly than is possible when u_τ has to be deduced from the velocity profile. As a further example of modern, high-Reynolds-number experiments in a large wind tunnel and using specialist instrumentation, figures 8.7(c) and 8.7(d) show data obtained using extremely small hot wires, allowing measurements down to $y^+ \approx 1$. The Re_θ values range from about 18000 to 60000. Note, incidentally, the slight overshoot at the top of the buffer layer, just visible in the DNS profile and symptomatic of a moderate-Reynolds-number boundary layer; this overshoot is captured in the full composite profiles that have

---

[2] https://http://www.mech.kth.se/pschlatt/DATA/

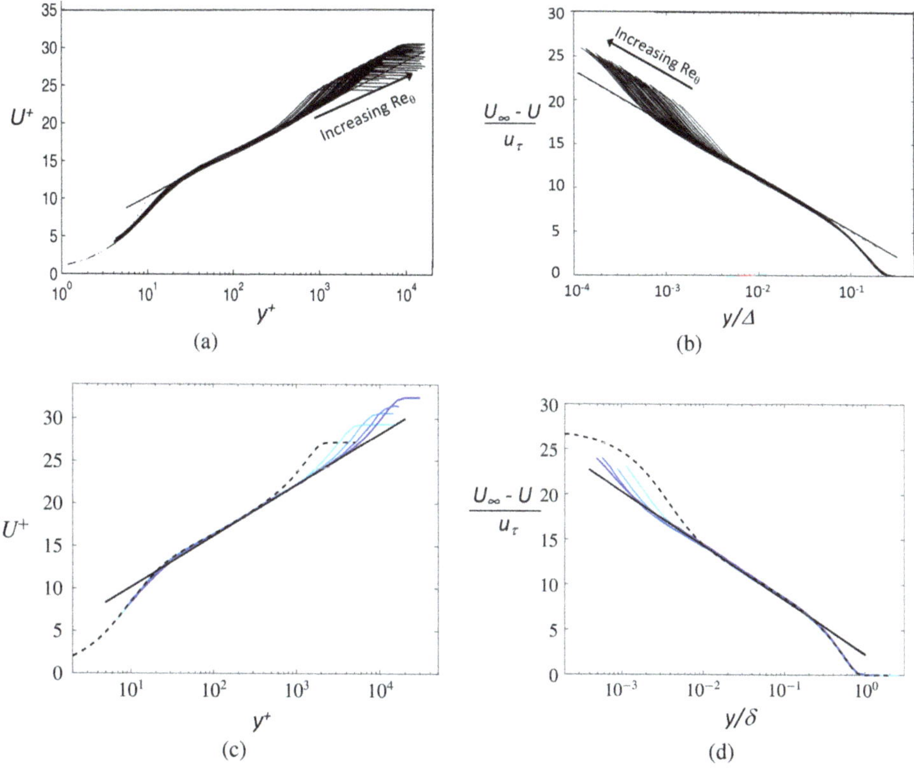

Figure 8.7. Mean velocity profile plotted using (a), (c) inner layer and (b), (d) outer layer scalings. For (a) and (b), from [71], u_τ was obtained from independent skin friction measurements. The Reynolds numbers are in the range $2530 < Re_\theta < 27300$. See also [72]. For (c) and (d), from [93][3], the Reynolds numbers are in the range $18000 < Re_\theta < 60000$. The measured profiles were fitted to a modern composite profile [20]. All the solid straight lines have $\kappa = 0.384$; in (a) and (b) they are the log law with $A = 4.173$. In (c), the straight line is the deficit log law, equation (8.26), with $B = 5.9$ and in (d) it is $(U_\infty - U)^+ = -\frac{1}{\kappa}\ln(y/\delta) + 2.3$. The dashed black lines in (c) and (d) are from a DNS at $Re_\theta \approx 7500$ [103].

been proposed [61]. The increasing extent of the logarithmic region with increasing Re_θ is very evident across all of figure 8.7.

A unified expression for the mean profile

Recall that our earlier expression for the full velocity profile (equation (8.27)) ignored the viscous sublayer and the buffer layer. In that sense, it does not represent the entire mean flow profile. Furthermore, recall the fact that the (Reynolds-number independent) logarithmic region only arises at large enough Reynolds numbers and appears as an overlap region between the inner and outer regions defined by f_i and f_o (equations (8.19) and (8.25)), within which $1 \ll y^+ \ll \delta^+$. Strictly, therefore, an appropriate expression for the complete velocity profile, all the way from the wall, can be written as a composite expansion:

---

[3] Reproduced with permission from [93], copyright Cambridge University Press.

$$U^+ = U_i^+(y^+) + L^+(y^+) + U_o^+(y/\Delta), \tag{8.30}$$

in which it is explicitly recognised that there is a logarithmic overlap layer (L^+) that joins the inner layer, $f_i = U_i^+$, and the outer layer, $U_o^+ = U_\infty^+ - W$ (where W is the f_o from equation (8.25)). It could be argued that a more fundamental way of assessing the classical scaling is to undertake Taylor series (or other kinds of) expansions of each term in the governing (Reynolds stress) equation (8.15), in order to see whether those scalings are asymptotically correct and to determine the extent to which modern data for the entire profile satisfy the results of the asymptotic analysis for U_i^+ and U_o^+ and lead to the expected logarithmic region. This all requires a rather detailed use of matched asymptotic expansions (MAE), which is beyond the scope of this book. It is not a particularly new approach, but there have been relatively recent efforts of this kind, often undertaken on the back of high-quality experimental and DNS data (in which u_τ is known independently of mean velocity data), e.g. [19, 61, 66, 74]. As well as helping to develop a composite profile of the entire boundary layer (which is algebraically much more complex than equation (8.27)), these have led, among other things, to the development of some criteria for assessing whether any particular set of experimental data is well-behaved, in the sense of being representative of the canonical equilibrium state [20]. In particular, it has been suggested that the profile's shape factor H (i.e. δ^*/θ) provides a useful test as to the state of a measured boundary layer. We return to this when considering the flow's integral parameters in section 8.3.4.

Overall, however, one of the major conclusions of all this work is that, in fact, the classical description of the mean flow profile, equation (8.27), is perfectly adequate for practical purposes [62], which is, perhaps, not a surprise. Some other comments on scaling analysis and its implications are given at the end of the following section, and further conclusions are of relevance in the context of the Reynolds stresses, which we briefly mention in section 8.4.

8.3.3 The skin friction

In the absence of a streamwise pressure gradient, the momentum integral equation (8.13) can be written as

$$\frac{c_f}{2} = \frac{\mathrm{d}\theta}{\mathrm{d}x} = \frac{u_\tau^2}{U_\infty^2} = \frac{1}{(U_\infty^+)^2}, \tag{8.31}$$

so, in addition to measuring the skin friction directly, it can, in principle, be deduced by measuring the variation of θ with x, but this is not easy to do accurately.

A consideration of equation (8.27) at $y = \delta$ leads directly to an (implicit) relationship between the local skin friction coefficient and the boundary layer thickness Reynolds number, $\mathrm{Re}_\delta = \delta U_\infty/\nu$:

$$\sqrt{\frac{2}{c_f}} = \frac{1}{\kappa} \ln\left(\mathrm{Re}_\delta \sqrt{\frac{c_f}{2}}\right) + A + \frac{2\Pi}{\kappa}. \tag{8.32}$$

Note, incidentally, that this can be written as

$$c_f = \frac{2\kappa^2}{\ln^2 \text{Re}_\tau} + \text{HOT},$$

where $\text{Re}_\tau = \delta u_\tau/\nu$ and HOT represents higher-order terms which become increasingly negligible relative to the first term as $\text{Re}_\tau \to \infty$. This means that as that limit is approached, both $u_\tau/U_\infty \to 0$ and $c_f \to 0$; the boundary layer effectively ceases to exist. Equation (8.32) can be used to compute data that show how c_f varies along the surface. An alternative (and explicit) expression, using Δ instead of δ, can be written as

$$U_\infty^+ \equiv \sqrt{\frac{2}{c_f}} = \frac{1}{\kappa} \ln(\text{Re}_{\delta^*}) + C^*, \tag{8.33}$$

which is directly obtained by adding equations (8.22) and (8.26), writing $C^* = A + B$, and using the following definition for Δ: $\Delta = \delta^* U_\infty^+$. A further alternative, often used in practice, is

$$U_\infty^+ \equiv \sqrt{\frac{2}{c_f}} = \frac{1}{\kappa} \ln(\text{Re}_\theta) + C. \tag{8.34}$$

where C^* and C must depend on the values used for κ, A, and B. (Strictly speaking, C is actually a weak function of Re_θ, given by $(\frac{1}{\kappa} \ln H + C^*)$.) These two latter relationships for c_f were used by Fernholz and Finley [27] in their compilation of data. Sometimes, power-law fits are used. For example, with standard values for κ, A, and Π, a power-law fit to data computed using equation (8.32) is just $c_f = 0.02\text{Re}_\delta^{-1/6}$ [119]. Given that the boundary layer growth is only a little slower than linear with x, this gives an immediate indication of how slowly c_f falls with x, compared to the $x^{-1/2}$ behaviour in a laminar boundary layer – equation (8.9). Power-law relationships, however, can only ever be appropriate for a relatively small Reynolds-number range. Numerous other relationships of the form $c_f = f_1(\text{Re}_x)$ are available [81, 98, 119].

Until fairly recently, given the absence of any independent, direct way of determining c_f, it was more common to obtain it by estimating u_τ from a best fit of the velocity profile data to the log law. The classical way is to use the 'Clauser chart' method, (e.g. [27]), but that requires assumptions about the values of κ and A (in equation (8.22)) and, in any case, also *assumes* the presence of a universal logarithmic region. The uncertainties are well known [34, 118]. The method can by extended by using a more comprehensive fit to the whole flow, but this also requires an assumption about the value of Π (as well as the wake function itself). Happily, as mentioned earlier, over the last couple of decades it has become possible to measure c_f directly, and this has made an assessment of the various expressions developed for c_f very much easier. Figure 8.8 shows two sets of modern data from experiments in facilities which allow larger ranges of Re_θ than is often possible. Also, and even more crucially, u_τ values were obtained directly, so no prior assumptions about the mean velocity profile were necessary. The data are compared with 13 of the classical

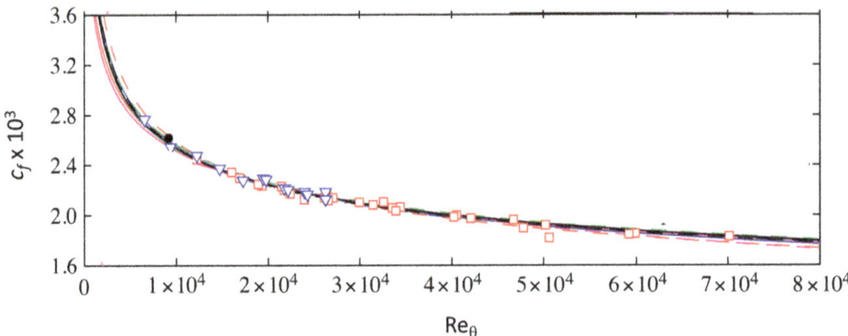

Figure 8.8. Skin friction variation with momentum thickness Reynolds number. The red squares represent data from the National Diagnostic Facility (NDF) at the Illinois Institute of Technology [67], and blue triangles denote data from the large-scale wind tunnel in Stockholm (KTH) [72][4]

relationships, each one reworked to embody the values of κ, A, and Π implied by the c_f and corresponding velocity data [66]. We do not identify all the relationships included in the figure, but point out that equations (8.32) and (8.34) are shown by the solid and dashed black lines, with $\kappa = 0.384$, $C^* = 3.354$, and $C = 4.127$, which collectively give the best fits of the equations to the experimental data. All the other lines come from various classical relationships for c_f in terms of Re_x, which thus require a relation between Re_θ and Re_x. This can be derived from the momentum integral equation (see [66], for example); a suitable power-law fit to the full result is

$$\mathrm{Re}_\theta = 0.01277(\mathrm{Re}_x)^{0.866}. \tag{8.35}$$

As an example, the classical Schultz-Grunow formula, $c_f = 0.37(\log \mathrm{Re}_x)^{-2.584}$ [81], then yields c_f in terms of Re_θ and this result (the dashed red line in figure 8.8), along with most others, fits the data remarkably well, particularly in the region of $\mathrm{Re}_\theta \approx 20000$ where, in fact, the values of c_f for all the relationships differ by less than 0.5%. We emphasise that these good fits *only* occur if (just one of) the numerical constants in the different relationships are adjusted in the light of the modern data, which (we recall) were obtained totally independently of the velocity profile. A further example is the classical c_f relationship, which can be easily deduced by assuming a one-seventh power law for the velocity profile; this is simply $c_f = 0.027\mathrm{Re}_x^{-1/7}$, and adjusting the constant from 0.027 to 0.0236 yields another of the curves in the figure (the solid red line). None of the relationships in their original form give good agreement with these modern data, largely because each was formulated on the basis of much smaller Reynolds-number ranges and indirect deductions about c_f (i.e. using the velocity profile). From a practical perspective, the use of either equation (8.33) or equation (8.34) with $\kappa = 0.384$, $C^* = 3.354$, and

---

[4] Republished with permission of the Royal Society from Nagib *et al* [66]; permission conveyed through Copyright Clearance Centre, Inc.

$C = 4.127$ provides accurate values of c_f at the Reynolds numbers commonly reached in laboratory facilities.

Note that even with the modifications implicit in figure 8.8, as Re_θ increases beyond about 10^5, all the relationships increasingly diverge from each other. It is therefore clear that none of them can be an asymptotically correct relationship. Indeed, it can be shown that asymptotically, i.e. as $\text{Re}_\theta \to \infty$, $H = \delta^*/\theta \to 1$ and thus (by equating equations (8.32) and (8.34)) that C must equal C^*, which is far from true in practical situations and thus demonstrates the distance between this limit and the Re_θ range covered by the data in figure 8.8. Given that over the practical Reynolds-number range, all the (appropriately modified) relationships yield good agreement with modern data to within experimental accuracy, how is one to choose between them? The answer can only be provided by theoretical arguments, and some work has recently taken place to address this. In particular, the classical boundary layer scaling has been reassessed, particularly as $\text{Re}_\theta \to \infty$, in the light of the newer measurements in experiments in which c_f is obtained directly. This has naturally involved careful asymptotic arguments, as mentioned in the previous section, because the classical log law strictly only appears in the overlap region between the wall region (f_i, equation (8.19)) and the outer region (f_o, equation (8.25)). The arguments have sometimes been driven by consideration of the way higher-order moments such as the Reynolds stresses (e.q. $\overline{u'^2}$) behave, because any self-consistent scaling analysis for the whole flow must be appropriate for these quantities as well as the mean velocity.

An alternative asymptotic approach to considering the entire boundary layer rests on using U_∞ rather than u_τ as the scaling velocity for the outer region. This leads to a power-law representation of U^+ rather than one which includes a central logarithmic region whose power exponent depends on the Reynolds number. This is, of course, unlike the central assumption that leads to the log law – i.e. that $y\partial U^+/\partial y$ is independent of U_∞, δ, and ν, so independent of Reynolds number and thus universal. The power-law approach, its assessment against the traditional log law, and the question of universality have been widely studied (e.g. [9, 15, 33, 34, 121]) in the context of both internal flows (recall our comments in section 3.2) and external flows. Regarding the debate surrounding the log law versus the power law, we quote a 2006 writer (W C George), who suggested that 'at least for the zero-pressure-gradient boundary layer, if the Österlund data had been available 15 years ago ... we probably would have never asked the questions we did about the log theory nor had reason to develop the power law alternative' [32]. (He was referring to the directly measured skin friction, which provides u_τ independently of any velocity profile assumptions [71], as discussed earlier.) This is revealing and tends to validate the later very positive conclusions about the log law, despite the same author concluding a year later that 'there is no justification $\cdots$ for a *universal* log law for wall-bounded flows' [33]. These issues can have significant practical importance, not least because most turbulence models used by the aircraft industry depend on the use of the log law and a 2% decrease in κ leads to a 1% reduction in the drag estimate for a large aircraft (e.g. an Airbus 350 or Boeing 787) [106] – a lot of money!

Before considering the turbulence field, in sections 8.4 and 8.5, which is naturally intimately linked with the mean velocity behaviour, we consider some of the important integral parameters of the latter.

8.3.4 Integral parameters

Recall the integral parameters defined by equations (8.10) and (8.11) and that the ratio of these two is the shape factor, $H = \delta^*/\theta$. In addition, recall the Rotta–Clauser length scale, $\Delta = \delta^* U_\infty/u_\tau$. We define an integral of the deficit velocity as follows:

$$I_1 = \int_0^\infty \frac{U_\infty - U}{u_\tau} \, d\left(\frac{y}{\delta}\right) = \frac{\Delta}{\delta}. \tag{8.36}$$

Both Clauser and Rotta [22, 90] defined a further parameter, which we will call I_2, given by

$$I_2 = \int_0^\infty \left(\frac{U_\infty - U}{u_\tau}\right)^2 d\left(\frac{y}{\Delta}\right). \tag{8.37}$$

Using the δ^* and θ definitions, it follows that

$$H = \left(1 - \frac{I_2}{U_\infty^+}\right)^{-1}. \tag{8.38}$$

I_2 (often called the Clauser parameter, which he called G) is a function only of Π, κ, and the wake profile shape w. In fact, it turns out that for most of the wake function shapes that have been used, including the Coles one, I_2 can be expressed as

$$I_2 = \frac{a + b\Pi + c\Pi^2}{\kappa(e + \Pi)}, \tag{8.39}$$

where a, b, c, and e are numerical constants that depend only on the specific profile shape.

It also follows by integrating the deficit profile, equation (8.29), to yield δ^*, that

$$I_1 \equiv \frac{\Delta}{\delta} \equiv \frac{\delta^*}{\delta}\frac{U_\infty}{u_\tau} \equiv \frac{\delta^*}{\delta}\sqrt{\frac{2}{c_f}} = \frac{1 + \Pi}{\kappa}, \tag{8.40}$$

showing that I_1, is *not* a function of the Reynolds number in this classical, two-parameter approach. These various relations (equations (8.36)–(8.40)) assume only that the two-parameter profile is an adequate representation of the mean flow and that the viscous sublayer and buffer layer can be ignored in integrations for δ^* and θ. (Incidentally, the relations are also independent of the nature of the surface, so they apply for rough walls also, with implications which we explore in section 8.7.)

We emphasise that the variation of H is described by just two parameters, U_∞^+ and I_2. Experiments have shown that I_2 tends asymptotically to a constant value at high Reynolds numbers, which is totally consistent with the fact that the $(U_\infty^+ - U^+)$ scales with y/Δ (except very close to the wall) – witness the excellent collapse we noted earlier,

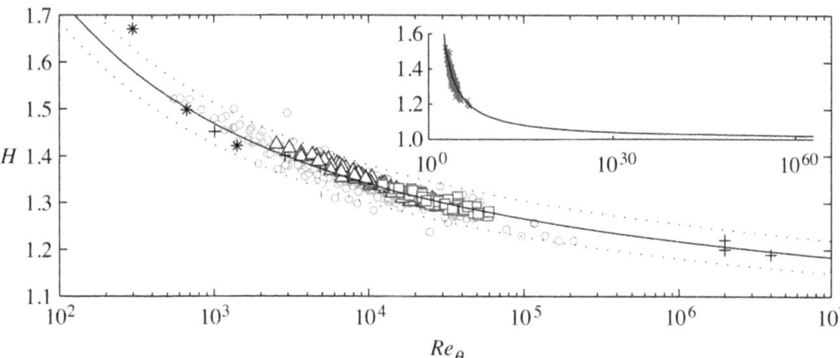

Figure 8.9. Variation of H with Re_θ. The squares represent data from the NDF [67] and triangles denote data from the large-scale wind tunnel in Stockholm (KTH) [72][5].

e.g. in figure 8.5. Figure 8.9 shows values of H from a very large number of experiments (see [66] for details) and, in particular, the relatively recent two sets of experiments highlighted in figure 8.8. Taking the value for I_2 in equation (8.38) to be 7.135 and using equation (8.34) to relate U_∞^+ to Re_θ yields good agreement (the solid black line) with the modern data over a reasonably wide Reynolds-number range. The inset shows how the result extends asymptotically to very large Re_θ, again emphasising how remote practical boundary layers are from the asymptotic limit, for which $H = 1$. The variation in H with Re_θ also emphasises the inadequacy of assuming a constant value for H; some aerodynamics textbooks give, for example, $H = 1.3$ (e.g. [119]), which is close to the value of 1.286 that arises if a one-seventh power law is assumed for the velocity profile. In any case, it is clear that in turbulent boundary layers H is much smaller than the Blasius value of 2.59 for a laminar boundary layer – section 8.2.3 – and, importantly, is a function of Re_θ.

The classical (constant) value of Δ/δ given by equation (8.40) is 4.036, using $\Pi = 0.55$ and $\kappa = 0.384$; this is a little lower than the ≈ 4.5 implied by the modern data (using δ as the boundary layer thickness defined by the $U = 0.99\,U_\infty$ point). These latter data are not inconsistent with the extremely slight variation of Δ/δ_{99} with Re_θ, implied by a more detailed (asymptotically driven) derivation of the complete velocity profile [66]. Sample exercise 8.2 explores the calculation of these integral parameters from DNS data.

By differentiating equation (8.33) with respect to x and using the momentum integral equation (8.31), it is straightforward to show that the growth of an equilibrium boundary layer (i.e. one which no longer depends on how it was initially generated other than on the resulting virtual origin) can be expressed as

$$\frac{\mathrm{d}\delta}{\mathrm{d}x} = \frac{1 + \kappa U_\infty^+}{I_2 + \kappa I_1 U_\infty^{+2} - \kappa I_2 U_\infty^+}, \tag{8.41}$$

---

[5] Republished with permission of the Royal Society from [66]; permission conveyed through Copyright Clearance Centre, Inc.

in which, as noted earlier, both I_1 and I_2 are constants. This shows immediately that the growth can only be linear if U_∞^+ (and thus the skin friction) is constant, which cannot occur in zero-pressure gradient-free stream conditions. Normally, we see that δ grows increasingly slowly as U_∞^+ increases. Choosing $\mathrm{Re}_\theta = 20000$ as typical for a moderately large Reynolds number, equation (8.34) gives $U_\infty^+ = 29.9$ at that local position, and from the above relation for $d\delta/dx$ we deduce that the local growth rate is only about 0.0115 – i.e. the angle of the 'wedge' defined by $y = \delta(x)$ is about 0.6°.

The growth of the boundary layer can also be successfully predicted through a full asymptotic analysis but the resulting expressions are long (see, e.g. [61]) and not very convenient. For $\mathrm{Re}_\theta > 10^5$, the results are actually very close to those obtained using equation (8.41) and the expressions that can be deduced from the momentum integral equation and using the 1/7th power law. For the record, these latter are

$$\frac{\theta}{x} = 0.015\,6\mathrm{Re}_x^{-1/7}, \quad \frac{\delta^*}{x} = 0.02\mathrm{Re}_x^{-1/7}, \quad \text{and} \quad \frac{\delta}{x} = 0.16\mathrm{Re}_x^{-1/7}, \tag{8.42}$$

in which the virtual origin, x_0, is implicit in Re_x [98].

We mention one other matter before moving to consider the turbulence statistics in section 8.4. First, recall that in section 7.3.3 we noted the ongoing discussion about the extent to which κ is universal. In the above discussion, we have generally used a value of around 0.384, as typical of most of the more recent boundary layer studies, but slightly different values were noted for channel and pipe flows. For internal flows, it has been argued that the Reynolds-number dependence on any quantity (in wall units) can be expressed mainly through an additional higher-order term of the order of Re_τ^{-1} so that, for example, the log law in the wall region at finite Reynolds numbers becomes $U^+(y^+, \mathrm{Re}_\tau) = f_\infty(y^+) + f_{\mathrm{Re}}(y^+)/\mathrm{Re}_\tau$, in which the first term is the regular log law – i.e. $f_\infty(y^+) = \frac{1}{\kappa}\ln(y^+) + A$; see, for example, [55]. This has made it possible to reconcile apparent differences in κ in the different internal geometries, by accounting for the differing effects of the axial pressure gradient on the velocity profile. However, it has been suggested that a suitable alternative to Re_τ is not obvious for external flows [107]. It thus remains a moot point whether κ is truly universal between internal and external flows and the reader is encouraged to watch for further developments in this area. One should also bear in mind that the uncertainty in the experimental determination of κ can never be less than the uncertainty in determining the friction velocity [102].

8.4 The turbulence properties

8.4.1 The Reynolds stresses

We begin by illustrating the behaviour of the Reynolds stresses. Figure 8.10 shows data obtained from a DNS at $\mathrm{Re}_\theta \approx 4060$, reported by Schlatter *et al* [96]. Note that the boundary layer thickness, δ_{99}, is defined as being where $U = 0.99U_\infty$, which is around $y^+ = \delta_{99}^+ = 1272$. Figure 8.10(a) shows that the viscous component of the shear stress is negligible above about $y^+ = 80$, and that below that point, the turbulence shear stress (the yellow line) slowly reduces to zero but is still noticeably

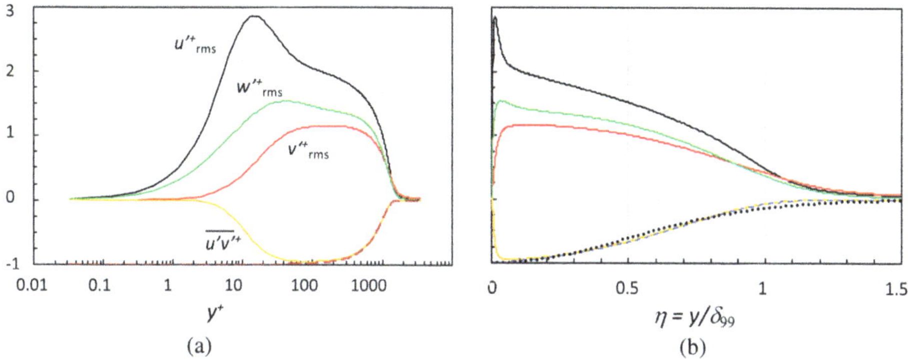

Figure 8.10. Turbulence fluctuations and shear stress in wall units (i.e. normalised using u_τ) in a turbulent boundary layer, plotted against y^+ (a) and y/δ_{99} (b); data obtained from a DNS at $\mathrm{Re}_\theta = 4060$. The dashed red line in (a) shows the *total* shear stress, with (minus) the viscous shear stress added to $\overline{u'v'}$. The data are in a database made available by Shatter[6], see [96] and also [95]. The four dashed and dotted lines in (a) are from laboratory data with sequentially increasing values of Re_θ up to about 48000 for the outermost one. From data made available by Marusic[7]; see [7]. The dotted black line in (b) represents $\overline{u'v'} = e^{-\eta^2/\alpha}$.

nonzero at $y^+ = 5$. It can be seen in figure 8.10(b) that the total shear stress is constant below about $y/\delta_{99} = 0.15$. A scaling analysis of the mean momentum equation (8.18), similar to that undertaken for free shear flows, shows that in the inner region, the terms on the left-hand side are small compared with the last term, which must therefore be zero. Thus, using equation (8.17), it follows by integration that $\nu \partial U/\partial y - \overline{u'v'} = u_\tau^2$. In the inertial sublayer, therefore, where viscous effects are negligible, it turns out that the Reynolds shear stress must be (reasonably) constant and equal to the wall stress.

In the outer region, the left-hand terms in the momentum equation cannot be ignored. Although it is possible, using similarity arguments, to relate the shear stress to the velocity through the momentum equation (just as we did for free shear flows in chapter 6), deducing a direct analytic relation between them using an eddy viscosity relation for closure is less useful, because the eddy viscosity ν_T is far from constant across the outer region. Nonetheless, if it *is* assumed to be constant, one can deduce that $\overline{u'v'}^+ \sim e^{-\eta^2/\alpha}$, where α is a fitting constant (see [109], for example). Figure 8.10(b) includes such a curve (the black dotted line). Adjusting α can improve the fit in one region to its detriment in another. In any case, since $\overline{u'v'}$ is constant in the inertial layer, we do not expect such an approach to be adequate below the outer layer.

The normal stress profiles are very similar to those in the plane channel, shown in figure 7.9(b) at a comparable Reynolds number, except that they naturally all decay to zero in the free stream. A peak in u'^+_{rms} at around $y^+ = 15$ is seen in figure 8.10(a) and its maximum value has also been found to vary with Reynolds number. This is

---

[6] https://www.mech.kth.se/~pschlatt/DATA/

[7] https://melbourne.figshare.com/articles/dataset/High-Reynolds-number-boundary-layer-statistics-stream-wise-spanwise-and-wall-normal-velocity-components/14141423

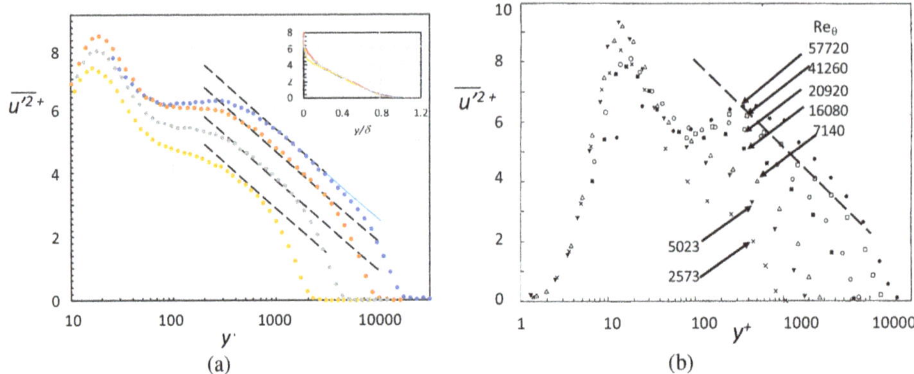

Figure 8.11. Streamwise Reynolds stress profiles. (a) Replotted data from the Melbourne set (shown as rms values in figure 8.10); $Re_\tau=$: 18390 (blue), 10493 (red), 5047 (grey), 2493 (yellow). The inset shows the same data plotted as lines and versus y/δ. (b) Redrawn from [27][8]. The dashed black lines in (a) and (b) represent $\overline{u'^2}^+ = -A\ln(Re_\tau) + B$, with $A = 1.25$, $B = 1.75$.

illustrated in figure 8.11(a), which shows the streamwise Reynolds stress profiles at various Re_τ. We noted just these features in internal flows (see figure 7.9(b) for channel flows and figure 7.14 for pipe flows), and it is worth emphasising again that there are clearly features in the inner layer which do not scale with wall units and are Reynolds number dependent. The variation in the inner peak is less obvious in figure 8.11(b), which shows sets of much older data from the compilation of Fernholz and Finley [27], but there were probably serious space and time measurement resolution issues even well away from the wall at the higher Reynolds numbers. This generates some doubt about the validity of the outer peaks visible in that figure (see [41] for an illuminating discussion); special measures were taken to acquire the more recent data shown in figure 8.11(a) – see, for example, [58]. Only at $Re_\tau = 18390$ (the blue symbols) does the beginning of a second (outer) peak seem to occur. There is some theoretical justification for an outer peak which increases in magnitude with increasing Re_τ [84], and it certainly occurs in pipe flows, recall figure 7.14.

Notice that when plotted in outer coordinates (see the inset of figure 8.11(a)) the profiles collapse in the outer layer (i.e. beyond about $y = 0.15\delta$) so there is, at most, only a very weak dependence on the Reynolds number, and the outer flow is essentially self-similar over a wide Reynolds-number range. (Note, however, that there have been some contrary findings [115].)

Now recall equation (7.43), which was suggested long ago by Townsend [114] on the basis of his attached eddy hypothesis. It can be written in a slightly different form as,

$$\overline{u'^2}^+ = -A_1\ln(Re_\tau) + B_1, \qquad (8.43)$$

and was anticipated to occur in the constant (shear) stress region – i.e. within the inertial subrange in which the mean velocity also has a logarithmic behaviour. A

---

[8] Reprinted from [27], copyright (1996) with permission from Elsevier.

similar variation was anticipated for the spanwise stress, $\overline{w'^2}^+$, whereas the vertical stress, $\overline{v'^2}^+$, was expected to be constant. Over the last two or three decades or so, this behaviour of the normal stresses, particularly $\overline{u'^2}^+$, has been widely examined, as typified by the early exposition of Perry *et al* [75], and more recently by, for example, Samie *et al* [93]. The Townsend behaviour appears to have been essentially substantiated, provided the Reynolds number is large enough. Figure 8.11 includes equation (8.42) (with $A_1 = 1.25$ and $B_1 = 1.75$) for a number of cases; there can be almost a decade of such behaviour once $\mathrm{Re}_\tau > 20000$ at least.

8.4.2 The turbulent kinetic energy balance

The turbulent kinetic energy equation for zero-pressure-gradient boundary layers is the same as that derived for free shear flows, equation (6.14), except for the addition of the viscous term, which is important in the near-wall region. We can therefore write it as

$$
\underbrace{U\frac{\partial k}{\partial x} + V\frac{\partial k}{\partial y}}_{I} = \underbrace{-\overline{u'v'}\frac{\partial U}{\partial y}}_{II} \; \underbrace{- \frac{\partial}{\partial y}\overline{v'\left(k + \frac{p'}{\rho}\right)}}_{IIIa + IIIb} \; \underbrace{-\; \epsilon}_{IV} \; \underbrace{+\; \nu\frac{\partial^2 k}{\partial y^2}}_{V}.
\tag{8.44}
$$

Figure 8.12 shows profiles of the various terms across a boundary layer at $\mathrm{Re}_\theta = 4061$. (Note that the relative importance of the terms as functions of y is highlighted by the normalisation used in figure 8.12(b)). The production (*II*) and dissipation (*IV*) terms roughly balance for values of y^+ above about 30 and y/δ below about 0.5, which is very similar to what is seen in internal flows (e.g. in a channel – figure 7.10). In that region, the energy equation thus approximately reduces to $-\overline{u'v'}\partial U/\partial y \approx \epsilon$, which becomes $\epsilon \approx u_\tau^3/(\kappa y)$ in the inertial region. (Incidentally, this provides a convenient way of obtaining a very rough estimate for ϵ without having to measure spectra.) Thus, in that region, production and dissipation decrease similarly to $1/y^+$ (since $\overline{u'v'}^+$ is unity and $\mathrm{d}U^+/\mathrm{d}y^+ = 1/(\kappa y^+)$). The production/dissipation balance is only approximate, since there is some turbulent transport of energy away from the region $0.2 < y/\delta < 0.6$ (see figure 8.12(b)). Furthermore, in the buffer layer, energy is transported away from the region $y^+ \approx 10$ to 35 and towards the wall below that region, see figure 8.12(a).

It is straightforward to show that the peak production occurs where $-\rho\overline{u'v'} = \tau_w/2$, which (from figure 8.12(a)) is at around $y^+ \approx 12$, agreeing with the location of peak production in figure 8.12(a). Very near the wall ($y^+ < 5$), the dissipation is increasingly balanced by viscous diffusion until, at $y = 0$, they are equal. Below $y^+ \approx 30$, both viscous diffusion (term *V*) and the turbulent transport (terms *IIIa* and *IIIb*) become significant, and the latter become important again in the outer region, above about $y/\delta = 0.5$, together increasingly balancing the advection, especially beyond the nominal edge of the boundary layer. Note that the advection (term *I*) is negligible in the inner region and is not shown in figure 8.12(a); it only becomes significant in the outer region above $y/\delta \approx 0.5$.

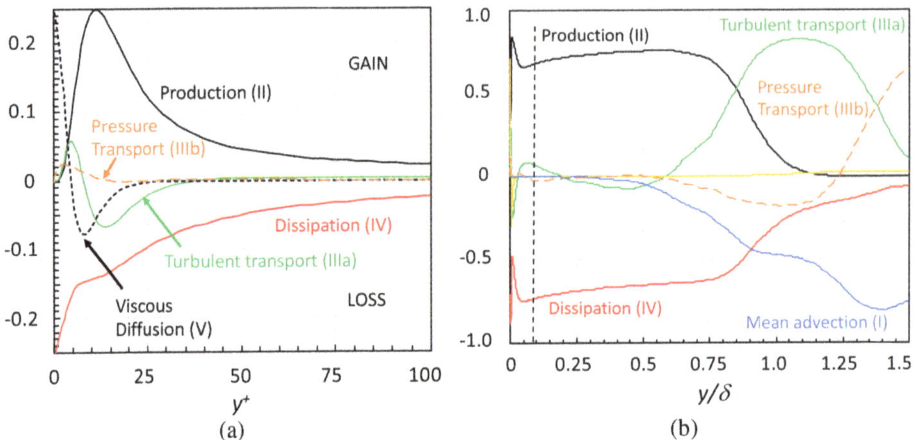

Figure 8.12. Turbulent kinetic energy balance; data from a DNS at $\mathrm{Re}_\theta = 4061$. (a) Terms normalised using viscous units; (b) terms normalised at each y so as to make the sum of the squares of all terms equal to unity. $y/\delta = 1$ is at $y^+ = 1272$, so that $y^+ = 100$ is at $y/\delta = 0.08$, as indicated by the vertical dashed line in (b). Drawn from data available at the KTH, Stockholm web site[9]. See also [95].

Similar balances can be constructed for the turbulence shear and normal stresses by considering each of the terms that contribute to their transport equations. Conclusions from such balances have similarities to some of those we discussed in the context of homogeneous shear flow (section 5.2), not least in the way in which the pressure strain terms operate and the fact that, for example, there is no direct production of $\overline{v'^2}$ and $\overline{w'^2}$ from the mean flow. In contrast, $\overline{u'^2}$ receives energy directly from the mean shear and the other two stresses receive energy via a process of redistribution from the streamwise stress. The interested reader could consult, for example, [13, 81], for further details.

8.4.3 The energy spectrum

Finally, we consider briefly how the turbulence energy is distributed across the wavenumber (or frequency) range at different locations in the turbulent boundary layer, by discussing typical energy spectra. Figure 8.13(a) shows some spectra of the streamwise component of the fluctuation velocity, using Kolmogorov scaling – i.e. $E_{11}(\kappa_1)(\epsilon\nu^5)^{1/4}$ versus $\kappa_1\eta$, with $\kappa_1 = 2\pi f/U$ and $\eta = (\nu^3/\epsilon)^{1/4}$. The reader will recall that energy spectra were introduced in section 3.5; a way in which they can be obtained experimentally was discussed in chapter 4 noting in particular, the common use of Taylor's hypothesis (section 4.5), which is how κ_1, the streamwise wavenumber, may be obtained from measured one-dimensional frequency spectra. Figure 3.5 showed a sequence of spectra measured in different kinds of flows and over a wide range of Reynolds numbers. The two spectra included there, which were

---

[9] https://www.mech.kth.se/~pschlatt/DATA

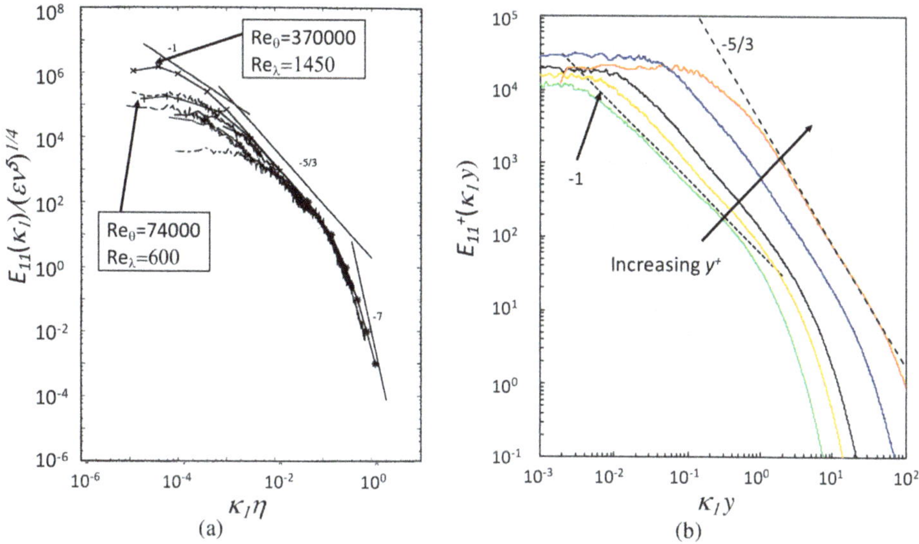

Figure 8.13. Energy spectra. (a) Plotted using Kolmogorov scaling. The curves with symbols are from [92[10]]. (b) Plotted as $E_{11}(\kappa_1 y)$ versus $\kappa_1 y$ (using viscous units). From the Melbourne data at $Re_\tau = 18390$ (as for figures 8.10 and 8.11). The y^+ (and y/δ) values are 53 (0.0029), 108 (0.00587), 192 (0.0104), 938 (0.051), and 3430 (0.19).

obtained in the world's largest wind tunnel and at around the middle of roughly 1 m-thick boundary layers [92], are included in figure 8.13(a). The other data come from other boundary layer experiments, as given by [27]. It is immediately apparent that the extent of the $-5/3$ inertial range increases with Reynolds number, as we noted in section 3.5. Since the data are plotted using Kolmogorov scaling, the spectra collapse in the inertial and dissipation ranges, as expected. The Taylor microscale Reynolds numbers (defined in section 3.2) for the two high-Reynolds-number boundary layer cases in the figure are 600 and 1450 and only the latter shows a reasonably extensive inertial region – perhaps about a decade. It has been argued that an Re_λ of around 1500 is required before a decade of inertial region can be expected (see, e.g. [92]).

Figure 8.13(b) shows selected spectra from the more recent experiments in Melbourne (e.g. [7, 8]) which, despite being relatively high-Reynolds-number cases (compared with most earlier laboratory experiments), have significantly lower Re_τ values than the two boundary layer cases highlighted in figure 8.13(a). Note first that a plausible inertial region with a $-5/3$ slope is only apparent at $y/\delta = 0.19$ and possibly at $y/\delta = 0.05$, but the extent of the region is not large. At these two points, the Re_λ values are 385 and 238, respectively, but nearer the wall the values are much lower (e.g. about 117 at $y/\delta = 0.01$), so that one would not expect to see a $-5/3$ region. Note that, as mentioned in section 4.5 (see figure 4.7), the best way of discerning the extent of any inertial ($-5/3$) region is to plot the spectrum as $\kappa_1^{5/3} E_{11}(\kappa_1)$ and look for the extent of the straight-line portion.

---

[10] Reprinted from [27], copyright (1996) with permission from Elsevier.

Second, it is evident in figure 8.13(b) that the spectrum at $y^+ = 53$ seems to have a significant region of κ_1^{-1} slope. There are sound theoretical arguments which suggest such behaviour in the near-wall region of boundary layers (e.g. [75]); these rely on the presence of a significant overlap region in which both inner scaling (i.e. using y and u_τ) and outer scaling (i.e. using δ and u_τ) of the wavenumber spectrum are simultaneously valid; this requires a sufficiently large Reynolds number and a location sufficiently close to the wall. For a decade of κ_1^{-1} behaviour, $y/\delta < 0.002$ is needed, whilst simultaneously, in order to avoid viscous effects, $y^+ > 100$ [69]; together, these imply $\mathrm{Re}_\tau > 52000$, which is beyond the capability of any current wind tunnel. The extent of the κ_1^{-1} region for $y/\delta = 0.0029$ in figure 8.13(b) is thus likely to be somewhat less than a decade. (Again, the best way to show this is to plot the compensated spectrum, $\kappa_1 E_{11}(\kappa_1)$.) We emphasise again the necessity for very specialised instrumentation to allow accurate measurements so close to the wall.

Third, notice that for all the spectra shown in figure 8.13(b), $E_{11}(\kappa_1 y)$ becomes constant at small κ_1. In fact, equation (4.49) states that this constant is linked directly to the integral length scale L_{11} – i.e.

$$L_{11}/y = \frac{\pi}{\overline{u'^2}} E_{11}(\kappa_1 y)|_{\kappa_1 = 0},$$

which allows the length scale to be deduced from the $E_{11}(\kappa_1 y)$ spectrum, provided the time periods of the u' records are sufficiently long to identify the spectral behaviour as $\kappa_1 \to 0$. It is also straightforward to deduce the Taylor microscale and thus its Reynolds number, Re_λ, from spectral data since, once ϵ has been determined, it follows from equations (4.23) and (5.7) that

$$\lambda^2 = \frac{30\nu \overline{u'^2}}{\epsilon}.$$

Note, however, that this assumes that the turbulence within the inertial range of wavenumbers (and beyond) is isotropic. There is plenty of evidence that this is true at high enough Reynolds numbers. Indeed, the experiments yielding the data in figures 8.13(b) and 8.14 (see below) also included measurements of the spectra for the other velocity components, i.e. E_{22} and E_{33}, from which it can be concluded that the expected constant factors between the three spectra in the inertial range, assuming isotropy (i.e. $E_{22} = E_{33} = \frac{4}{3}E_{11}$; see section 3.5) are valid. Within the log-law region, the dissipation can be estimated by assuming that it balances energy production, so that $\epsilon = u_\tau^3/\kappa y$, from which it follows that $\mathrm{Re}_\lambda^2 = 30\kappa y^+ \overline{u'^{2+}}$, which provides an alternative way of estimating Re_λ.

A convenient way of assessing the spread of energy across the wavenumber range in the whole boundary layer is to construct contours of constant (compensated) energy in the λ/y plane. Figure 8.14(a) shows just such a plot, with contour lines replaced by a continuous colour range. It is immediately apparent that near the wall, up to a little below the start of the log-law region, energy is concentrated at wavelengths of around 0.1δ, whereas within the log-law region, energy is mostly

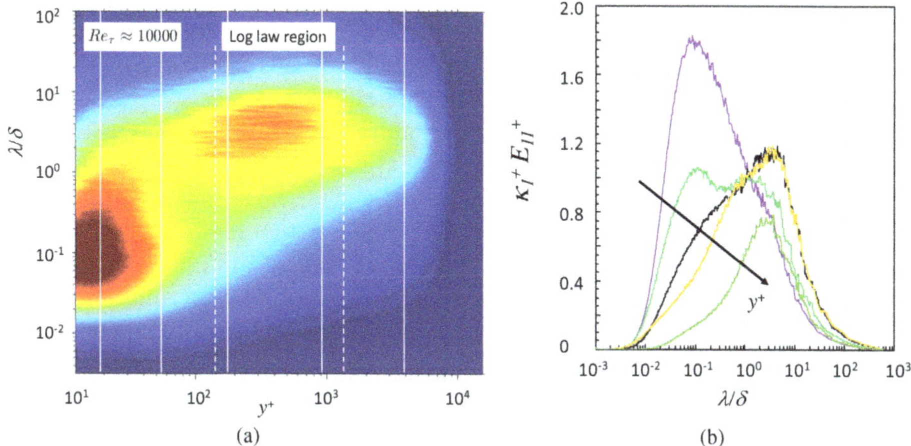

(a)

(b)

Figure 8.14. (a) Values of $\kappa_1^+ E_{11}^+(\kappa_1 y)$ throughout a boundary layer at $Re_\tau = 10100$, over all wavelengths. The extent of the log-law region lies between the vertical dashed white lines. (b) Plotted as $E_{11}(\kappa_1 y)$ versus λ/δ. y^+=19, 55, 187, 938, and 3948; these locations are marked by the vertical white lines in (a). (a) and (b) were both prepared using data made available at by Marusic[11] (as given in figure 8.10).

concentrated in a wavelength range of a few boundary layer thicknesses. Slices of this plot at fixed y^+ are shown in figure 8.14(b).

We emphasise that all the experimental spectra illustrated in the above discussion required the use of Taylor's hypothesis (section 4.5) in order to convert the measured frequency spectra to wavenumber spectra. This was done in the usual way by taking the convection velocity as the local velocity. Whilst this is undoubtedly adequate for the lower wavenumber (long wavelength) parts, it is inherently less likely to be true for the high wavenumber motions – indeed, it is known that the convection velocity is scale dependent [4, 64], but this does not seriously impair the validity of the above discussion.

The reader should by now be aware that there are various different ways of plotting spectra. The particular ways chosen depend on the specific aspect of the spectral structure that is of interest. In switching between different representations, it is helpful to bear in mind that:

$$\overline{u'^2} = \int_0^\infty E_{11}(f)\,\mathrm{d}f = \int_0^\infty E_{11}(\kappa_1)\,\mathrm{d}\kappa_1 = \int_0^\infty E_{11}(\kappa_1 y)\,\mathrm{d}(\kappa_1 y) \tag{8.45}$$

with

$$\kappa_1 = \frac{2\pi f}{U_c}, \qquad E_{11}(\kappa_1 y) = \frac{E_{11}(\kappa_1)}{y}. \tag{8.46}$$

<hr>

[11] https://melbourne.figshare.com/articles/dataset/High-Reynolds-number_boundary_layer_statistics_-_streamwise_spanwise_and_wall-normal_velocity_components/14141423

8.5 The turbulence structure

Nothing we have discussed in the earlier sections of this chapter has given any information about the dynamics, or even the presence, of specific structural features (whether coherent or not) within the boundary layer. Even the spectra, with associated spatial correlations, only provide time-averaged information, albeit showing the likely presence of motions across a wide spatial scale. Two of the earliest features of the dynamics to be identified, in the 1940s–1960s, were the presence in the viscous sublayer ($y^+ < 50$, say) of 'streaky' motions, elongated in the streamwise direction, e.g. [45, 48], and, because of the intermittency in the boundary layer's outer regions, the presence of identifiable large-scale eddies which 'roll' along, engulfing the outer inviscid fluid in a time-dependent manner, e.g. [26, 36, 113]. This early work sparked an explosion of interest in the flow's dynamics, with numerous studies of many kinds (including the early efforts using DNS). A useful review of the state of understanding up to the late 1980s is provided by Robinson [88]. Further advances in instrumentation, with developments not just in multipoint measurement techniques but also full-field techniques, such as particle image velocimetry (PIV), has led to yet more work, much of which has solidified the often qualitative understanding previously gained. We can summarise the basic features as follows:

1. Low-speed streaks very near the wall
2. Ejections of low-speed fluid outward from the wall
3. Sweeps of higher-speed fluid towards the wall
4. Vortical structures of various kinds
5. Large-scale motions in the outer region, which can be very long in the streamwise direction

Some of the motions near the wall occur in 'packets', as we will see later. Useful descriptions of the major details of these structures are provided by Pope [81] and Robinson [88].

We illustrate some of the features with reference to relatively recent visualisations and videos from DNS studies. Snapshots from available videos are shown in figures 8.15 and 8.17, but the reader is encouraged to view the videos in their entirety. Firstly, notice how the surface skin friction patterns shown in figure 8.15 reveal that there are instantaneously long thin regions of high c_f (in red) at different locations across the span, which are interspersed with long low c_f regions (in blue). These are necessarily the imprint of the motions above, typified by the classical hairpin-type vortices explored long ago, e.g. [38]. Such structures have been widely identified, both singly and in packets, and seem to be prevalent in wall turbulence generally. Close to the wall, their spanwise spacing is typically about 100 viscous units. Figure 8.16(a) is a cartoon of a single hairpin structure, showing the motions it induces because of its inherent vorticity. It was shown analytically long ago how spanwise vortex lines could be stretched by the shear into hairpin loops [110], but this model has to be modified in a near-wall region to include long quasi-streamwise vortices (the 'legs' in figure 8.16(a)) connected to the hairpin's head by vortex necks

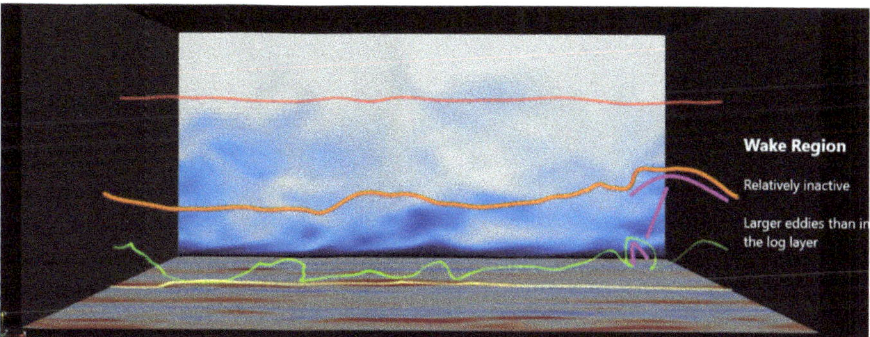

Figure 8.15. Snapshot from a video by Ruan and Blanquart[12] [91] rather charmingly called 'The Hitchhiker's Guide to Flat Plate Boundary Layers'. The colours on the backplane indicate the mean velocity, ranging from deep blue near the surface to white in the free stream. Instantaneous c_f contours are shown on the surface, from around zero (dark blue) up to about 0.005 (dark red).

inclined at around 45° to the wall [88]. To quote Adrian [3], '*With this simple model, the low-speed streaks are explained as the viscous, low-speed fluid that is induced to move up from the wall by the quasi-streamwise vortices*'. Fluid is induced to move up and through the hairpin loop, so creating the ejection (Q2 is a second-quadrant event – i.e. one that has positive v' and negative u'). This event encounters an incoming sweep (Q4 is a fourth-quadrant event – i.e. one that has negative v' and positive u') and thus creates a stagnation point flow which generates an inclined region of shear. The associated hairpin loops make significant contributions to the Reynolds shear stress.

Hairpins normally seem to occur in packets, and the three-dimensional conceptual picture in figure 8.16(c) suggests that they are more common in the lower half of the boundary layer, especially within the log-law region, although the larger ones can extend to the edge of the flow. Their legs usually originate very near to the wall. A complete description of the hairpin packet model and how it accounts for a number of the observed structural features in turbulent boundary layers has been provided by Adrian *et al* [3], where it is suggested that the average angle, γ, of the hairpin 'ramps' (seen in figure 8.16(b)) is around 12°. The near-wall hairpins may either grow to form larger vortices, creating a hierarchy of such eddies throughout the boundary layer, or grow by extracting vorticity from the mean flow; both scenarios can lead to a logarithmic mean velocity profile [77]. Such structures are coherent in the sense that they are sufficiently long-lived to be observable with appropriate processing of the velocity and/or vorticity fields. There are various ways of detecting vortex structures, and a common technique is to use what is called the Q-criterion, where Q is defined as the second invariant of the velocity gradient tensor. This can be expressed by $Q = \frac{1}{2}(||\Omega^2|| - ||S^2||)$, with S_{ij} and Ω_{ij} as defined by equations (2.9) and (2.10). It is thus a local measure of the excess of rotation rate

---

[12] https://gfm.aps.org/meetings/dfd-2020/5f6012cd199e4c091e67c029

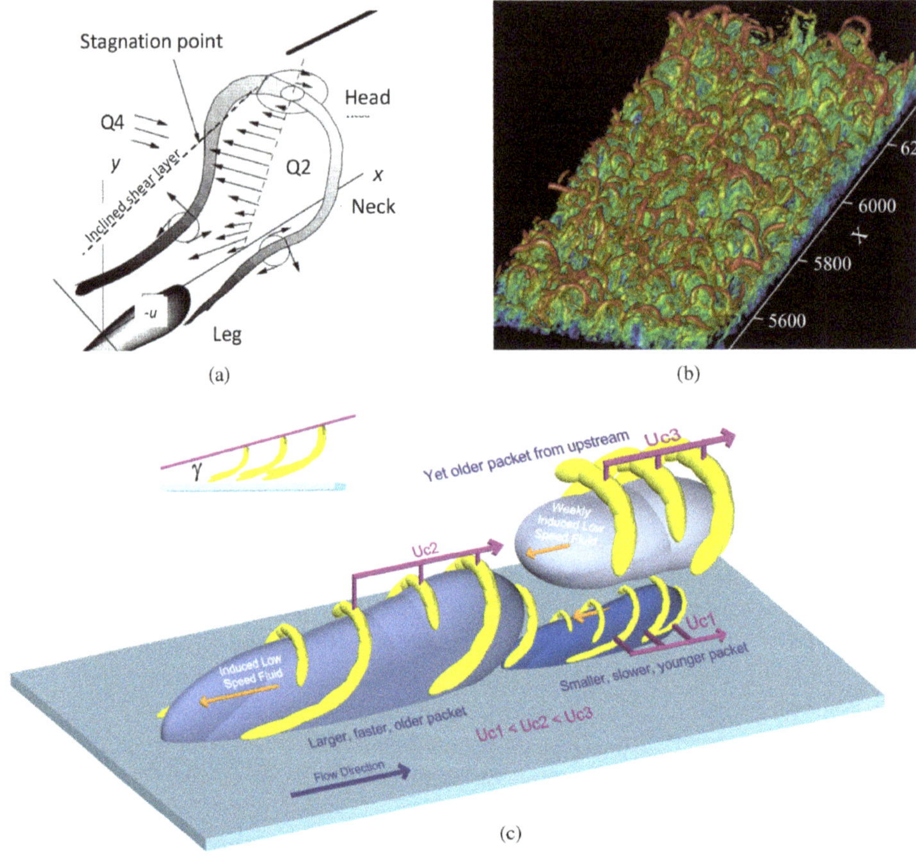

Figure 8.16. (a) Schematic of an individual hairpin vortex attached to the wall, with its induced motions[13]. (b) Visualisation of hairpins from a DNS of a low-Reynolds-number boundary layer[14]. X refers to the distance downstream normalised by the initial momentum thickness. At the farthest downstream location, $\mathrm{Re}_\theta \approx 1000$. (c) A 3D conceptual visualisation of a packet of the hairpins shown in (a)[13], copyright Cambridge University Press.

relative to the strain rate and has been shown to be a particularly good way of detecting vortex filaments [18, 39]. DNS allows a much clearer picture to emerge than would be possible to obtain from laboratory studies, since full-field data are available at all times. Figure 8.16(b) illustrates what is now possible; it shows the isocontours of a Q value close to zero. The prevalence of hairpins is very clear, although it should be noted that the boundary layer is at a relatively low Reynolds number.

Hairpins may provide a fundamental building block of the mechanism that sustains near-wall turbulence, see e.g. [2, 37]. Alternatively, the basic mechanism for the latter may be one of instability and transient growth. This latter view has recently been expounded in terms of some exact coherent structures (coherent in

---

[13] Reproduced with permission from [3], copyright Cambridge University Press.
[14] Reproduced with permission from [120], copyright Cambridge University Press.

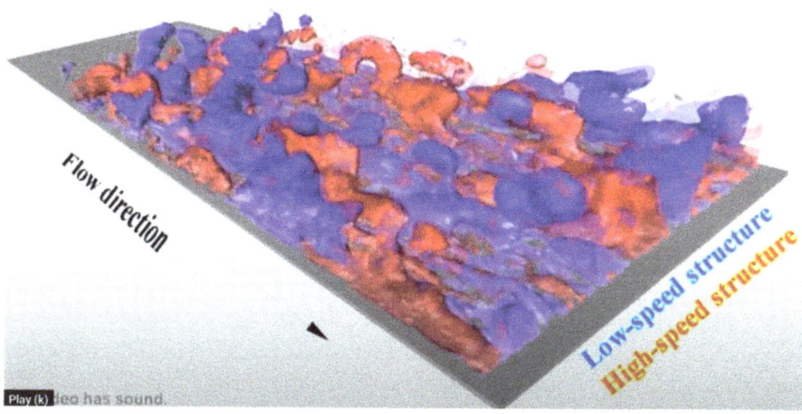

Figure 8.17. Snapshot from a video by Lee and Zaki[15] [52] from the DNS of a boundary layer at $Re_\tau = 1000$, highlighting the large-scale structures, with low-speed ones shown in blue and high-speed ones in red.

time and space) which can be shown to be exact solutions of the Navier–Stokes equations (for parallel flows) and seem to encompass many features of near-wall turbulence; the structures arise from a three-dimensional instability due to the superposition of streamwise vortices and the mean shear [35]. There are various other scenarios, however [73], and the study of the precise way in which turbulence is generated and maintained by these structures will undoubtedly continue for some time.

The much-larger-scale outer flow structures can be seen in the visualisation of figure 8.17. This large-scale motion (often called LSM), like the near-wall hairpin structure, has been studied experimentally, e.g. [3, 31], as has its impact on the turbulent/non-turbulent interface at the outer part of the flow, e.g. [52]. The convoluted boundary between these turbulent bulges and the vorticity-free external flow is not sharp; there is a 'superlayer' (as discussed in section 6.6.2 in the context of free shear flows), within which viscosity is important and across which the entrainment of the external flow into the turbulent region occurs. The LSMs have a scale of $\mathcal{O}(\delta)$ but there are also very much larger – more elongated – structures (very-large-scale motions (VLSMs)) with streamwise scales of around $\mathcal{O}(10\delta)$. It is these, in particular, which make adequate DNSs more expensive than would be the case in their absence (because long domain lengths are needed to capture them properly). Similar 'superstructures' also occur in internal flows, although it is currently rather unclear how similar they are to those in boundary layers. The way in which the various outer layer, larger-scale structures interact with the near-wall flow and its energy production cycle has been of interest for a long time (see, e.g. [25, 86, 117]) and more recently has been the subject of further experimental and DNS studies (e.g. [1, 40, 65, 105, 111]). It is evident that the interactions increase in importance as the Reynolds number increases.

---

[15] https://gfm.aps.org/meetings/dfd-2017/59bbd22bb8ac316d38841fe1

How do these various structural features of boundary layer turbulence differ from those in internal flows? Recall that we concluded chapter 7 by stating (at the end of section 7.4) that the major structural features in internal flows are, in many respects, essentially the same as those in external flows. This is most evidently true for the processes in the inner region, items (1)–(4) in the list enumerated above. These are not very dependent on the nature of the outer boundary, whether it be solid (or – more pertinently – a centreline symmetry boundary, for internal flows) or a genuine free stream (for external flows). What is less obvious, however, is the extent of the similarities in the large-scale motions, particularly the VLSMs (superstructures) that have a scale of $\mathcal{O}(10\delta)$. It has been found that the energy contained within them is much the same in channels and pipes, but it exists at larger wavelengths and further from the wall than in boundary layers [63]; the latter, at least, is perhaps unsurprising given the encroaching effects of the entrainment processes in the outer region of boundary layers. Some quantitative differences would be expected because of the influence of the fully turbulent symmetry centreline in pipes and channels (across which turbulent fluid from above is intermittently transported), contrasted with the entrainment processes. The topic continues to be studied; Marusic [57] has provided a relatively recent review.

We have hardly mentioned atmospheric boundary layer (ABL) flows. Most of these occur over some kind of rough surface (and will thus be discussed further in section 8.7), but it is worth noting here that some relatively recent measurements on a practically smooth desert floor have been helpful in solidifying, not least, the fact that the mean velocity profile in the wall region is essentially independent of the Reynolds number (e.g. [60]). Such ABL data have also provided evidence that the near-wall turbulence can be scaled similarly to the scaling at much lower Reynolds numbers and has structural features very similar to those discussed above [47, 56, 83].

There are two major reasons for all the attention being paid in the literature to the eddy structures and their dynamics. First, they determine the form of the mean velocity and the Reynolds stress profiles (and spectra). Townsend [112, 114] first made this link specific through his 'attached eddy hypothesis', which has been very influential over nearly three-quarters of a century. He postulated in 1951 that *the bulk of the energy containing eddies are, in a sense, attached to the wall, and the dependence of scale on distance from the wall is not a local effect but due to the attachment of most of the eddies.* He modelled the flow in terms of these large, energy-containing, attached eddies, and thus explained his correlation measurements, showing also that the hypothesis leads not only to the mean velocity log law but also to the logarithmic behaviour of the two normal stresses, $\overline{u'^2}$ and $\overline{w'^2}$ (e.g. equation (8.43)). These ideas have since been significantly extended and shown to also lead to the κ_1^{-1} behaviour of the u' energy spectrum in the near-wall region – indeed, to the complete spectral behaviour across all wavenumbers in the logarithmic layer, e.g. [75, 76]. Townsend's attached eddy hypothesis will undoubtedly continue to be a source of inspiration for the foreseeable future.

The second reason for all the attention paid to the eddy dynamics near walls is that understanding the structures and their influence on the turbulence generation and maintenance processes promises the hope of being able to develop successful control strategies to minimise the surface drag, with all the obvious practical advantages that could give [46, 59].

8.6 Effects of pressure gradients

In practice, turbulent boundary layers are very often subjected to streamwise pressure gradients, whose effects are qualitatively similar to those that occur if the boundary layer is laminar. They depend crucially on the sign of the pressure gradient. If it is favourable, that is, if the pressure decreases in the streamwise direction, the free-stream velocity increases, which tends to accelerate the boundary layer – usually leading to δ decreasing with x or, perhaps, remaining constant. This latter case is a special one, yielding a truly self-similar, equilibrium flow [109, 114]. If the pressure gradient is adverse, i.e. the pressure increases with x, the boundary layer is decelerated and may eventually separate from the surface.

We may define a parameter Ψ, a measure of the importance of the pressure gradient, as

$$\Psi = \frac{\delta^*}{\tau_w} \frac{\mathrm{d}P_\infty}{\mathrm{d}x} = -\frac{\delta^* U_\infty}{u_\tau^2} \frac{\mathrm{d}U_\infty}{\mathrm{d}x}, \tag{8.47}$$

which allows the momentum integral equation (8.13) to be rewritten as

$$\frac{\mathrm{d}\theta}{\mathrm{d}x} = \frac{u_\tau^2}{U_\infty^2}\left[1 + \Psi\left(1 + \frac{2\theta}{\delta^*}\right)\right]. \tag{8.48}$$

Notice that for a constant (positive) pressure gradient, increasing δ^* and decreasing u_τ both increase the growth rate (because of the falling velocities) – a sort of positive feedback. Figure 8.18 shows how a favourable pressure gradient ($\Psi < 0$) accelerates the flow, making the boundary layer fuller. Eventually, relaminarisation may occur. On the other hand, as Ψ increases from zero, the velocity profile becomes less full and the shear stress profile develops a peak, with a lower value near the wall. This value gradually decreases with increasing Ψ until separation is reached, at which point the velocity profile has a zero slope at the wall and the surface stress is zero. Downstream from that location, the surface stress has the opposite sign and the mean velocity profile contains a region of negative velocity. For favourable pressure gradients, the shape factor H gradually decreases (recall that it is larger in turbulent boundary layers than in laminar ones) and it increases for $\Psi > 0$.

With the usual kind of scaling analysis, it can be shown formally that in the inner region Ψ has very little influence; along with the left-hand side's mean advective terms, the pressure gradient term in the momentum equation (8.18) is of a lower order than the final stress term. Thus, as we noted in chapter 7, conventionally the inner-region analysis is valid for nearly all kinds of wall flow. This is not true, however, in the outer layer, in which the pressure gradient term cannot be neglected any more than can the advective terms. The outer layer is affected first and has the

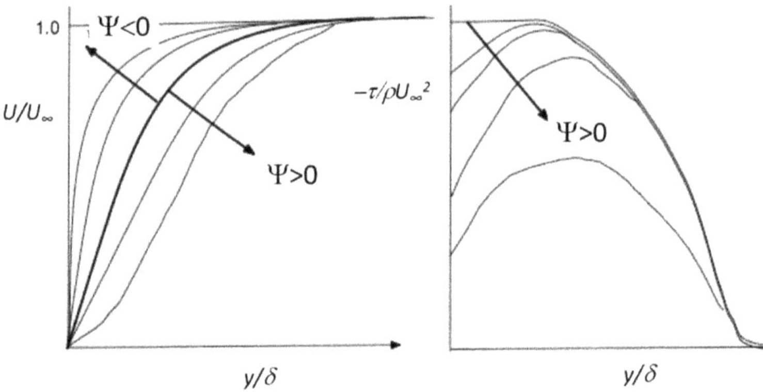

Figure 8.18. A sketch of the effects of the pressure gradient parameter, Ψ, on the profiles of mean velocity (left) and shear stress (right).

initial effect of merely changing the surface stress; the inner layer is practically unchanged, although it has a smaller log-law extent in y/δ.

This is all illustrated in figure 8.19, which shows some relatively early measurements of the mean velocity profile at a series of stations (1–12) in an increasingly adverse pressure gradient. It is immediately evident that the outer layer becomes more and more prominent; the shape factor rises from 1.39 at the first station to 1.61 at the last, and, correspondingly, the Coles' wake factor Π rises from about 0.6 to 2.9. This flow is still far from separating, but it is clearly the case that, nonetheless, the outer layer is strongly affected. The inner layer, however, appears to conform well to the classical log law. As noted above, this was the conventional wisdom for many decades [5, 24, 78]. It is perhaps intuitive, however, that if the outer region is so strongly affected, it would seem unlikely that the inner region could remain the same, particularly in the light of the now well-known structural interactions between the inner and outer regions, as discussed in the previous section, 8.5. More recent studies, some of which include (unlike [94]) direct measurements of c_f, have shown that there *are* changes in the inner region; these include changes in κ and A, e.g. [68]. For a very recent, helpful review and a discussion of the new measurements, see Knopp *et al* [49].

It is worth emphasising what happens to the shear stress profile at the wall. From the momentum equation (8.15), this must be given by

$$\left.\frac{\partial \tau}{\partial y}\right|_0 = \frac{\mathrm{d}P_\infty}{\mathrm{d}x}$$

so the shear stress increases with y for $\Psi > 0$, as illustrated in figure 8.18. Of course, once separation is reached, the velocity profile has a zero slope at the wall and $c_f = 0$ there. Even before that point, the boundary layer equations become invalid, not least because the flow is no longer 'thin', in the sense explained in section 6.2.2 for thin shear flows. Full analysis of flows near the separation region is therefore complex, but there have been some similarity-type (scaling) analyses for the flow further upstream, where the boundary layer can still be considered as thin, starting with the

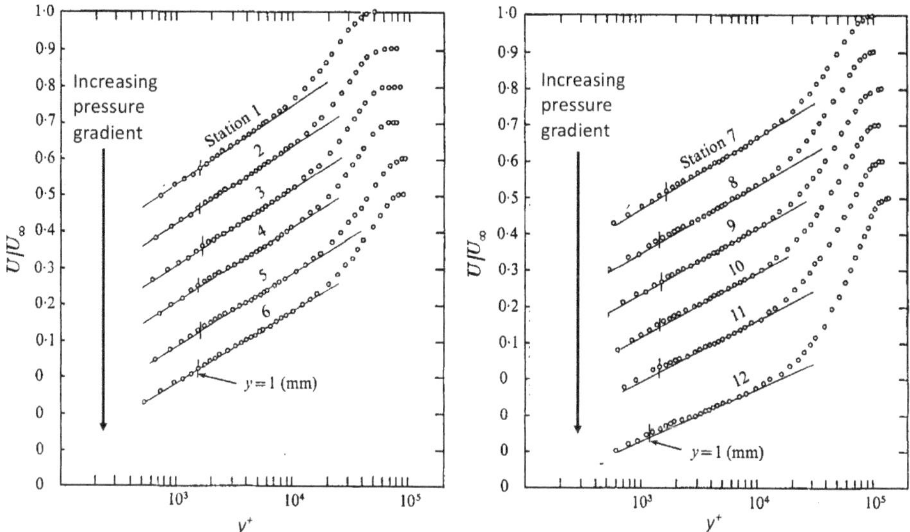

Figure 8.19. The effect of increasing adverse pressure gradient on mean velocity profiles. The data are from 12 consecutive downstream locations after the imposition of the pressure gradient. Reproduced with permission from [94], copyright Cambridge University Press.

early classical approaches of Clauser and Rotta [22–24, 90], and more recently by, for example, [16, 80]. The available data sets for boundary layers in pressure gradients are relatively sparse and, given the practical importance of flows in pressure gradients, we expect significantly more effort on this topic.

8.7 Effects of roughness

Throughout our discussion of wall flows, both internal and external, the walls have been assumed to be completely smooth. In practice, this is often not the case, so the study of the influence of surface roughness has a long history. For external flows, the most obvious case is perhaps the ABL, which can only be considered to be a smooth-wall flow under very unusual geomorphological circumstances. Often, in this case, the roughness elements on the surface (i.e. trees, buildings, other man-made objects, etc.) may be large – perhaps up to 10%–15% of the boundary layer depth. Of course, there are other important cases – consider, for example, ocean-going liners and cargo transporters. These do not have to be in service for very long before their hulls collect marine and/or chemical accretions of many kinds, leading to significantly increased drag (and thus energy consumption).

Figure 8.20 shows the various regions of the flow above the $y = 0$ surface when there is a rough region of depth h on it. In the classical engineering context, this layer may consist of tightly packed objects (like the sand grains used by Nikuradse – see below). There will often be very little flow within it. In the meteorological literature, this layer is commonly called the 'canopy layer' and in that context it may well consist of inhomogeneous arrays of large roughness 'elements' (e.g. buildings). There will almost always be significant flow within it – responsible for the immediate

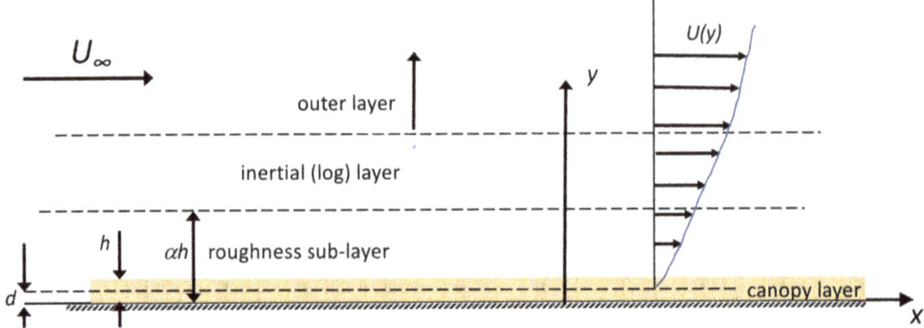

Figure 8.20. Sketch of the various regions in a rough-wall boundary. The yellow-patterned layer represents the roughness of height h, which constitutes the 'canopy' layer.

dispersion of pollutants and/or heat which may be generated therein. Any flow in this canopy layer must be strongly inhomogeneous, even if the roughness morphology is not. So there will inevitably be a region above $y = h$ up to $y = \alpha h$, say, in which the flow slowly loses its inhomogeneity. This may be quite thick – α may perhaps be as much as five. Provided the overall boundary layer thickness is sufficiently large compared with h, there could be the usual inertial layer above, in which the flow is no longer spatially variant in the horizontal.

We will briefly discuss some major points about rough-wall flows in the present context of boundary layers, although much of it is relevant to pipe and channel flows as well (as we will indicate). Indeed, the early work on roughness was largely generated by the industrial importance of rough-wall pipe flows [70]. Classical thinking suggested that the roughness merely acts to increase the surface stress, without causing any important structural changes in the flow, i.e. the *Townsend hypothesis* [114]. This assumes, crucially, that for a given roughness height, characterised by its Reynolds number $h^+ = hu_\tau/\nu$, the flow Reynolds number δ^+ is sufficiently large that h/δ remains small – no more than 2%–3%, say, as argued convincingly by Jiménez in his useful review of rough-wall flows [43]. This ensures that the roughness affects less than half of the thickness of the logarithmic layer. Since the most energetic part of a smooth-wall boundary layer, including the region in which the energy production and self-generation processes occur, lies below about $y^+ \approx 100$, and typical air or water flows in engineering have viscous length scales of $\mathcal{O}(10)$ μm, it is clear that quite small h ($\mathcal{O}(1)$ mm) would be enough to obliterate this region. Even smaller h must effectively remove the viscosity-dominated sublayer. The surface stress in such circumstances will (usually) be significantly larger than the regular skin friction and dominated by pressure drag generated by the individual roughness elements. (There are roughness morphologies which are specifically designed to *reduce* the surface drag – the well-known 'riblets', for example – but we do not consider this specialist topic here. References [11] and [85] provide an entrance to the relevant literature.)

In the engineering context, the mean velocity profile for the log-law region is classically modified to become

$$U^+ = \frac{1}{\kappa} \ln(y^+) + A - \Delta U^+(h^+,,,), \tag{8.49}$$

where ΔU^+ is the *roughness function*, a function of h^+ but also of the geometrical parameters defining the roughness elements and their layout (the roughness morphology). Notice that the addition of surface roughness causes a downward shift of the log law, illustrated in figure 8.21(a). Nikuradse began a very long-standing trend by writing this equation using a 'sand grain roughness length', k_s, as

$$U^+ = \frac{1}{\kappa} \ln(y/k_s) + A' - \Delta U^+, \tag{8.50}$$

because his experiments on pipes used closely packed sand grains as the roughness. This relation is effectively a definition of k_s. Much of the literature has thus expressed other kinds of roughness in terms of 'equivalent sand grain roughness' of height k_s, with h some multiple of k_s.

For many practical cases, the surfaces are 'fully rough', in the sense that the viscous drag is negligible compared with the (pressure) drag generated directly by the roughness elements. Certainly, this is almost always the case for the ABL, so that meteorologists have long used the alternative log law expressed by

$$U^+ = \frac{1}{\kappa} \ln(y/y_o), \tag{8.51}$$

where y_o is the *roughness length*, which implicitly embodies the effect of the roughness function and is determined by h and the roughness morphology alone. Typically, $y_o = 0.033k_s$ (for $A' = 8.5$ and $\kappa = 0.4$). We can define an alternative roughness Reynolds number by $\mathrm{Re}^* = y_o u_\tau/\nu$. Note, incidentally, that for a smooth wall, $y_o/h = e^{-A\kappa}/h^+$, so $\mathrm{Re}^* = 0.135$, for $\kappa = 0.4$ and $A = 5$. It is generally reckoned that a hydraulically smooth flow requires $h^+ < 5$, but this limit must depend somewhat on the morphology of the surface roughness.

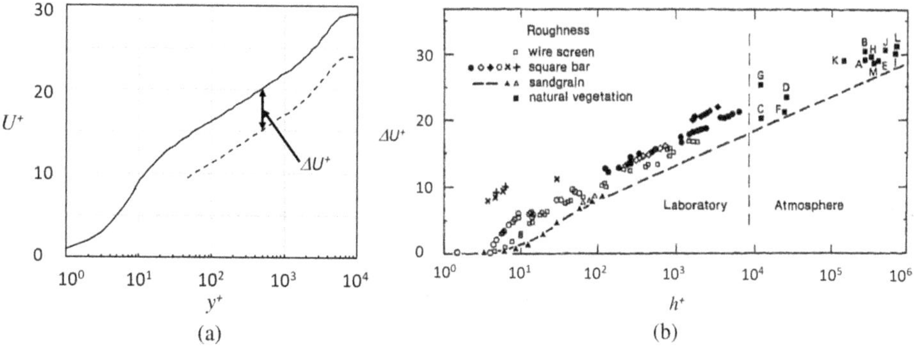

Figure 8.21. (a) A sketch showing a typical rough-wall profile offset of about $\Delta U^+ = 5$ below the usual log law. (b) Variation of ΔU^+ with h^+ for numerous surfaces, both in the laboratory and in the atmosphere; from [87] with permission from The American Society of Mechanical Engineers, which gives the origin of all the different experiments.

For a fully rough surface, $y_o/h \neq f(h^+, \mathrm{Re}^*)$, although it remains dependent on the surface morphology; y_o and ΔU^+ are then entirely equivalent measures of the effect of the roughness on the mean profile, and equations (8.49) and (8.51) together imply that

$$\Delta U^+ = A + \frac{1}{\kappa}\ln(\mathrm{Re}^*) \equiv A + \frac{1}{\kappa}\ln(h^+) + \frac{1}{\kappa}\ln\left(\frac{y_o}{h}\right). \tag{8.52}$$

It is usual to consider that fully rough conditions require $k_s \geqslant 80$, i.e. $\mathrm{Re}^* \geqslant 2$, but the precise value must depend on the nature of the roughness. Appropriately spaced square flat plates, for example, are more 'efficient' at generating surface drag, and fully rough conditions can exist for $\mathrm{Re}^* = 1$ or even less [104]. Equation (8.52) shows immediately that since y_o/h is a constant for any particular surface (but different for different surfaces), the roughness function increases logarithmically with h^+, and the precise value is dependent on y_o/h. This is all illustrated in figure 8.21(b), which shows a large collection of ΔU^+ data from different experiments using different kinds of rough surface. The dashed line represents classical sand grain roughness [98], but all the other surfaces, despite following the logarithmic behaviour individually (for high enough values of h^+), do not collapse together because of the dependence of ΔU^+ on the roughness morphology. At low h^+, the flows are transitionally rough – i.e. viscosity plays a part in setting the surface drag (so equation (8.51) is arguably not such an appropriate form of the velocity profile).

We must emphasise at this point that in all the equations above, y should be understood as $y - d$, where y is measured from the bottom of the roughness and d is known as the *zero plane displacement* [87], as illustrated in figure 8.20. (Note that $y = d$ is *not* exactly the location where $U = 0$, as seen immediately from equation (8.51).) The fact that for many roughness morphologies, d cannot be considered to be zero, but lies somewhere below $y = h$, makes an appropriate analysis of experimental data more difficult than for smooth-wall flows, because d is usually unknown. This difficulty is exacerbated by the fact that the surface stress cannot easily be determined independently – it requires a specially designed 'floating balance' (e.g. [28, 50]) or pressure tapping of the roughness elements (e.g. [21, 79]). These techniques can also allow d to be determined independently if it is assumed to coincide with the plane at which the surface drag appears to act; there is some theoretical justification for that assumption [42]. If these techniques are unavailable or impractical, one has to resort to the somewhat ill-conditioned process of fitting the mean velocity profile to the data, maximising the quality of fit by judicious adjustments to d, y_o, and u_τ, although measurement of the turbulence shear stress profile may provide an estimate of the latter (with some arguably dubious assumptions). This also, of course, requires an assumption for the values of κ and, if the outer layer is included in the fitting process, Π. These ideas are explored in sample exercise 8.3.

Recall now our earlier point that classical thinking suggests that the roughness merely increases the surface stress, but causes no structural changes to the flow. Very close to the roughness (i.e. in the roughness sublayer, perhaps up to $5h$ above $y = 0$)

this obviously cannot ever be true; the roughness generates small-scale turbulent motions which are spatially inhomogeneous. This inevitably *does* lead to near-wall changes which must be very specific to the particular roughness type. Nonetheless, appropriate modifications of the classical correlations of, for example, skin friction and shape factor versus Re_θ have been found to remain accurate, which is certainly the case for $h/\delta \lesssim 0.02$. We pointed out in section 8.3.4 that the integral and associated relationships, equations (8.36)–(8.40), are also true for rough surfaces. The correlation for c_f versus Re_θ originally calculated by Rotta [90] can thus be reworked for rough surfaces, for which Re_θ may be replaced by θ/y_o, and provides a good fit to data even for roughnesses significantly larger than 0.02δ, as shown in figure 8.22(a). The roughness types vary from sandpapers, to wire meshes, to rectangular blocks and the values of Re_θ range up to about 47800. (Note that, because of the very different roughness types used, the variation in θ/y_o is not monotonic with Re_θ – for the case when the latter was 47800, for example, the former was only about 100.) In figure 8.22(b), a parameter related directly to I_1 in equation (8.40) is plotted for the same data and, within experimental uncertainties, has the same constant value anticipated by the classical two-parameter analysis – with $\kappa = 0.41$ and $\Pi = 0.55$, one expects that $c_f^{1/2}\delta/\delta^* = \sqrt{2}/I_1 = \sqrt{2}\kappa/(1 + \Pi) \approx 0.35$. (Note that the Coles' curve plotted in figure 8.22(b) used $\kappa = 0.41$.) There is a hint in figure 8.22(a) that the value of Π may be a little higher for very rough surfaces (see also, for example, [108]), but it should be recognised that Π is difficult to determine accurately, depending as it does on the values of u_τ, κ, y_o, and d used for the profile fit.

However, whilst all this suggests that the mean flow may be universal even beyond the $h/\delta \lesssim 0.02$ limit, and well described by the classical two-parameter boundary layer analysis, this is not always much help from a practical perspective, because it requires knowledge of y_o before c_f can be deduced from the mean velocity profile. There have been numerous attempts, particularly in the meteorological context (but beginning with Schlichting [97]), to produce correlations for y_o (or k_s) in terms of morphological parameters (such as λ_f, the ratio of the frontal area of the

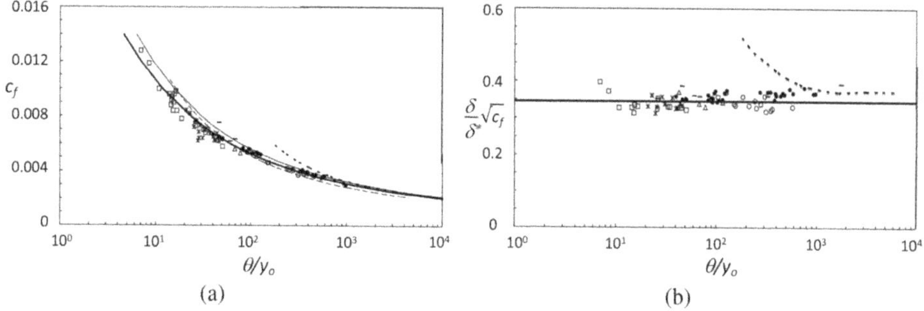

Figure 8.22. (a) c_f and (b) $c_f^{1/2}\delta/\delta^*$ variations for rough-wall boundary layers. The various symbols refer to different kinds of roughness. The dotted lines are obtained from the Coles' relations and the solid lines in (a) have $\Pi = 0.55$ (upper) and 0.7 (lower). From the authors' laboratory, along with data from [12] and [29]. Reproduced with permission from data from [17], copyright Cambridge University Press.

elements to the plan area on which they sit), but these do not always help for roughness types very different from the ones used to generate the correlation.

The extent to which a rough surface, even in cases in which h/δ is very small, affects the turbulence processes beyond the near-wall region is, as yet, not entirely clear. Some have found significant changes all the way to the outer layer, but usually in the probably rather special case of two-dimensional spanwise rib roughness (see [51, 101], for example). This type of surface was first distinguished from the more usual 'k'-type roughness, comprising three-dimensional elements, by Perry $et\ al$ [79]. They called it 'd'-type roughness: 'depressions or narrow lateral grooves in the wall'. The canopy flow is then typified by spanwise-coherent vortex motions trapped (intermittently) in the cavities. What does seem clear, however, is that roughness promotes a rise in the streamwise stress within the logarithmic region – enhancing the 'hump' which appears in the smooth-wall case at high Reynolds numbers (recall figure 8.11). This may be an indication of broader changes which, in any case, one might expect given the increasingly clear links between the motions of the near-wall structures and those in the outer flow in the smooth-wall case; roughness seriously inhibits or even destroys some of the former. Nonetheless, there is considerable evidence that structural changes in the outer layer are, if they exist at all, very minor provided that h/δ does not exceed a few percent [29, 116], or perhaps even more [6], so that the original Townsend hypothesis that all roughness does is alter the surface stress seems fairly secure. Except for the enhanced skin friction, fully rough turbulent boundary layers over arbitrary three-dimensional roughness of all kinds may thus be considered as equivalent in many ways to very-high-Reynolds-number smooth-wall boundary layers.

8.8 Subject giants: L Prandtl

Ludwig Prandtl (1875–1953) was a German scientist and was the founding co-director (together with Carl Runge) of the Institute for Applied Mathematics and Mechanics at Göttingen University. During his nearly 50 years at Göttingen, the institute became a leading place for turbulence research and among Prandtl's 85 doctoral students are the famous names of Heinrich Blasius, Theodore von Kármán, Max Munk, Johann Nikuradse, Walter Tollmien, Hermann Schlichting, and Karl Wieghardt [14].

In 1904, Prandtl made a remarkable breakthrough in aerodynamics when he proposed the idea of the boundary layer. The Navier–Stokes equations had already been developed and the importance of viscosity was well known, but little progress had been made in how to solve the equations in order, for example, to predict the drag of an aerofoil. This was particularly relevant at the time, as the Wright Brothers had just completed the first powered flight in 1903. While the Euler equations could be used to predict the lift of an aerofoil assuming inviscid flow, the effects of viscosity could not be ignored in estimating the drag.

In 1914, Prandtl reviewed Eiffel's 1912 experiments on the drag of a sphere and correctly showed how the drag on a sphere is a function of the Reynolds number. More importantly, he published work during 1918–1919 describing a 'thin-airfoil

Figure 8.23. Ludwig Prandtl (1875–1953) at his water tunnel in the mid-to-late 1930s. Image reproduced with permission from [14], copyright Cambridge University Press.

theory', which remains important to this day. It included work on induced drag and showed that a finite wing with an elliptical planform minimised the induced drag and was thus the most efficient. The designer of the Spitfire fighter used the theory to great effect!

Much of Prandtl's research overlapped and complemented that of G. I. Taylor and the two of them wrote several letters back and forth before WWII. However, during the war, Prandtl was unable to collaborate internationally, and when the war ended, he was forced to stop all research. After some time, he was permitted to work and to teach again, but was by then reaching the end of his career.

Prandtl's work greatly shaped the development of modern fluid dynamics through his research and his teaching. Various editions of his lecture notes and textbooks on the essentials of fluid mechanics are still in publication and capture his typical visual approach and style.

Sample exercises

8.1. The file 'TBLData.txt'[16] contains the velocity profile of a smooth-wall turbulent boundary layer at $Re_\theta = 4060$ obtained using DNS by Schlatter and Orlu [95]. This exercise explores how the velocity profile compares with the expected viscous sublayer, log law, and wake profiles.

(a) Plot the velocity profile U^+ versus y^+ in linear axes. Zoom in to the near-wall region and compare the profile with $U^+ = y^+$, which one expects to see in the viscous sublayer. To what value of y^+ is this

---

[16] https://github.com/cvanderwel/TurbulentFlows/blob/main/data/TBLData.txt

valid? What percentage of the boundary layer depth (in terms of δ) does this region cover?

(b) Zoom out and adjust the axes to display this plot of U^+ versus y^+ in log-linear axes. Fit the log-law equation (8.22) using $\kappa = 0.384$ and $A = 4.173$.

(c) Plot the velocity profile against outer units (i.e. U^+ versus y/δ). Plot the log law with the wake function (i.e. equation (8.27)) with $\Pi = 0.55$. Note how well this fits in the outer region (and clearly does not apply beyond $y/\delta = 1$).

8.2. This exercise considers the integral parameters that describe the velocity profile data given in the file 'TBLData.txt'[17]. The file contains data for a smooth-wall turbulent boundary layer at $Re_\theta = 4060$ obtained using DNS by Schlatter and Orlu [95].

(a) Given that $c_f = 0.002971$ for this flow (and by definition, $u_\tau/U_\infty = \sqrt{c_f/2}$), integrate the velocity profile to confirm that the normalised displacement thickness is $Re_{\delta^*} = \int_0^\infty \left(\frac{U_\infty}{u_\tau} - U^+ \right) dy^+ = 5633$, that the normalised momentum thickness is $Re_\theta = \frac{U_\infty}{u_\tau} \int_0^\infty \left(\frac{U_\infty}{u_\tau} - U^+ \right) dy^+ = 4061$, and that $H = 1.387$.

(b) Calculate the normalised value of the Rotta–Clauser integral thickness $Re_\Delta = U_\infty \Delta/\nu = (U_\infty/\nu)\delta^*(U_\infty/u_\tau) = Re_{\delta^*} U_\infty/u_\tau$. Determine its ratio to the normalised boundary layer thickness $Re_{\delta_{99}} = U_\infty \delta_{99}/\nu = \delta_{99}^+ U_\infty/u_\tau$ and compare this with the classical (constant) value of Δ/δ from the text.

8.3. This exercise explores the velocity profile over a rough surface and the challenges in determining u_τ and ΔU^+. The file 'RoughWallData.txt'[18] contains measurements of the velocity and Reynolds shear stress over a surface covered with 320 grit sandpaper in channel flow, made by Flack, Schultz, Barros, and Kim [30].

(a) Try to fit a log law in the form of equation (8.49) to the measurements of U versus y by adjusting the values of u_τ, d, and $C = (A - \Delta U^+)$, assuming u_τ is unknown.

(b) Plot the Reynolds stress and get a second estimate of u_τ from the peak of $-\overline{u'v'}$ in the inertial (log) layer as $-\overline{u'v'}_{peak} = u_\tau^2$. How does this compare with the previous estimate from the log-law fit?

(c) In these experiments, the wall shear stress was measured directly from the pressure drop in the channel and u_τ was found to be 0.155 m/s. How different were the previous estimates? Replot the log law in inner units and compare the results with a smooth-wall channel flow (with $\kappa = 0.387$, $A = 4.5$; exercise 7.4) to determine ΔU^+.

---

[17] https://github.com/cvanderwel/TurbulentFlows/blob/main/data/TBLData.txt
[18] https://github.com/cvanderwel/TurbulentFlows/blob/main/data/RoughWallData.txt

References

[1] Abe H, Kawamura H and Choi H 2004 Very large-scale structures and their effects on the wall shear-stress fluctuations in a turbulent channel flow up to $Re_\tau = 640$ *J. Fluids. Eng.* **126** 835–43

[2] Adrian R 2007 Hairpin vortex organization in wall turbulence *Phys. Fluids.* **19** 041–301

[3] Adrian R J, Meinhart C D and Tomkins C D 2000 Vortex organisation in the outer region of the turbulent boundary layer *J. Fluid. Mech.* **422**

[4] Álamo J C D and Jiménez J 2009 Estimation of turbulent convection velocities and corrections to Taylor's approximation *J. Fluid. Mech.* **640** 5–26

[5] Alving A E and Fernholz H H 1995 Mean velocity scaling around a mild, turbulent separation bubble *Phys. Fluids.* **7** 1956–69

[6] Amir M and Castro I P 2011 Turbulence in rough-wall boundary layers *Exp. Fluids.* **51** 313–26

[7] Baidya R, Philip J, Hutchins N, Monty J P and Marusic I 2017 Distance-from-the-wall scaling of turbulent motions in wall-bounded flows *Phys. Fluids.* **29** 1–10

[8] Baidya R, Philip J, Hutchins N, Monty J P and Marusic I 2021 Spanwise velocity statistics in high-Reynolds-number turbulent boundary layers *J. Fluid. Mech.* **913** 1–25

[9] Barenblatt G I 1993 Scaling laws for fully developed turbulent shear flows. Part I. Basic hypotheses and analysis *J. Fluid. Mech.* **248** 513–20

[10] Batchelor G K 1967 *An Introduction to Fluid Dynamics* (Cambridge University Press: Cambridge)

[11] Bechert D W, Bruse M, Hage W, de Hoeven J G T and Hoppe G 1997 Experiments on drag-reducing surfaces and their optimisation with an adjustable geometry *Exp. Fluids.* **338** 59–87

[12] Bergstrom D J, Akinlade O G and Tachie M F 2005 Skin friction correlation for smooth and rough wall turbulent boundary layers *Trans. ASMEI: J. Fluids. Eng.* **127** 1146–53

[13] Bernard P S and Wallace J M 2002 *Turbulent Flow. Analysis, Measurement and Prediction* (Wiley: New York)

[14] Bodenschatz E and Michael E 2011 Prandtl and the Göttingen school *A Voyage Through Turbulence* ed P A Davidson, Y Kaneda, K Moffatt and K R Sreenivasan (Cambridge: Cambridge University Press) pp 40–100

[15] Buschmann M and el Hak M G 2003 Generalized logarithmic law and its consequences *AIAA J* **41** 565–72

[16] Castillo L and George W K 2001 Similarity analysis for turbulent boundary layer with pressure gradient: outer flow *AIAA J* **39** 41–7

[17] Castro I P 2007 Rough-wall boundary layers: mean flow universality *J. Fluid. Mech.* **585** 469–85

[18] Chakraborty P, Balachandar S and Adrian R J 2005 On the relationships between local vortex identification schemes *J. Fluid. Mech.* **535** 189–214

[19] Chauhan K A, Nagib H M and Monkewitz P A 2007 On the composite logarithmic profile in zero pressure gradient turbulent boundary layers *45th AIAA Aerospace Sciences Meeting and Exhibit (Reno, Nevada, 8–11 January 2007)*

[20] Chauhan K A, Monkewitz P A and Nagib H M 2009 Criteria for assessing experiments in zero pressure gradient boundary layers *Fluid. Dyn. Res.* **41** 1–23

[21] Cheng H and Castro I P 2002 Near-wall flow over urban type roughness *Bound. Layer. Meteorol.* **104** 229–59

[22] Clauser F H 1954 Turbulent boundary layers in adverse pressure gradients *J. Aerosp. Sci.* **2** 91–108

[23] Clauser F H 1956 The turbulent boundary layer *Adv. Appl. Mech.* **4** 1–51

[24] Coles D 1956 The law of the wake in the turbulent boundary layer *J. Fluid. Mech.* **1** 191–226

[25] Coles D E 1987 Coherent structures in turbulent boundary layers *Perspectives in Turbulence* ed H U Meier and P Bradshaw (Berlin: Springer) pp 93–114

[26] Corrsin S and Kistler A L 1954 The free-stream boundaries of turbulent flows *NACA* Tech Rep TN-3133

[27] Fernholz H H and Finley P J 1996 The incompressible zero-pressure-gradient turbulent boundary layer: an assessment of the data *Prog. Aerospace. Sci.* **32** 245–311

[28] Ferreira M A, Rodriguez-Lopez E and Ganapathisubramani B 2018 An alternative floating element design for skin-friction measurement of turbulent wall flows *Exp. Fluids.* **59** 1–15

[29] Flack K A, Schultz M P and Connelly J S 2007 Examination of a critical roughness height for outer layer similarity *Phys. Fluids.* **19** 1–9

[30] Flack K A, Schultz M P, Barros J M and Kim Y C 2016 Skin-friction behavior in the transitionally-rough regime *Int. J. Heat. Fluid. Flow.* **61** 21–30

[31] Ganapathisubramani B, Longmire E K and Marusic I 2003 Characteristics of vortex packets in turbulent boundary layers *J. Fluid. Mech.* **478** 35–46

[32] George W K 2006 Recent advancements toward understanding the of turbulent boundary layers *AIAA J* **44** 2435–49

[33] George W K 2007 Is there a universal log law for turbulent wall-bounded flows? *Phil. Trans. R. Soc.* A. **365** 789–806

[34] George W K and Castillo L 1997 Zero-pressure gradiennt turbulent boundary layer *Appl. Mech. Rev.* **50** 689–729

[35] Graham M D and Floryan D 2021 Exact coherent states and the non-linear dynamics of wall-bounded turbulent flows *Ann. Rev. Fluid. Mech.* **53** 227–53

[36] Grant H L 1958 The large eddies of turbulent motion *J. Fluid. Mech.* **4** 149–90

[37] Hamilton J M, Kim J and Waleffe F 1995 Regeneration mechanisms of near-wall turbulence *J. Fluid. Mech.* **287** 317–48

[38] Head M R and Bandyopadhyay P 1981 New aspects of turbulent boundary layer structure *J. Fluid. Mech.* **107** 297–338

[39] Hunt J C R, Wray A A and Moin P 1988 Eddies, streams and convergent zones in turbulent flows *Studying Turbulence Using Numerical Simulation Databases, 2. Proceedings of the 1988 Summer Program* CTR-S88 Center for Turbulence Research 193–208

[40] Hutchins N and Marusic I 2007 Large-scale influences in near-wall turbulence *Phil. Trans. R. Soc.* A **365** 647–64

[41] Hutchins N, Nickels T B, Marusic I and Chong M S 2009 Hot-wire spatial resolution issues in wall-bounded turbulence *J. Fluid. Mech.* **635** 103–36

[42] Jackson P S 1981 On the displacement height in the logarithmic velocity profile *J. Fluid. Mech.* **111** 15–25

[43] Jiménez J 2004 Turbulent flow over rough walls *Annu. Rev. Fluid. Mech.* **36** 173–96

[44] Kachanov Y F 1994 Physical mechanisms of laminar boundary layer transition *Annu. Rev. Fluid. Mech.* **26** 411–82

[45] Kim H T, Kline S J and Reynolds W C 1971 The production of turbulence near a smooth wall in a turbulent boundary layer *J. Fluid. Mech.* **50** 133–60

[46] Kim J 2011 Physics and control of wall turbulence for drag reduction *Phil. Trans. R. Soc.* A **369** 1396–411

[47] Klewicki J C, Priyadarshana P J A and Metzger M M 2008 Statistical structure of the fluctuating wall pressure and its in-plane gradients at high Reynolds number *J. Fluid. Mech.* **609** 195–220

[48] Kline S J, Reynolds W C, Schraub F A and Runstadler P W 1967 The structure of turbulent boundary layers *J. Fluid. Mech.* **30** 741–73

[49] Knopp T, Reuther N, Novara M, Schanz D, Schülein E, Schröder A and Kähler C J 2021 Experimental analysis of the log law at adverse pressure gradient *J. Fluid. Mech.* **918** 1–32

[50] Å Krogstad P, Efros V 2016 Rough wall skin friction measurements using a high resolution surface balance *Int. J. Hear. Fluid. Flow.* **31** 429–33

[51] Krogstad P-å, Antonia R A, Browne L W B 1992 Comparison between rough- and smooth-wall turbulent boundary layers *J. Fluid. Mech.* **245** 599–617

[52] Lee J, Sung H J and Zaki T A 2017 Signature of large-scale motions on turbulent/non-turbulent interfaces in boundary layers *J. Fluid. Mech.* **819**

[53] Lee J H, Kwon Y S, Monty J P and Hutchins N 2012 Tow-tank investigation of the developing zero-pressure-gradiennt turbulent boundary layer *18th Australasian Fluid Mechanics Conf. (Launceston, Australia)*

[54] Lee J H, Kwon Y S, Hutchins N and Monty J P 2013 Spatially developing turbulent boundary layer on a flat plate arXiv:1210.3881 Entry V84181 to the Gallery of Fluid Motion.

[55] Luchini P 2017 Universality of the turbulent velocity profile *Phys. Rev. Lett.* **188** 1–4

[56] Marusic I and Hutchins N 2008 Study of the log-layer structure in wall turbulence over a very large range of Reynolds number *Flow. Turb. Comb.* **81** 115–30

[57] Marusic I, McKeon B J, Monkewitz P A, Nagib H M, Smits A J and Sreenivasan K R 2010 Wall-bounded turbulent flows at high Reynolds numbers: recent advances and key issues *Phys. Fluids.* **22** 103

[58] Marusic I, Baars W and Hutchins N 2017 Scaling of the streamwise turbulence intensity in the context of inner-outer interactions in wall turbulence *Phys. Rev. Fluids.* **2** 1–22

[59] McKeon B J, Sharma A S and Jacobi I 2013 Experimental manipulation of wall turbulence: a systems approach *Phys. Fluids.* **25** 1–34

[60] Metzger M M and Klewicki J C 2001 A comparitive study of near-wall turbulence in high and low Reynolds number boundary layers *Phys. Fluids.* **13** 692–701

[61] Monkewitz P A, Chauhan K A and Nagib H M 2007 Self-consistent high-Reynolds-number asymptotics for zero-pressure-gradient turbulent boundary layers *Phys. Fluids.* **19** 1–12

[62] Monkewitz P A, Chauhan K A and Nagib H N 2008 Comparison of mean flow similarity laws in zero pressure gradient boundary layers *Phys. Fluids.* **20** 1–16

[63] Monty J, Hutchins N, Ng H, Marusic I and Chong M 2009 A comparison of turbulent pipe, channel and boundary layers *J. Fluid. Mech.* **632** 432–42

[64] Monty J P and Chong M S 2009 Turbulent channel flow: comparison of streamwise velocity data from experiments and direct numerical simulation *J. Fluid. Mech.* **633** 461–74

[65] Morrison J F 2007 The interaction between inner and outer regions of turbulent wall-bounded flow *Phil. Trans. R. Soc.* A **365** 683–98

[66] Nagib H M, Chauhan K A and Monkewitz P A 2007 Approach to an asymptotic state for zero pressure gradient turbulent boundary layers *Phil. Trans. R. Soc.* A **365** 755–70

[67] Nagib H N, Christiphorou C, Ruedi J-D, Monkewitz P, Österlund J, Gravante S, Chauhan K and Pelivan I 2004 Can we ever rely on results from wall-bounded turbulent flows without direct measurements of wall shear stress? *24th AIAA Aerodynamics Measurement Technology and Ground Testing Conf.* 2392 *(Portland, Oregon, 28 June–1 July 2004)*

[68] Nickels T B 2004 Inner scaling for wall-bounded flows subject to large pressure gradients *J. Fluid. Mech.* **521** 217–39

[69] Nickels T B, Marusic I, Hafez S and Chong M S 2005 Evidence of the k_1^{-1} law in a high-Reynolds number turbulent boundary layer *Phys. Rev. Lett.* **95** 1–4

[70] Nikuradse J 1932 Strömungsgesetze in rauhen Rohren *Forschung auf dem Gebiete des Ingenieurwesens* Supplement **4**(B) (English Transl. NACA TT F-10, 359.) VDI research booklet 361. Supplement to Research on the Areas of Engineering Issue B Volume 4.

[71] Österlund J M 1999 Experimental studies of zero pressure gradient turbulent boundary layer flow *PhD thesis* (Stockholm: Kungl Tekniska Högskolan)

[72] Österlund J M, Johansson A V, Nagib H M and Hites M H 1999 Wall shear stress measurement in high Reynolds number boundary layers from two facilities *30th AIAA Fluid Dynamics Conf.* 3814 *(Norfolk, Virginia, 28 June 1999 – 01 July 1999)* 1–7

[73] Panton R L 2001 Overview of the self-sustaining mechanisms of wall turbulence *Prog. Aerospace. Sci.* **37** 341–83

[74] Panton R L 2007 Composite asymptotic expansions and scaling wall turbulence *Phil. Trans. R. Soc.* A **365** 733–54

[75] Perry A, Henbest S and Chong M 1986 A theoretical and experimental study of wall turbulence *J. Fluid. Mech.* **165** 163–99

[76] Perry A E and Abell C J 1977 Asymptotic similarity of turbulence structures in smooth- and rough-wall pipes *J. Fluid. Mech.* **79** 785–99

[77] Perry A E and Chong M S 1982 On the mechanism of wall turbulence *J. Fluid. Mech.* **119** 173–217

[78] Perry A E, Bell J B and Joubert P N 1966 Velocity and temperature profiles in adverse pressure gradient boundary layers *J. Fluid. Mech.* **25** 299–320

[79] Perry A E, Schofield W H and Joubert P N 1969 Rough wall turbulent boundary layers *J. Fluid. Mech.* **37** 383–413

[80] Perry A E, Marusic I and Jones M B 2002 On the streamwise evolution of turbulent boundary layers in arbitrary pressure gradients *J. Fluid. Mech.* **461** 61–91

[81] Pope S B 2000 *Turbulent Flows* (Cambridge: Cambridge University Press)

[82] Prandtl L 1932 Zur turbulenten Strömung in Rohren und längs Platten *Ergeb Aerodyn Versuchsanst Göttingen* **4** 18–29

[83] Priyadarshana P J A, Klewicki J C, Treat S and Foss J F 2007 Statistical structure of turbulent-boundary-layer velocity-vorticity products at high and low Reynolds numbers *J. Fluid. Mech.* **570** 307–46

[84] Pullin D I, Inoue M and Saito N 2013 On the asymptotic state of high Reynolds number, smooth-wall turbulent flows *Phys. Fluids.* **25** 1–9

[85] Ran W, Zare A and Jovanović M R 2021 Model-based design of riblets for turbulent drag reduction *J. Fluid. Mech.* **906** 1–38

[86] Rao K N, Narasimha R and Narayanan M A B 1971 The 'bursting' phenomenon inn a turbulent boundary layer *J. Fluid. Mech.* **48** 339–52

[87] Raupach M R, Antonia R A and Rajagopalan S 1991 Rough-wall turbulent boundary layers *Appl. Mech. Rev.* **44** 1–25

[88] Robinson S K 1991 Coherent motions in the turbulent boundary layer *Ann. Rev. Fluid. Mech.* **23** 601–39

[89] Rotta J C 1950 *Über die Theorie der turbulenten Grenzschichten* Max-Planck-Institut für Strömungsforschung, Göttingen Available as NACA TM 1344, 1953, Tech Rep 1

[90] Rotta J C 1962 The calculation of the turbulent boundary layer *Prog. Aerospace. Sci.* **2** 1–219

[91] Ruan J and Blanquart G 2021 Direct numerical simulations of a statistically stationary streamwise periodic boundary layer via the homogenized Navier-Stokes equations *Phys. Rev. Fluids.* **6** 1–20

[92] Saddoughi S G and Veeravalli S V 1994 Local isotropy in turbulent boundary layers at high Reynolds number *J. Fluid. Mech.* **268** 333–72

[93] Samie M, Marusic I, Hutchins N, Fu M K, Fan Y, Hultmark M and Smits A J 2018 Fully resolved measurements of turbulent boundary layer flows up to $Re_\tau = 20000$ *J. Fluid. Mech.* **851** 391–415

[94] Samuel A E and Joubert P N 1974 A boundary layer developing in an increasingly adverse pressure gradient *J. Fluid. Mech.* **66** 481–505

[95] Schlatter P and Örlü R 2010 Assessment of direct numerical simulation data of turbulent boundary layers *J. Fluid. Mech.* **659** 116–26

[96] Schlatter P, Örlü R, Li Q, Brethouwer G, Fransson J H M, Johansson A V, Alfredsson P H and Henningson D S 2009 Turbulent boundary layers up to $Re_\theta = 2500$ studied through simulation and experiment *Phys. Fluids.* **21** 1–4

[97] Schlichting H 1936 Experimentelle untersuchungen zum rauhugkeitsproblem *Ing. Arch.* **7** 1–34

[98] Schlichting H 1979 *Boundary Layer Theory* 7th edn (New York: McGraw-Hill)

[99] Schlichting H and Gersten K 2017 *Boundary Layer Theory* 9th edn (Berlin: Springer)

[100] Schmid P J and Henningson D S 2001 *Stability and Transition in Shear Flows.* (Berlin: Springer)

[101] Schultz M P, Volino R J and Flack K A 2010 Boundary layer structure over a two-dimensional rough wall *Proc. of IUTAM Symposium on the Physics of Flow Over Rough Walls* ed T Nickels (Berlin: Springer)

[102] Segalini A, Örlü R and Alfredsson P H 2013 Uncertainty analysis of the von Kármán constant *Exp. Fluids.* **54** 1–9

[103] Sillero J, Jiménez J and Moser R 2013 One-point statistics for turbulent wall-bounded flows at Reynolds numbers up to $\delta^+ \approx 2000$ *Phys. Fluids.* **25** 1–15

[104] Snyder W H and Castro I P 2002 The critical reynolds number for rough-wall boundary layers *J. Wind. Eng. Ind. Aero.* **90** 41–54

[105] Spalart P R 1988 Direct simulation of a turbulent boundary layer up to $R_\theta=1410$ *J. Fluid. Mech.* **187** 61–98

[106] Spalart P R 2006 *Turbulence are we getting smarter? 36th Fluid Dynamics Conf. and Exhibit* (Fluid Dynamics Award Lecture)

[107] Spalart P R and Abe H 2021 Empirical scaling laws for wall-bounded turbulence deduced from direct numerical simulations *Phys. Rev. Fluids.* **6** 1–17

[108] Tani I 1987 Turbulent boundary laer development over rough surfaces *Perspectives in Turbulence Studies* ed H U Meier and P Bradshaw (Berlin: Springer) 239–49

[109] Tennekes H and Lumley J L 1972 *A First Course in Turbulence* (Cambridge, MA: MIT Press)

[110] Theordorsen T 1952 Mechanism of turbulence *Proc. 2nd Midwestern Conf. on Fluid Mech.* (Columbus, Ohio: Ohio State University) pp 1–19

[111] Toh S and Itano T 2005 Interaction between a large-scale structure and near-wall structures in channel flow *J. Fluid. Mech.* **524** 249–62

[112] Townsend A A 1951 The structure of the turbulent boundary layer *Proc. Camb. Phil. Soc.* **47** 375–95

[113] Townsend A A 1956 *Structure of Turbulent Shear Flow* 1st edn (Cambridge: Cambridge University Press)

[114] Townsend A A 1976 *Structure of Turbulent Shear Flow* 2nd edn (Cambridge: Cambridge University Press)

[115] Vallikivi M, Hultmark M and Smits A 2015 Turbulent boundary layer statistics at very high Reynolds number *J. Fluid. Mech.* **779** 371–89

[116] Volino R J, Schultz M P and Flack K A 2007 Turbulence structure in rough- and smooth-wall boundary layers *J. Fluid. Mech.* **592** 263–93

[117] Wark C E and Nagib H M 1991 Experimental investigation of coherent structures in turbulent boundary layers *J. Fluid. Mech.* **230** 183–208

[118] Wei T, Schmidt R and McMurtry P 2005 Comment on the clauser chart method for determining the friction velocity *Exp. Fluids.* **38** 695–9

[119] White F M 2005 *Viscous Fluid Flow* 3rd edn (New York: McGraw Hill)

[120] Wu X and Moin P 2009 Direct numerical simulation of turbulence in a nominally zero-pressure-gradient flat-plate boundary layer *J. Fluid. Mech.* **630** 5–41

[121] Zagarola M V and Smits A J 1998 A new mean velocity scaling for turbulent boundary layer *Proc. FEDMS'98, ASME Fluids Enginneering Division Summer Meeting* FEDSM98-4950

IOP Publishing

Turbulent Flows: an Introduction

Ian P Castro and Christina Vanderwel

Chapter 9

Turbulent mixing

Turbulence can greatly enhance mixing and diffusion processes, and this is important for a range of engineering and natural processes. In this chapter, we discuss how scalar fields are mixed in the presence of turbulent flows, limiting our scope to passive scalar fields, examples of which can include concentration or temperature. In section 9.1, we introduce the basic processes involved in scalar mixing. Section 9.2 describes Fick's law governing molecular diffusivity. This is used in section 9.3, which presents the scalar transport equation for laminar flow and simple solutions to this equation for a variety of scenarios involving point-source releases, which are typical of many environmental applications. Section 9.4 presents the scalar transport equation for turbulent flows (produced using Reynolds averaging) and then shows how an apparent turbulent diffusivity model can account for the added mixing associated with the presence of turbulence. There are some practical limitations to this approach; thus, in section 9.5, two contrasting alternative approaches to quantifying turbulent dispersion, proposed by Taylor and Richardson, are presented. Sections 9.6 and 9.7 consider practical applications of turbulent mixing in confined mixing processes and free-shear flows. Finally, in section 9.8, we discuss the scalar spectrum and how the important length scales of the scalar field compare with the eddy sizes in the velocity field.

9.1 Initial remarks

As we discussed in section 1.3, one of the important characteristics of turbulent flows is their diffusive nature. Applications in which this is important include chemical mixing, combustion, water contamination, and the transport of atmospheric pollutants. Anyone who has done any sort of flow visualisation will appreciate how the introduction of some sort of visible substance can reveal the impressive intricate features of turbulent flows, as seen in figures 1.1 and 8.1, for example.

We consider the motion and mixing of a passive scalar field superimposed on, or contained within, a turbulent flow. Examples of scalar fields include the concentration of a substance mixed with the fluid or the temperature of the fluid, which may both be considered passive if they do not affect the velocity field of the flow (recall section 2.2). Reacting flows, such as chemical processes and combustion, are beyond our scope and require additional equations governing the reaction processes, further adding to the complexity of the mixing processes involved.

To introduce the basic processes involved in passive scalar mixing, consider a blob of fluid mixing into another fluid, as illustrated in figure 9.1. In the absence of turbulence, the mixing of two fluids would occur solely through molecular actions (figure 9.1(a)), as molecules in regions of high concentrations are transported down local gradients towards regions of low concentrations; the rate at which this happens will be defined in section 9.2. It is a relatively slow process, which gets slower and slower in time as the concentration gradients become less steep, and it can take quite a while for any scalar to reach the boundaries of the domain. In the presence of turbulence, eddies of different sizes can transport large portions of the blob of fluid

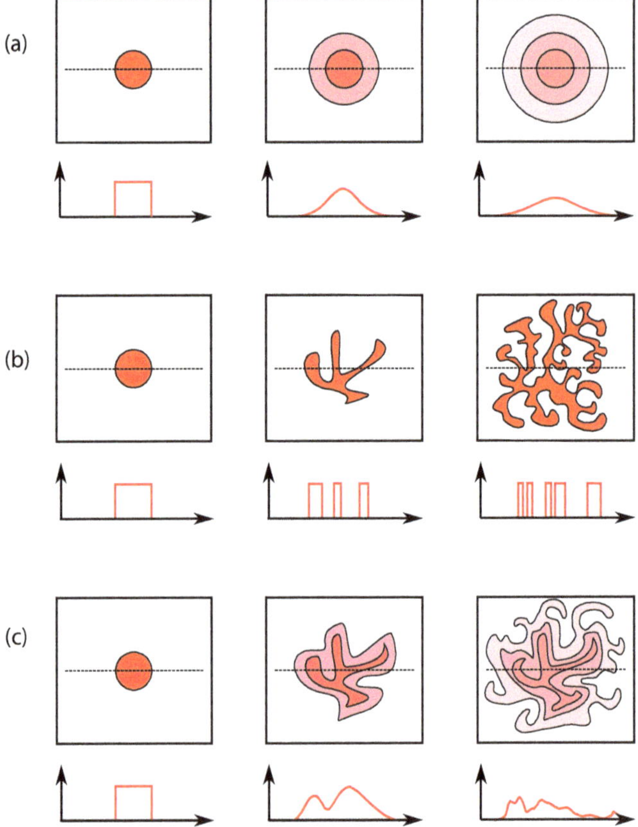

Figure 9.1. Three cases demonstrating the diffusion of an initially circular blob of fluid mixing into another fluid (a) due solely to molecular diffusion, (b) due solely to turbulent diffusion, and (c) due to both molecular and turbulent diffusion. Cross-sectional maps of the concentration field develop in time from left to right and the concentration signal taken through the centreline is shown beneath each map. (Inspired by [7].)

around the domain. If one could completely remove the effect of molecular diffusion, the result would look like figure 9.1(b). Eddies larger than the initial blob are able to transport the blob around in its entirety, whereas eddies smaller might stretch and fold the contours of the blob, spreading it around the domain. The result, however, is a very intermittent concentration field, where the concentration signal would be binary, as illustrated in the cross-sections plotted underneath each map. But in practice, of course, molecular and turbulent effects coexist and the result is very efficient mixing (figure 9.1(c)). The turbulence acts to elongate the contours of the blob of fluid, creating a larger interface with steep gradients over which molecular diffusion acts.

We discuss these different mechanisms by considering different physical interpretations of the turbulent mixing processes and ways to describe the interaction between the different scales of eddies and the scalar field. First, we consider the basic molecular processes which operate whether the containing flow is laminar or turbulent.

9.2 Molecular diffusion

The process of diffusion describes the transport of a scalar property in a fluid from a region of high concentration to a region of low concentration. In the absence of any flow at all, substances are only able to diffuse towards regions of lower concentration through molecular motions. In the case of temperature, the heat flux, q_i, in a direction x_i, is controlled purely by conduction and is described by Fourier's law,

$$\frac{q_i}{A} = -\kappa \frac{\mathrm{d}T}{\mathrm{d}x_i}, \tag{9.1}$$

where κ is the thermal conductivity. Similarly, the mass flux, $\dot{m}_i$, of a substance with a concentration, c, in the direction x_i, by molecular diffusion is directly proportional to the concentration gradient and is described by Fick's law,

$$\frac{\dot{m}_i}{A} = -\gamma \frac{\mathrm{d}c}{\mathrm{d}x_i}, \tag{9.2}$$

where γ is the molecular diffusivity. The similarities between equations (9.1) and (9.2) are clear; both temperature and concentration fluxes are transported via molecular diffusion, acting to reduce gradients in the flow. Both temperature and concentration can be considered as *passive scalar* quantities so long as they do not interact with the velocity field, i.e. there are no buoyancy effects (in the temperature case) or reactions between the scalar and the background fluid. In this section, we will assume the scalar of interest is an added substance whose local concentration is c; however, everything we discuss is equally relevant to the understanding of temperature fields in turbulent flows, when one replaces c with T, provided that the temperature differences are small and thus do not produce significant buoyancy effects (usually $\Delta T \lesssim 5°$).

Typical thermal and molecular diffusivity constants are presented in tables 9.1 and 9.2 for various fluids. These constants are often presented nondimensionally in terms of the Prandtl number and the Schmidt number, respectively, as

Table 9.1. Typical thermal diffusivity constants for various fluids.

fluid	thermal diffusivity ($\kappa c_p^{-1} \rho^{-1}$) ($m^2\ s^{-1}$)	Pr
Air	2.22×10^{-5}	0.708
CO_2	1.06×10^{-7}	0.770
Water vapour	2.04×10^{-5}	1.06
Glycerol	9.47×10^{-8}	12500
Engine oil	8.72×10^{-8}	10400
Mercury	4.60×10^{-6}	0.0249

Table 9.2. Typical molecular diffusivity constants for various fluids.

fluids	diffusivity γ ($m^2\ s^{-1}$)	Sc
O_2 in water	1.80×10^{-9}	558
CO_2 in water	1.60×10^{-9}	628
NaCl in water	1.35×10^{-9}	744
Sucrose in water	0.45×10^{-9}	2230
Rhodamine 6G dye in water	4.0×10^{-10}	2500
O_2 in air	0.206×10^{-4}	0.748
CO_2 in air	0.164×10^{-4}	0.940
Ethyl alcohol in air	0.119×10^{-4}	1.29
Water vapour in air	0.246×10^{-4}	0.612

$$\mathrm{Pr} = \frac{\nu}{\kappa/(c_p\rho)} \quad \text{and} \quad \mathrm{Sc} = \frac{\nu}{\gamma}, \tag{9.3}$$

where c_p is the specific heat of the fluid. The Prandtl number essentially represents the ratio of the momentum diffusivity (the numerator) to the thermal diffusivity (the denominator). Likewise, the Schmidt number represents the ratio of the momentum diffusivity to the mass diffusivity.

The diffusion of molecules dissolved in liquids is typically very slow over everyday length scales (much slower than diffusion in air) and is almost always dominated by advection. (Recall the simple example given in section 1.3.) Sample problem 9.1 demonstrates that if molecular diffusion were the only mechanism of mixing, it would take on the order of a day for milk to mix in a cup coffee, and surely it would be too cold to drink by then!

9.3 The scalar transport equation

The scalar transport equation (2.7) was presented in chapter 2 and we repeat it here:

$$\frac{\partial c}{\partial t} + u_j \frac{\partial c}{\partial x_j} = \gamma \frac{\partial^2 c}{\partial x_j \partial x_j}. \tag{9.4}$$

The first term, $\partial c/\partial t$, is the time rate of change of the concentration, the second term, $u_j \partial c/\partial x_j$, represents the rate of transport of the species by advection with the flow, and the third term represents the rate of transport of the species through molecular diffusion. This equation is sometimes known as the advection–diffusion equation, as these are the two processes that govern any change in the scalar concentration. For passive scalars, it is decoupled from the Navier–Stokes equations, which means that once the velocity field is known, the equation can be solved for the motion of the scalar. Note that the equation is linear in c, whether or not there is any advection, which makes it relatively easy to solve. Analytical solutions are available for the concentration in simplified scenarios, and, as an introduction to the more complex turbulent situations, we discuss a few of these in this section.

The simplest case is that of a single instantaneous point source spreading into an ambient fluid in which there is no advection – i.e. $u_j = 0$. Equation (9.4) then becomes just the simple diffusion equation, and its solutions for how the concentration of this puff of mass, M, located at the origin, changes in time and expands in one, two, or three dimensions are straightforward to obtain and are presented in table 9.3. An example of when the 1D solution would apply is the mixing of a slug of contaminant introduced across the whole cross-section of a thin pipe; it can spread only axially. The 2D solution might describe the case of an oil droplet spreading on the surface of some water; this can only spread within that 2D surface plane (assuming the diffusion of oil into water is essentially negligible). More commonly, the 3D solution might be appropriate for describing the rate of expansion of a puff of smoke in the atmosphere. In each case, the concentration field exhibits a Gaussian profile. The maximum concentration of the puff always remains at the origin, and can therefore be determined from the coefficient to the exponential in the solutions for the concentration. It is important to note that the decay rates of the maximum concentration are different for the three cases, and the decay rate is fastest for the case in three dimensions, simply because there are more directions into which the mass can diffuse. The characteristic width of the puffs can be defined as the standard deviation, σ, of the normal distribution. Comparing the solutions provided in table 9.3 with the typical form of a normal distribution, which has $\exp(-x^2/2\sigma^2)$, it is clear that the puff width always grows in time according to $\sigma(t) = \sqrt{2\gamma t}$, no

Table 9.3. Solutions of the scalar transport equation for an instantaneous point source in static ambient conditions

	Concentration equation	Max concentration	Puff width
1D:	$c(x, t) = \dfrac{M}{A_{yz}(4\pi\gamma t)^{1/2}} \exp\left(-\dfrac{x^2}{4\gamma t}\right)$	$c_{\max}(t) = \dfrac{M}{A_{yz}(4\pi\gamma t)^{1/2}}$	$\sigma(t) = \sqrt{2\gamma t}$
2D:	$c(x, y, t) = \dfrac{M}{L_z(4\pi\gamma t)} \exp\left(-\dfrac{x^2}{4\gamma t} - \dfrac{y^2}{4\gamma t}\right)$	$c_{\max}(t) = \dfrac{M}{L_z(4\pi\gamma t)}$	$\sigma(t) = \sqrt{2\gamma t}$
3D:	$c(x, y, z, t) = \dfrac{M}{(4\pi\gamma t)^{3/2}} \exp\left(-\dfrac{x^2}{4\gamma t} - \dfrac{y^2}{4\gamma t} - \dfrac{z^2}{4\gamma t}\right)$	$c_{\max}(t) = \dfrac{M}{(4\pi\gamma t)^{3/2}}$	$\sigma(t) = \sqrt{2\gamma t}$

matter how many dimensions are considered. This is because the diffusion rate is governed by the concentration gradient (in equation (9.2)) in every direction. In some cases, it is preferable to define the plume width as the 'full-width half-maximum' (FWHM), which is related to the half width as FWHM $=2\sqrt{2\ln 2}\,\sigma \approx 2.355\sigma$.

It is not difficult to adapt these solutions to describe the evolution of an instantaneous point source in a fluid with a uniform (laminar) flow along the x-axis. Considering a frame of reference moving with the centre of the puff at the convection velocity of the flow, U, the solutions provided in table 9.3 can be adapted by replacing x with $(x - Ut)$. The previous relations for $c_{\max}$ are also still valid, but the peak concentration no longer stays at the origin, but rather remains at the centre of the puff as it is advected along, as seen in figure 9.2(a). When transport of the mass occurs through both advection and diffusion, it becomes important to understand the significance of each mechanism. If diffusion is very slow, advection dominates, and the puff can preserve its peak concentration over quite a distance. On the other hand, if advection is very slow, the puff spreads very quickly and might even transport some concentration upstream. The Peclet number is a nondimensional parameter that represents the balance between the advection and diffusion processes and is defined as

$$\mathrm{Pe} = \frac{LU}{\gamma}. \tag{9.5}$$

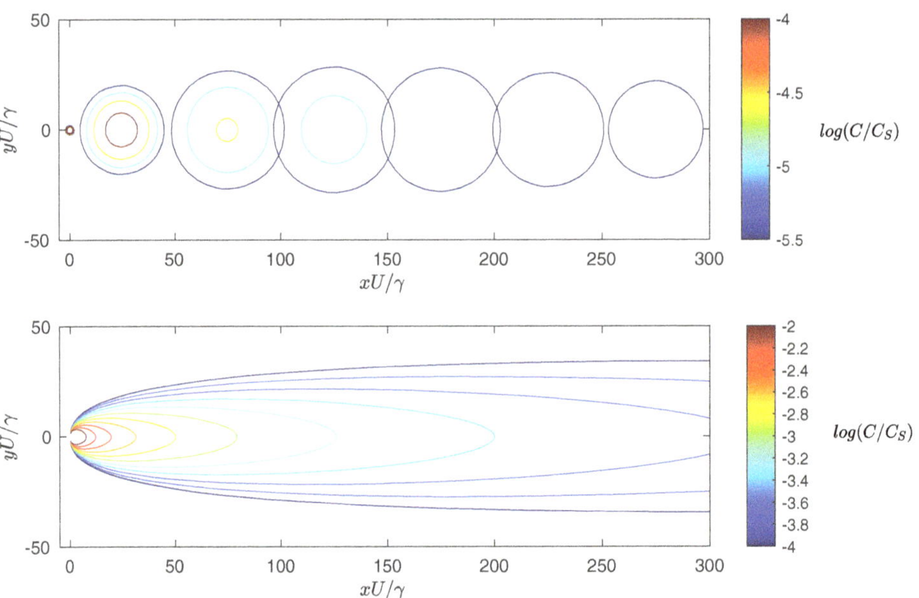

Figure 9.2. Plots of the analytical solutions for the mean concentrations of single puffs at various instances following their release (top) and of a continuous plume (bottom). The solid lines show contours of constant concentration in the central plane (i.e. at $z = 0$, say), with magnitudes matching the labels on the colourbar. Note that we expect the Gaussian plume model (bottom plot) to predict the concentration well for locations beyond $xU/\gamma > 8$ (i.e. Pe > 8).

The relevance of this parameter becomes clear if one considers the timescales of the different processes. The timescale of advection can be estimated as the time required to travel some distance L at the speed U,

$$t_{\text{adv}} = \frac{L}{U}. \qquad (9.6)$$

The timescale of diffusion can be estimated as the time it would take the edge of a diffusing cloud centred at the origin to reach a distance of $x = L$. If the edge of the diffusing cloud is taken to be at 2σ (where $c \approx 0.13c_{\max}$), then equating the two as $2\sigma = 2\sqrt{2\gamma t} = L$ leads to a timescale for diffusion,

$$t_{\text{diff}} = \frac{L^2}{8\gamma}. \qquad (9.7)$$

The balance of these two timescales can be expressed by their ratio:

$$\frac{t_{\text{diff}}}{t_{\text{adv}}} = \frac{L^2/8\gamma}{L/U} = \frac{LU}{8\gamma} = \frac{1}{8}\text{Pe}. \qquad (9.8)$$

When this ratio is equal to one, advection and diffusion are equally strong. In other words, when $\text{Pe} \gg 8$, diffusion acts very slowly and advection dominates, and when $\text{Pe} \ll 8$, advection acts very slowly and diffusion dominates. In example problem 9.3, the evolution of a puff is explored under different conditions.

The case is slightly more complicated for a continuous source. This situation can be thought of as the superposition of infinitely closely spaced individual releases of puffs, as illustrated in figure 9.2(b). An approximate (three-dimensional) solution for the concentration of a continuous point source in a uniform flow, with a mass flow rate of $\dot{m}$, assuming that streamwise diffusion is negligible compared with the advection (i.e. $\text{Pe} \gg 8$), is known as the *Gaussian plume model* [3, 25]:

$$c = \frac{\dot{m}}{4\pi\gamma x}\exp\left[-U\left(\frac{y^2 + z^2}{4\gamma x}\right)\right]. \qquad (9.9)$$

The maximum concentration of the plume decays according to x^{-1} and the plume width grows according to $x^{0.5}$.

For typical situations in atmospheric dispersion and air quality modelling, for example, the prediction of the impact of industrial chimney exhausts and other sources of pollution that are in proximity to the ground, the Gaussian plume model can be adapted to account for the effect of a solid surface. The form typically used in meteorology, when the source is located at an elevation h, is [3, 25]

$$c = \frac{\dot{m}}{2\pi U\sigma_y\sigma_z}\exp\left(\frac{-y^2}{2\sigma_y{}^2}\right)\left[\exp\left(\frac{-(z-h)^2}{2\sigma_z{}^2}\right) + \underbrace{\exp\left(\frac{-(z+h)^2}{2\sigma_z{}^2}\right)}_{\text{added ground reflection}}\right], \qquad (9.10)$$

which has an added source term to account for reflections by the ground. Note also that the term $4\pi\gamma x$ in equation (9.9) has been replaced by $2\pi U \sigma_y \sigma_z$, recalling the relationship $\sigma = \sqrt{2\gamma x / U}$. In the meteorological context, the observed plume spreading rates are heavily influenced by atmospheric turbulence and may vary in the vertical and spanwise directions; hence, two separate variables are used, σ_y and σ_z. Alternatively, this solution can be expressed by replacing the molecular diffusivity γ by an appropriate turbulent diffusivity γ_T, as we discuss below.

9.4 Turbulent transport equations

For turbulent flows, using Reynolds decomposition (see section 2.4), we separate the turbulent variables into their mean and fluctuating components, i.e. $u_i = U_i + u_i'$ and $c = C + c'$. Applying this decomposition to equation (9.4) produces the equation of the mean scalar, introduced in chapter 2 and repeated here:

$$\frac{\partial C}{\partial t} + U_j \frac{\partial C}{\partial x_j} = \frac{\partial}{\partial x_j}\left(\gamma \frac{\partial C}{\partial x_j} - \overline{c'u_j'} \right). \tag{9.11}$$

The term $-\overline{c'u_j'}$ is called the turbulent scalar flux vector and it constitutes three additional unknowns in the mean scalar equation: $\overline{c'u_1'}$, $\overline{c'u_2'}$, and $\overline{c'u_3'}$. Just as described in section 2.4 for the Reynolds shear stress, it is straightforward to derive an equation expressing the transport of $-\overline{c'u_j'}$. Likewise, the transport equation for the scalar variance $\overline{c'^2}$ can be derived and written as

$$\frac{\partial \overline{c'^2}}{\partial t} + U_j \frac{\partial \overline{c'^2}}{\partial x_j} = \underbrace{-2\overline{c'u_j'}\frac{\partial C}{\partial x_j}}_{I} - \underbrace{\frac{\partial \overline{c'^2 u_j}}{\partial x_j}}_{II} + \underbrace{\gamma \frac{\partial^2 \overline{c^2}}{\partial x_j \partial x_j}}_{III} - \underbrace{2\gamma \overline{\frac{\partial c'}{\partial x_j}\frac{\partial c'}{\partial x_j}}}_{IV}. \tag{9.12}$$

Term *I* represents the rate of production of scalar fluctuations. Term *IV* represents the rate of destruction of scalar fluctuations by molecular motions and is commonly written as $-2\epsilon_c$, where ϵ_c is referred to as the scalar dissipation rate. The remaining terms on the right-hand side of the equation, *II* and *III*, represent the rate of transport of scalar fluctuations by turbulence and molecular diffusion, respectively. The values of the scalar variance $\overline{c'^2}$, the turbulent flux vector $\overline{c'u_j'}$, and the mean scalar dissipation rate ϵ_c, are of particular importance in many problems [51]. Measurements of the various terms in the governing equations, their trends, and their relative importance provide insight into the fundamental mechanisms of turbulent transport. Such measurements can also be used to develop and test theoretical and numerical models and to validate computer simulations.

The additional unknown in equation (9.11), i.e. $\overline{c'u_j'}$, means that the turbulent scalar transport equation cannot be solved analytically without introducing an additional closure model for the turbulent mass flux vector, mirroring the closure problem discussed in the context of the momentum equation (2.16) in section 2.4. Similarly, the solution of the variance equation (9.12) requires further modelling (e.g. for the term containing $\overline{c'^2 u_j'}$). Apart from the introduction of the turbulent eddy viscosity in section 2.4 (relating the shear stress to the mean velocity gradient) and its use in providing analytic solutions for free-shear flows in chapter 6, we have avoided discussing turbulence modelling in this book. It is appropriate here, however, to discuss the corresponding model for the mass flux vector. The most commonly used approach is the first-order gradient transport model, expressed by [3, 25, 47]

$$-\overline{c' u_i'} = \gamma_{T_{ij}} \frac{\partial C}{\partial x_j}, \tag{9.13}$$

where $\gamma_{T_{ij}}$ is the turbulent, or 'eddy', diffusivity tensor, which has nine components, the values of which depend on the turbulent field. This model is based on the assumptions that the macroscopic characteristic length scale of the scalar concentration is much greater than that of the transporting mechanism (i.e. the turbulence) and that the flow properties are approximately homogeneous over the turbulence length scale [8]. Despite the fact that these assumptions are not satisfied in most cases of interest, the gradient transport model is known to provide fairly accurate predictions in a variety of situations [40]. Measurements of all nine components of $\gamma_{T_{ij}}$ are quite scarce, so a common practice is to assume that the turbulent diffusivity is isotropic and can be represented by a single constant, γ_T.

The advantage of this model arises because, typically, the turbulent diffusivity far exceeds the magnitude of the molecular diffusivity, so molecular processes can be ignored and the turbulent scalar transport equation becomes

$$\frac{\partial C}{\partial t} + U_j \frac{\partial C}{\partial x_j} = \frac{\partial}{\partial x_j}\left(\gamma_T \frac{\partial C}{\partial x_j}\right). \tag{9.14}$$

The turbulent diffusivity effectively replaces the molecular diffusivity and this equation is of the same form as the (laminar flow) advection–diffusion equation (9.4). It follows that the previously discussed solutions to this equation may also be used (provided γ_T is assumed to be constant), with various levels of accuracy that depend not least on the Peclet number.

The main challenge with this approach is the determination of a reasonable value for the turbulent diffusivity, as this is not a material property, but rather depends on the specific conditions of the flow (just as the eddy viscosity is a property of the flow, not the fluid). A classical approach for measuring turbulent diffusion in the laboratory or in the environment is to relate the rate of growth of puffs and plumes generated in the turbulent flow to an apparent turbulent diffusivity. Recall that the solutions for instantaneous point-source releases (table 9.3) predict that the characteristic cloud width evolves in time according to $\sigma(t) = \sqrt{2\gamma t}$ for laminar cases. Likewise, for turbulent cases, the width might be expected to grow according to $\sigma(t) = \sqrt{2\gamma_T t}$. Rearranging this relationship leads to an estimate of the apparent turbulent diffusivity as

$$\gamma_T = \frac{1}{2}\frac{\mathrm{d}\sigma^2}{\mathrm{d}t}. \tag{9.15}$$

This is exactly the approach used in the early investigations by Richardson described in section 9.9.

Examples of investigations of turbulent diffusion in the environment include early observations of anti-aircraft shell bursts [33, 42], plumes of lycopodium spores released in the atmosphere [14], and dyes released in a lake [9]. Examples of relevant laboratory studies include investigations of diffusion behind a point and line sources in grid turbulence [1, 41, 50], in uniformly sheared flow [17, 21, 44, 49], in channel flows [19, 27, 52], and in boundary layers [12, 43]. In these turbulent flows, the turbulent kinetic energy and length scale generally evolve downstream and so do the diffusivities of superimposed scalar plumes. As a result, equation (9.14) with a constant γ_T is not applicable, and the plume growth rate deviates from the classical power laws discussed above; even more fundamentally, the assumptions embodied in equation (9.13) are invalid. Experimental studies of plumes in grid turbulence and shear flows have found that the plume widths do follow power laws, but with powers different from the theoretical (constant diffusivity) value of 0.5. Insights into this discrepancy between the simplified solutions and actual turbulent flows can be found through Taylor's alternative analysis of diffusion in homogeneous turbulence, discussed in the following section.

9.5 Two perspectives on point-source dispersion

9.5.1 Taylor's dispersion theory

In 1922, Taylor made a departure from the classical gradient transport approach and considered the problem of turbulent diffusion in terms of the motion of individual fluid particles released from a fixed point in stationary isotropic turbulent flow [46]. This perspective requires thought about the motion of the particle from the point of view of the particle as it travels through the flow (a *Lagrangian* approach) rather than in an absolute stationary frame of reference (a *Eulerian* approach).

Imagine releasing a large number of particles from a fixed point in space and watching them disperse into a cloud; or, equivalently, releasing a single particle over and over and tracking the positions of each one. Taylor used the variance of the ensemble of all the particle displacements, $\overline{X^2}$, relative to the centre of gravity of the cloud, following time t from their release, essentially as a measure of the width of the cloud, σ. Taylor demonstrated that this variance should depend only on the net effect of the Lagrangian turbulence properties that the particles encountered since their release, according to

$$\overline{X^2}(t) = 2\overline{v^2} \int_0^t (t - \xi)R(\xi)d\xi, \tag{9.16}$$

where v represents the Lagrangian velocity fluctuations and $R(\xi)$ is the Lagrangian correlation coefficient.

Taylor's analysis led to the identification of three regimes of diffusion known as the *molecular diffusion, turbulent convection* and *turbulent diffusion* regimes [1, 50]. In these regimes, equation (9.16) can be simplified to, respectively,

$$\overline{X^2}(t) \approx \sqrt{2\gamma t} \ \ \text{for} \ \ t \ll \gamma/\overline{v^2}, \tag{9.17}$$

$$\overline{X^2}(t) \approx \overline{v^2}t^2 \ \ \text{for} \ \ \gamma/\overline{v^2} \ll t \ll \mathscr{T}, \tag{9.18}$$

$$\overline{X^2}(t) \approx 2\overline{v^2}\mathscr{T}t \ \ \text{for} \ \ \mathscr{T} \ll t, \tag{9.19}$$

where $\mathscr{T}$ is the Lagrangian integral timescale of the turbulence. Taylor defined the effective rate of dispersion as

$$\frac{1}{2}\frac{\mathrm{d}\overline{X^2}}{\mathrm{d}t} = \overline{v^2}\int_0^t R(\xi)d\xi. \tag{9.20}$$

Recognising that the variance of the particle displacements $\overline{X^2}$ should be related to the width of a point-source release puff σ, a comparison with equation (9.15) illustrates that equation (9.20) is effectively a measure of the turbulent diffusivity γ_T. It follows that the rate of dispersion has the following limits for these three regimes

$$\frac{1}{2}\frac{\mathrm{d}\overline{X^2}}{\mathrm{d}t} = \gamma \ \ \text{for} \ \ t \ll \gamma/\overline{v^2}, \tag{9.21}$$

$$\frac{1}{2}\frac{\mathrm{d}\overline{X^2}}{\mathrm{d}t} \approx \overline{v^2}t \ \ \text{for} \ \ \gamma/\overline{v^2} \ll t \ll \mathscr{T}, \tag{9.22}$$

$$\frac{1}{2}\frac{\mathrm{d}\overline{X^2}}{\mathrm{d}t} \approx \overline{v^2}\mathscr{T} \ \ \text{for} \ \ \mathscr{T} \ll t. \tag{9.23}$$

These relations imply that at the moment of particle release, the rate of dispersion is purely due to molecular actions. After a short time, turbulence takes over and the rate of dispersion then increases linearly with the dispersion time. Eventually, the rate of dispersion reaches an asymptote which depends only on the Lagrangian integral timescale of the turbulence and the variance of the Lagrangian velocity fluctuations. Note also that Taylor's analysis assumes homogeneous isotropic conditions so, in this case, it is correct to represent γ_T as a single value rather than a tensor.

Batchelor later extended this theoretical analysis to three dimensions and defined a three-dimensional diffusion coefficient tensor $\frac{1}{2}\mathrm{d}\overline{X_iX_j}/\mathrm{d}t$ in terms of the mean Lagrangian displacement tensor $\overline{X_iX_j}(t)$ of a particle transported by homogeneous turbulence [4]. He further demonstrated that in homogeneous, non-sheared turbulence and for a Gaussian particle displacement distribution, $\frac{1}{2}\mathrm{d}\overline{X_iX_j}/\mathrm{d}t$ would be equal to the turbulent diffusivity tensor $\gamma_{T_{ij}}$, as defined by equation (9.13). As in Taylor's theory, the magnitudes of the components of $\gamma_{T_{ij}}$ would increase with dispersion time and would eventually reach asymptotes that depend only on the Lagrangian integral timescales of the turbulence $\mathscr{T}_{ij}$ and the intensities of the Lagrangian velocity fluctuations. Subsequently, it was suggested that the asymptotic values of the turbulent diffusivities could be estimated from Eulerian properties, as

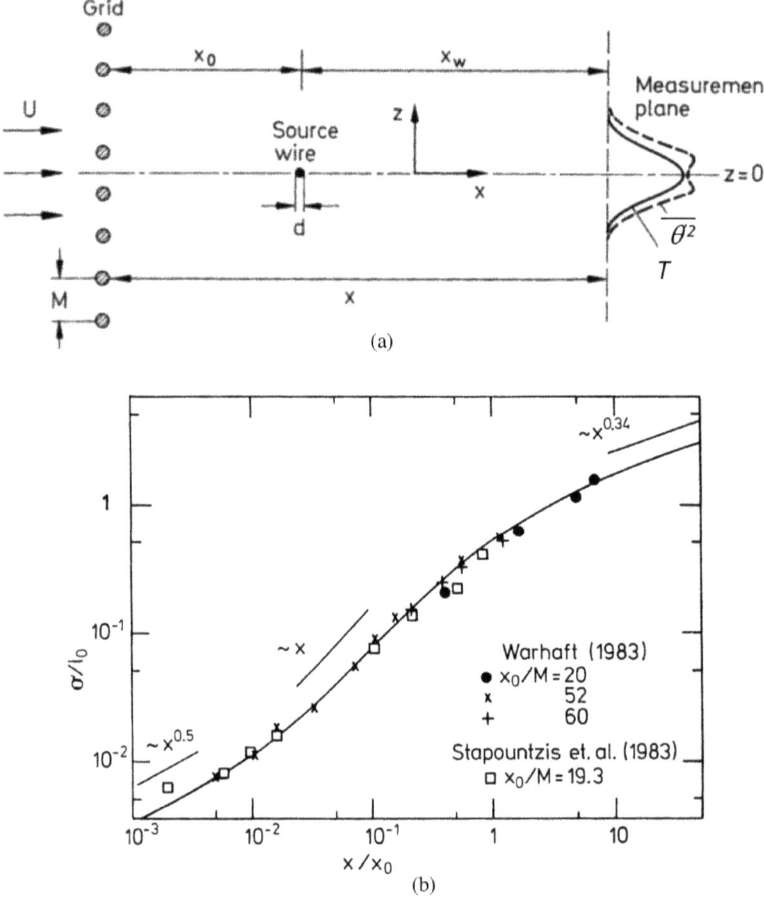

Figure 9.3. Measurements of the transport of (scalar) temperature behind a heated line source in grid turbulence. (a) The experimental setup, with profiles of mean temperature (T) and temperature variance $\overline{\theta'^2}$ at a distance x_w from the heat source. (b) The growth of the thermal wake thickness σ (normalised by the turbulence length scale at the wire l_0) follows the power-law predictions for the molecular diffusion, turbulent convection, and turbulent diffusion regimes[1].

surrogates for their Lagrangian counterparts, for example as $\gamma_{T_{22}} \approx \overline{u_2'^2} T_{11}$ [8]. Various authors (including [32, 34, 45, 49, 55]) have since expanded Batchelor's analysis for homogeneous turbulent shear flow and noted that the normal diffusivities are sometimes unequal and that some of the cross-diffusivities are non-zero. These studies highlight the limitations of assuming that turbulent diffusivity is isotropic.

These predictions of the rate of dispersion embodied in equations (9.21)–(9.23) have been tested extensively in grid turbulence. Figure 9.3(a) illustrates the classical experiments of [41, 50], which measured the diffusion of temperature behind a

---

[1] Reproduced from [1], copyright (1985) with permission from Springer. With data from [41, 50].

heated wire in grid turbulence. Although the wire is hot, the magnitude of the temperature fluctuations is only a few degrees and buoyancy effects are negligible. While the mean temperature profiles collapse to the expected Gaussian shape, the temperature variance does not and is sometimes observed to have a double peak. The production of scalar variance peaks at the locations of the greatest mean scalar gradient on either side of the plume centreline, producing this double peak. The growth of the plume width follows a power law; however, the three regimes of diffusion can be observed as the effective turbulent diffusivity varies downstream. First, equation (9.21) predicts that very close to the source, only molecular diffusion acts to spread the plume and that $\sigma \propto x^{0.5}$. Farther downstream, in the turbulent convective regime, equation (9.22) predicts that the turbulent diffusivity will grow linearly, causing the plume width to grow proportionally to x^1. In the far field, equation (9.23) predicts that the turbulent diffusivity will be proportional to $\overline{v^2}\mathscr{T}$. For grid turbulence, it is known that the Lagrangian timescale evolves according to $\mathscr{T} \propto x$ and that the velocity variance decays according to $\overline{v^2} \propto x^{-n}$, where typically $n \approx 1.3$ (see section 5.1). We therefore expect the turbulent diffusivity to vary according to $\gamma_T \propto x^{1-n}$ and the width of a plume to grow according to $\sigma \propto x^{\frac{1}{2}(2-n)}$. This is all exactly as observed in figure 9.3(b).

9.5.2 Richardson's relative dispersion theory

One aspect that Taylor's dispersion theory fails to capture is the importance of the particular, dynamic, eddy structure of the flow to the resulting mixing of the point-source plume. While the previously discussed solutions focus on the properties of the mean concentration of the plume, an instantaneous snapshot of the concentration field might have a very different shape. In many meteorological applications, the time-averaged plume might be much wider than the instantaneous plume. In engineering applications, for example, for toxic chemical releases, predicting the peak concentration downstream from a point source is more important than knowing the mean.

Figure 9.4 illustrates how different scales of eddies in the flow have different consequences on the spread of a point-source plume. If eddies are present that are larger than the plume, they can create time-dependent transverse shifts of the entire plume, a process commonly referred to as meandering. When a plume is meandering, the measured mean concentration is typically much smaller than the magnitude of the instantaneous peaks. If only small-scale eddies are present, they tend to act at the plume boundary, making small distortions that help mix the scalar into the external environment. In many turbulent flows, and especially in the atmosphere, a range of eddy scales are present and so both these processes occur simultaneously. In these situations, it can be helpful to consider the rate at which the boundaries of a plume spread relative to its cross-sectional centre of mass. This approach removes the effects of motions with scales larger than the plume width, essentially removing the effect of meandering, and focusses on the effect of the smaller eddies in the flow. The 'relative' width, σ_r, of the puff is then defined as the standard deviation of the concentration relative to the instantaneous centre of mass, y_c, of the cloud rather

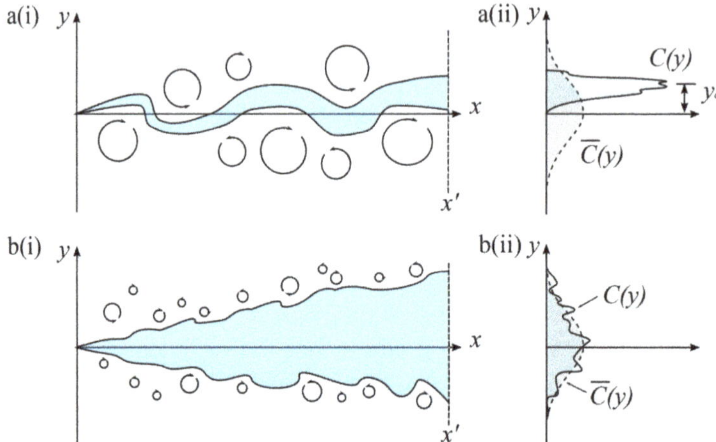

Figure 9.4. (a(i)) When the scalar field is thin compared with the size of the eddies in the flow, the resulting plume exhibits meandering; (a(ii)) in this case, profiles of the time-averaged concentration significantly underpredict the peak instantaneous concentration. (b(i)) When the scalar field is wide compared with the eddies in the flow, the turbulence acts to distort the instantaneous boundary of the plume, but there is not much meandering; (b(ii)) the profiles of the mean and instantaneous concentrations are not too dissimilar in this case.

than the fixed coordinate system. The intensity of meandering of the centre of mass, y_c, can be expressed by its standard deviation, $\sigma_m = \sqrt{\overline{y_c^2}}$. With these definitions, it can be shown that the 'absolute' dispersion in a fixed frame of reference is the sum of the relative dispersion and the meandering as [10]

$$\sigma^2 = \sigma_r^{\,2} + \sigma_m^{\,2}. \tag{9.24}$$

Figure 9.5 illustrates the balance of relative dispersion and meandering in the typical time history of a concentrated puff. In the initial phases when the puff is still very small ($\sigma_r < \eta$), relative dispersion is limited to molecular diffusion, and meandering motions are the dominant mechanism of dispersion. At large diffusion times, meandering tends to a constant asymptotic value on the order of the eddy length scale L, and the rate of relative dispersion essentially coincides with the rate of absolute dispersion (i.e. $(1/2)(d\sigma_r^2/dt) \approx \overline{v^2}\,\mathscr{T}$, equation (9.23)). However, at intermediate times, a phase exists in which the relative width σ_r is in the inertial subrange ($\eta \ll \sigma_r \ll L$), where

$$\frac{1}{2}\frac{d\sigma_r^2}{dt} = \frac{1}{2}(g\epsilon)^{1/3}\sigma_r^{4/3}. \tag{9.25}$$

This relationship was first discovered by Richardson [28] and is one of the few empirically derived turbulence relationships that has stood the test of time. The constant g is known as 'Richardson's constant' and typically has a value of approximately 0.5.

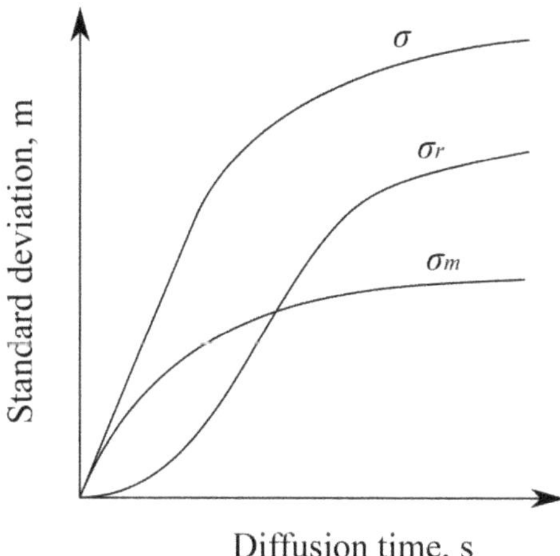

Figure 9.5. The typical development of the absolute plume width, relative plume width, and standard deviation of meandering of a concentrated puff (inspired by [10]).

This approach to the problem of turbulent dispersion is known as 'relative' dispersion or 'two-particle dispersion', i.e. the evolution of the separation distance between a pair of particles in the flow. For further details of this topic, we refer the reader to recent reviews [36, 37].

9.6 Optimising mixing and stirring in industrial flows

The efficient mixing of scalars in flows is important for various industrial processes; fuel and air mixtures should be evenly mixed for complete combustion, and many industrial chemical processes require efficient stirring of the component species. We have already discussed how molecular diffusion only acts where there are gradients in concentration, mostly at the interfaces between substances. As discussed in section 9.1, adding turbulence helps to stretch and distort these interfaces; however, different sizes of eddies relative to the size of the scalar have different effects. In this section, we will discuss how the particular characteristics of different turbulent flows influence the transport of scalars seeded within them.

In industrial contexts, where there is some control over the mixing process and the sizes of eddies in the turbulence, clever work has been done to use these ideas to optimise mixing processes. A typical application in chemical mixing requires that the mixture ratio of two solutions equalises as quickly as possible. The initial state consists of two distinct solutions and then their concentrations mix over time. Assuming that these solutions are nonreactive and passive, either the mixture ratio or the concentration of one of the solutions can be monitored as a passive scalar. In these sorts of chemical mixing process, any optimisation of the mixing process can reduce power requirements and improve product quality.

There are many ways of defining the 'mixedness' of the system; one indicator is the variance of the concentration field. For a time-varying system, assuming a scalar field whose spatial average is zero, it has been demonstrated that a differential equation describing the time evolution of the scalar variance can be derived from the advection–diffusion equation (9.4) as follows [13]:

$$\frac{1}{2}\frac{\mathrm{d}}{\mathrm{d}t}||c||^2 = -\mathrm{Pe}^{-1}||\nabla c||^2, \tag{9.26}$$

regardless of whether the flow is laminar or turbulent and where the variance is expressed as $||c||^2$, the Euclidean (L^2) norm of the c-field, which represents an average over the whole domain, and similarly for the scalar gradient, $||\nabla c||^2$. Equation (9.26) reveals that for a finite Peclet number, the rate at which the variance decreases depends on the magnitude of the concentration gradients in the flow. This is consistent with the idea that molecular diffusion acts at the scalar interfaces where large gradients occur and is why, in the absence of stirring, two fluids that are separated will remain separated for a long time. As they are divided by a thin interface, although the local concentration gradient is high at the interface, it is zero everywhere else, so the average gradient is very small. Once 'stirring' begins, advection quickly extends the interfacial regions between the two fluids. The longer the interface is stretched and contorted, the greater the average value of the gradient becomes and the more quickly the fluids are mixed. Equation (9.26) also highlights the theoretical case of the absence of molecular diffusion, i.e. $\gamma = 0$ (Pe $\to \infty$); in this case, the variance is unchanging. Again, this is consistent with the picture we initially drew in figure 9.1, illustrating that molecular diffusion is the necessary mechanism to equalise concentrations.

Based on this strategy, Foures *et al* [13] mathematically optimised the mixing of two fluids in Poiseuille flow by first energising the flow, then allowing the perturbations to stretch and distort the concentration interfaces, increasing the spatially-averaged concentration gradient, and then letting the energy decay and allowing molecular actions to take over as the flow relaxed and diffused. Aref *et al* [2] reviewed various other applications of similar chaotic advection methods for the optimisation of active and passive mixing in various kinds of flows including bounded and unbounded flows, stirred flows, cavity flows, as well as microfluidics applications. In many natural flows, buoyancy-driven instabilities are also very efficient drivers for advective mixing, such as in the ocean [39].

So does laminar or turbulent flow produce better mixing? In the absence of buoyancy, in many cases, it is still favourable for the flow to remain laminar to avoid the increased energy consumption associated with turbulent flows (e.g. for many internal flows). Laminar mixing can be optimised to produce very effective mixing, as has been done in figure 9.6(a) for a two-rod stirrer [11]. However, recent studies also demonstrate that increasing the Reynolds number can increase the mixing efficiency, justifying the increased energy cost, because the initial vorticity created by stirrers and boundaries can detach and be advected into the entire flow, as illustrated in figure 9.6(b) [16]. In either case, the optimal approach clearly requires breaking the initial interface and making use of the advective motions of eddies to allow molecular diffusion to take hold.

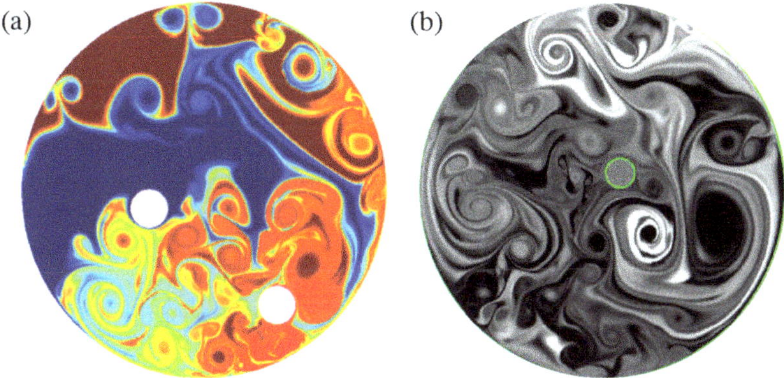

Figure 9.6. (a) Optimised laminar mixing performed by two stirrers[2], also available as a video[3]. (b) Optimised turbulent mixing by a single stirrer[4].

9.7 The behaviour of scalar fields in free-shear flows

In addition to all the situations previously discussed, which were mostly for cases in which there is a localised source of scalar within the flow, scalar fields completely filling free-shear flows, such as jets and wakes, are also commonly encountered in engineering. As examples, consider the turbulent wake behind a heated cylinder, or a heated jet issuing into a cooler ambient. In these canonical free-shear flows, the mean scalar equation can be simplified following the same procedure as that used for the velocity field (section 6.2.3). The *thin shear layer equation* for the scalar field for planar flows is

$$U\frac{\partial C}{\partial x} + V\frac{\partial C}{\partial y} = -\frac{\partial \overline{c'v'}}{\partial y} + \gamma\frac{\partial^2 C}{\partial y^2}, \tag{9.27}$$

which has clear similarities with equation (6.10). Just as before, if molecular transport is negligible (in this case Sc ⩾ 1), then the final term can be neglected, as this term becomes small under the assumption that the Peclet number is large. The resulting *thin shear layer equation* for the scalar field applies for planar jets, wakes, and mixing layers, and the same procedures that were used to develop solutions for the velocity field for each of the flows can be used to develop solutions for the scalar field. One notable difference in obtaining such solutions is that in place of the eddy viscosity model, the eddy diffusivity model, equation (9.13), with an isotopic γ_T, is assumed in the replacement of the turbulent scalar flux term $\overline{c'v'}$. This leads to the result that the mean concentration field C scales in the same way as U_S in all these flows (recall the useful summary in table 6.1).

---

[2] Reprinted with permission from [11], copyright Cambridge University Press.
[3] https://www.cambridge.org/core/journals/journal-of-fluid-mechanics/article/mixing-enhancement-in-binary-fluids-using-optimised-stirring-strategies/9AE72F972881686E217DAA36E1F79201#supplementary-materials
[4] Reprinted with permission from Kadoch *et al* [16]. Copyright (2020) by the American Physical Society.

Scalar diffusion in jets, in particular, has been studied extensively [23, 24, 53]. Figure 9.7 presents some measurements comparing the velocity and concentration fields in a typical turbulent jet. The concentration profiles all collapse, demonstrating self-similarity, and they follow the same predicted shape as the velocity profile but with a notably wider spread. Theory predicts that both the velocity and concentration profiles will decay linearly with downstream distance in a turbulent jet, and measurements report that typically the concentration decreases faster. Overall, this indicates that turbulent scalar transport is stronger than the turbulent transport of momentum. Typically, the shape of the scalar fluxes closely follows the shape of the Reynolds stresses, changing sign on either side of the jet, indicating net transport away from the centreline [53, 54].

The fact that the behaviour of a passive scalar field closely mimics the properties of the velocity field in these flows is not too surprising, as the characteristics of the turbulence that transport both the momentum and the scalar are determined by the conditions that created the jet/wake itself. This is a recurrent theme and is commonly referred to as the *Reynolds analogy*. The idea is that the same turbulent mechanisms that transport momentum (and hence determine the velocity field) also cause turbulent diffusion and therefore determine the scalar field. Of course, these mechanisms do not necessarily act with equal magnitudes. Building on the gradient transport approach, the relative balance of these actions is often described by the turbulent Prandtl number for temperature or the turbulent Schmidt number for concentration, defined respectively as

$$\mathrm{Pr}_T = \frac{\nu_T}{\kappa_T/(c_p\rho)} \quad \text{and} \quad \mathrm{Sc}_T = \frac{\nu_T}{\gamma_T}, \tag{9.28}$$

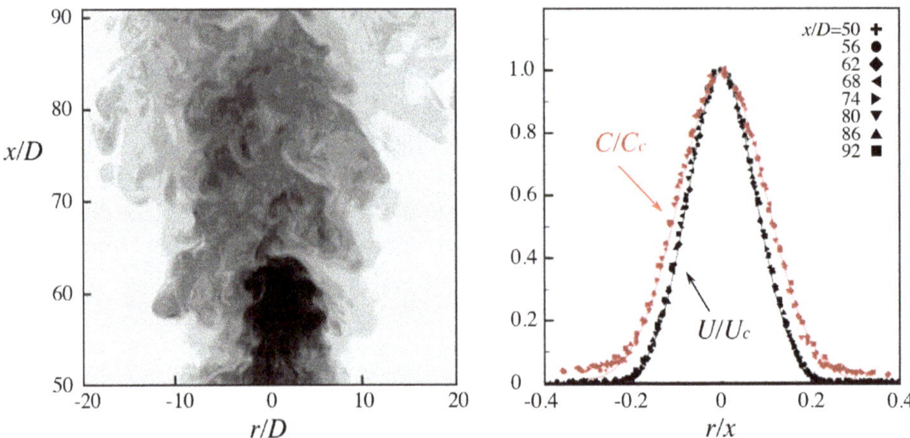

Figure 9.7. Measurements of scalar diffusion in a turbulent jet at a Reynolds number of 3000. (a) An example of the instantaneous concentration field of the turbulent jet. (b) Concentration profiles (in red) and velocity profiles (in black) at various locations downstream (*x/D*) from the origin. Reprinted from [53], copyright (2001) with permission from Springer.

where κ_T is the turbulent thermal conductivity, akin to the turbulent diffusivity γ_T for concentrations, and ν_T is the eddy viscosity. In the jet example shown in figure 9.7, the relative behaviour of the momentum and scalar profiles implies that $Sc_T < 1$. Just as γ_T is not necessarily uniform or isotropic and is expected to vary depending on the flow conditions, the assumption that a single value describes Pr_T or Sc_T (as is often assumed in computational fluid dynamics) is somewhat tenuous. Nonetheless, in some flows, such as free-shear flows, these parameters can be useful for relating the width of the scalar field with that of the velocity field, for example.

9.8 The scales of turbulent mixing

We have already discussed the fact that turbulence consists of eddies at different scales and that these eddies transport the passive scalar in different ways. The result is that the scalar concentration is also characterized by a range of scales. The three-dimensional spectral density function, $E_c(\kappa)$, of the scalar fluctuation c', describes this distribution of scales, and its integral over all wavenumbers is related to the scalar variance,

$$2 \int_0^\infty E_c(\kappa) \, d\kappa = \overline{c'^2}. \tag{9.29}$$

The largest scale of the scalar spectrum, L_c, say, will be determined from the initial conditions as the scale at which the scalar is introduced into the flow. This might be different from the largest scales of eddies in the flow, so L_c and L will not necessarily be the same.

The smallest scale of the scalar field, representing the scale at which diffusion is dominated by molecular processes, is represented by the Batchelor scale η_B. This is the smallest scale at which concentration gradients occur, and it depends on the molecular diffusivity, γ, of the fluid. If $\gamma < \nu$, the diffusion of the concentration field occurs at smaller scales than the dissipation of turbulent eddies. The ratio of η_B to the Kolmogorov length scale η, which represents the smallest scales of the velocity field, is equal to the square root of the Schmidt number (or the Prandtl number) as

$$\frac{\eta}{\eta_B} = \left(\frac{\nu}{\gamma}\right)^{1/2} = Sc^{1/2}. \tag{9.30}$$

Because $Sc > 1$ for most substances ($\mathcal{O}(10^3)$ for the common combination of salt in water), the scalar field typically exhibits much finer fluctuations than the size of the smallest eddies in the flow.

If the Reynolds number is large enough, the scalar spectra exhibit a so-called *inertial–convective subrange*, which occurs somewhere between L_c and η_c, as sketched in figure 9.8. This inertial–convective subrange is akin to the inertial subrange in the velocity spectra and should similarly scale solely by the dissipation rate of the velocity field, ϵ, and that of the scalar field, ϵ_c. The dissipation rate of the scalar field can be defined as

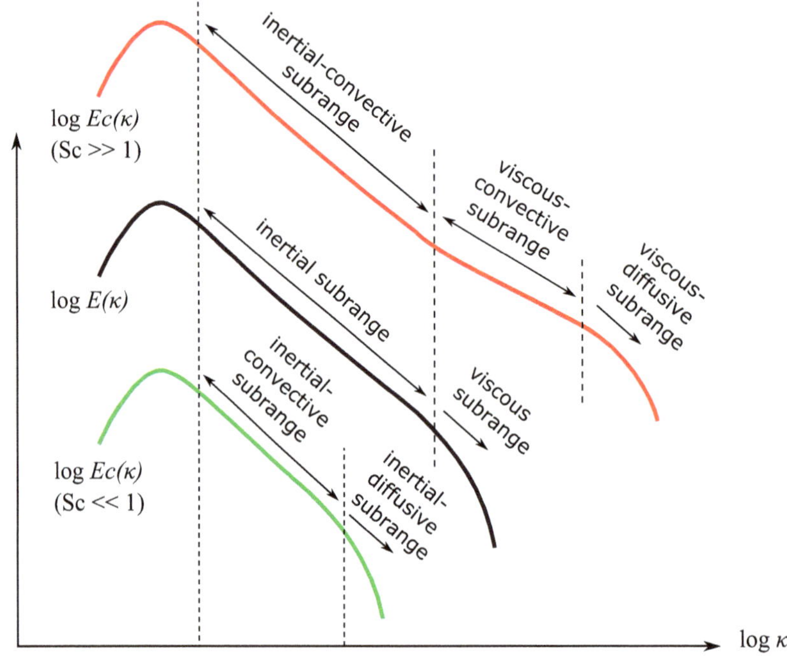

Figure 9.8. Sketch of the velocity and scalar spectra in fluids with large and small Schmidt numbers. (Adapted from [48].)

$$\epsilon_c = \gamma \overline{\frac{\partial c'}{\partial x_i}\frac{\partial c'}{\partial x_i}}. \tag{9.31}$$

Corrsin [6] and Obukhov [22] both first suggested this dependence, which leads to the relation,

$$E_c(\kappa) = C_c\epsilon_c\epsilon^{-1/3}\kappa_1^{-5/3}. \tag{9.32}$$

The constant C_c is known as the Corrsin–Obukhov constant, and has been shown to have a value of approximately 0.5 [48].

At very large wavenumbers – the scales at which molecular diffusion begins to occur – the shape of the scalar spectra begins to depend on γ. This region is known as the *viscous–diffusive subrange* and it ends at wavelengths around η_B, the Batchelor scale.

If $\gamma > \nu$ ($Sc < 1$), molecular diffusivity becomes important at larger scales and the inertial–convective subrange might be cut short. This deviation from equation (9.32), illustrated in figure 9.8, is referred to as the *inertial–diffusive subrange*. It is fairly rare that this occurs, because typically the Prandtl number or the Schmidt number of the fluid is greater than one; however, it occurs in mercury, for example.

If $\gamma < \nu$ ($Sc > 1$), molecular diffusivity occurs at scales even smaller than the smallest eddies, and the range of scales between the inertial–convective subrange and the viscous–diffusive subrange is known as the *viscous–convective subrange*. In this range, the scalar field is influenced by eddies of size η and strength $(\nu/\epsilon)^{-1/2}$. It follows that the form of the scalar spectra in this range must have the form

$$E_c(\kappa) = A\epsilon_c(\nu/\epsilon)^{-1/2}\kappa^{-1}, \tag{9.33}$$

where A is a nondimensional coefficient. The fact that most liquids of interest have a Schmidt number greater than one (see, e.g. Tables 9.1 and 9.2) means that computational methods (particularly direct numerical simulation (DNS)) struggle to simulate turbulent mixing in these cases, as they would require the numerical mesh to be much finer than the one needed to capture the velocity field to resolve all the fine scales in this extended scalar spectrum. Furthermore, it is at these scales that molecular diffusion occurs, so, if the mesh is not sufficiently fine, some form of modelling is required to parameterise these fine-scale mechanisms.

The physical mechanisms with which the turbulent eddies transport the concentration is still an ongoing area of research. The structure and shape of eddies can clearly be visualised by passive scalar tracers in the flow, but the fact that a swirl is present in the scalar field does not necessarily mean that an eddy is present at that moment, and, likewise, the presence of an eddy is not always reflected in the scalar field. The flow field can carry these vortex signatures away from their origin as shadows of previous events, resulting in the fact that the velocity and concentration fluctuations are typically not directly correlated. An example of this is a contrail left behind by an aeroplane on a clear day, which may have originated in a swirling exhaust or a wingtip vortex, but is clearly no longer swirling (or indeed turbulent) far downstream.

The effects of large coherent eddies on scalar transport are particularly important in the context of environmental flows. It has been noted that in the atmospheric surface layer, the ejection and sweep motions that are typically associated with coherent structures are responsible for much of the land-surface evaporation, heat, and momentum fluxes [18, 20]. The impact of coherent vortices in quasigeostrophic flows in the ocean and the atmosphere has been usefully discussed [26], along with a description of how scalars seeded within vortices experience reduced dispersion compared to the free stream and become, in a sense, 'islands of stability'. Investigations of particle dispersion have also demonstrated that coherent structures are responsible for particle clustering, which has direct ramifications for reactive flows, such as combustion, as well as to droplet coalescence and the formation of clouds in the atmosphere [35, 38]. A better understanding of the role of the different scales of eddies in scalar transport is therefore important for improving our understanding and modelling of turbulent diffusion processes.

9.9 Subject giants: L F Richardson

Lewis Fry Richardson (11 October 1881–30 September 1953) was an English mathematician with a background in meteorology. Richardson worked for several years for the Meteorological Office in the UK, and it was during this time that he published his book *Weather Prediction by Arithmetical Finite Differences* [30], which is regarded as the very first attempt at computational fluid dynamics [5]. He also had

Figure 9.9. Lewis Fry Richardson (11 October 1881–30 September 1953) pictured in the early 1930s. Credit: NOAA.

an interest in the motion of water in clouds in the atmosphere [29], which may have inspired his further research on atmospheric dispersion. In addition to these studies, he was one of the first researchers to recognise the importance of the reflection of radiation by the Earth and to measure the albedos of different landscapes [31].

Richardson and Taylor developed their theories of dispersion not long after Fick's model of diffusion (equation (9.1)) was formulated in 1855. With the advent of Einstein's atomic theory around 1905, the focus at the time was to relate diffusivity to molecular properties [5]. However, Richardson noted that in turbulent flows, the diffusivity constant was able to range over several orders of magnitude and could not depend on the material properties alone, but had to depend also on the turbulence properties. In order to investigate the diffusion properties of turbulence, Richardson focussed on the separation of different particles. He hypothesised that the separation of two particles, l, is only influenced by eddies of a size smaller than l; eddies larger than l would transport the particles equally and thus would not affect their relative separation. Therefore, he concluded that the diffusivity must depend on the scale of the phenomenon. Richardson's analysis that led him to the 'four-thirds law' (equation (9.25)) was based on experimental observations, several of which he conducted himself. One of these experiments involved tracking the paths of balloons released in a balloon competition on Brighton Beach [15]. These balloons carried labels that were returned to Richardson by people who found them after the balloons had landed. The trajectories of the balloons had a greater variability the farther the balloons travelled. Although Richardson's law was based on relatively simple experimental observation, it has never been disproven and studies are continuing to confirm and build upon Richardson's original theory.

It is remarkable that Richardson and Taylor both published their theories on diffusion nearly a century ago, and yet the significance of their work has stood the test of time, as our current framework for the understanding and analysis of turbulent mixing is still based on their contributions.

Sample exercises

9.1 Estimate the diffusion timescale $t_{\text{diff}} = L^2/(8\gamma)$ (equation (9.7)), representing the time required for the diffusion of milk in a cup of coffee in the absence of advection, choosing appropriate values for L and γ.

9.2 Plot the concentration profiles $c(x)$ for a single point-source puff of mass $M = 1$ kg released in an ambient fluid without any advection, assuming $\gamma = 1\,\mathrm{m^2\,s^{-1}}$. Compare the rate of decay of the maximum concentration for 1D, 2D, and 3D scenarios.

9.3 Plot the evolution of an instantaneous 1 kg point source released at the origin in a uniform flow with $U = 1\,\mathrm{m\,s^{-1}}$ and $\gamma = 1\,\mathrm{m^2\,s^{-1}}$ at several time steps. Compare the concentration level with the scenario of a continuous $1\,\mathrm{kg\,s^{-1}}$ point source released at the origin under the same conditions.

9.4 The files 'PlumeData1.txt'[5] and 'PlumeData2.txt'[6] each contain a dataset containing 1000 samples of all three velocity components, U, V, and W (in mm/s), and concentration, C (normalised by the source concentration, so it is unitless), all measured at 2 Hz (i.e. not time-resolved). The two datasets were collected from two points at the edge of a thin continuous point-source plume at locations of $y_1 = 16.1$ mm and $y_2 = 18.5$ mm measured from the centreline of the plume, respectively.

   (a) Calculate the mean concentration at each point. From this, determine the vertical gradient of the mean concentration between these two points.

   (b) Make a scatter plot of the concentration fluctuations versus fluctuations in the vertical velocity component at both points. Does this indicate a positive, negative, or neutral correlation?

   (c) Calculate the vertical turbulent concentration flux $\overline{cv}$ at both points and report the average. Explain the significance of this term.

   (d) Using the first-order gradient transport model, determine the turbulent diffusivity, γ_T (including units). The intermittency of the concentration signal creates some uncertainty in the result – estimate the range of confidence in this result.

---

[5] https://github.com/cvanderwel/TurbulentFlows/blob/main/data/PlumeData1.txt
[6] https://github.com/cvanderwel/TurbulentFlows/blob/main/data/PlumeData2.txt

References

[1] Anand M and Pope S 1985 Diffusion behind a line source in grid turbulence *Turbulent Shear Flows* vol 4 (Berlin: Springer) pp 46–61

[2] Aref H, Blake J R, Budišić M, Cardoso S S, Cartwright J H, Clercx H J, El Omari K, Feudel U, Golestanian R and Gouillart E *et al* 2017 Frontiers of chaotic advection *Rev. Mod. Phys.* **89** 007–25

[3] Arya S P 1999 *Air Pollution Meteorology and Dispersion* (Oxford: Oxford University Press)

[4] Batchelor G 1949 Diffusion in a field of homogeneous turbulence. I. Eulerian analysis *Aust. J. Chem.* **2** 437–50

[5] Benzi R 2011 Lewis Fry Richardson *A Voyage Through Turbulence* ed P A Davidson, Y Kaneda, K Moffatt and K R Sreenivasan (Cambridge: Cambridge University Press)

[6] Corrsin S 1951 On the spectrum of isotropic temperature fluctuations in an isotropic turbulence *J. Appl. Phys.* **22** 469–73

[7] Corrsin S 1961 Turbulent flow *Am. Sci.* **49**(3) 300–25

[8] Corrsin S 1975 Limitations of gradient transport models in random walks and in turbulence *Adv. Geophys.* **18** 25-60

[9] Csanady G 1963 Turbulent diffusion in Lake Huron *J. Fluid. Mech.* **17** 360–84

[10] Csanady G T 1973 *Turbulent Diffusion in the Environment* (Dordrecht: D. Reidel)

[11] Eggl M F and Schmid P J 2020 Mixing enhancement in binary fluids using optimised stirring strategies *J. Fluid. Mech.* **899** A24

[12] Fackrell J and Robins A 1982 Concentration fluctuations and fluxes in plumes from point sources in a turbulent boundary layer *J. Fluid. Mech.* **117** 1–26

[13] Foures D P, Caulfield C P and Schmid P J 2014 Optimal mixing in two-dimensional plane Poiseuille flow at finite Péclet number *J. Fluid. Mech.* **748** 241–77

[14] Hay J S and Pasquill F 1959 Diffusion from a continuous source in relation to the spectrum and scale of turbulence *Adv. Geophys.* **6** 345–65

[15] Hunt J C R 1998 Lewis Fry Richardson and his contribution to mathematics, meteorology, and models of conflict *Annu. Rev. Fluid. Mech.* 30 (Fry Richardson: Lewis) xiii–xxxvi

[16] Kadoch B, Bos W J and Schneider K 2020 Efficiency of laminar and turbulent mixing in wall-bounded flows *Phys. Rev.* E **101** 043–104

[17] Karnik U and Tavoularis S 1989 Measurements of heat diffusion from a continuous line source in a uniformly sheared turbulent flow *J. Fluid. Mech.* **202** 233–61

[18] Katul G, Kuhn G, Schieldge J and Hsieh C-I 1997 The ejection-sweep character of scalar fluxes in the unstable surface layer *Bound. Layer. Meteorol.* **83** 1–26

[19] Lepore J and Mydlarski L 2011 Lateral dispersion from a concentrated line source in turbulent channel flow *J. Fluid. Mech.* **678** 417–50

[20] Li D and Bou-Zeid E 2011 Coherent structures and the dissimilarity of turbulent transport of momentum and scalars in the unstable atmospheric surface layer *Bound. Layer. Meteorol.* **140** 243–62

[21] Nakamura I, Sakai Y, Miyata M and Tsunoda H 1986 Diffusion of matter from a continuous point source in uniform mean shear flows (1st report) *B JSME* **29** 1141–8

[22] Obukhov A 1949 The structure of the temperature field in a turbulent flow *Izv Akad Nauk SSSR, Ser Geogr And Geophys* **13** 58–69

[23] Panchapakesan N R and Lumley J L 1993 Turbulence measurements in axisymmetric jets of air and helium. part 1. air jet *J. Fluid. Mech.* **246** 197–223

[24] Papanicolaou P N and List E J 1988 Investigations of round vertical turbulent buoyant jets *J. Fluid. Mech.* **195** 341–91

[25] Pasquill F 1962 *Atmospheri Diffusion.* D (Van Nostrand)

[26] Provenzale A 1999 Transport by coherent barotropic vortices *Annu. Rev. Fluid. Mech.* **31** 55–93

[27] Rahman S and Webster D 2005 The effect of bed roughness on scalar fluctuations in turbulent boundary layers *Exp. Fluids.* **38** 372–84

[28] Richardson L 1926 Atmospheric diffusion shown on a distance-neighbour graph *Proc. R. Soc. Lond.* A **110** 709–37

[29] Richardson L F 1919 Measurement of water in clouds *Proc. R. Soc. Lond.* A **96** 19–31

[30] Richardson L F 1922 *Weather Prediction by Numerical Processes* (Cambridge: Cambridge University Press)

[31] Richardson L F 1930 The reflectivity of woodland, fields and suburbs between London and St Albans *Q J R Met Soc* **56** 31–8

[32] Riley J and Corrsin S 1974 The relation of turbulent diffusivities to Lagrangian velocity statistics for the simplest shear flow *J Geophys Res* **79** 1768–71

[33] Roberts O 1923 The theoretical scattering of smoke in a turbulent atmosphere *Proc. R. Soc. Lond.* A **104** 640–54

[34] Rogers M, Mansour N and Reynolds W 1989 An algebraic model for the turbulent flux of a passive scalar *J. Fluid. Mech.* **203** 77–101

[35] Rouson D and Eaton J 2001 On the preferential concentration of solid particles in turbulent channel flow *J. Fluid. Mech.* **428** 149–69

[36] Salazar J and Collins L 2009 Two-particle dispersion in isotropic turbulent flows *Annu. Rev. Fluid. Mech.* **41** 405–32

[37] Sawford B 2001 Turbulent relative dispersion *Annu. Rev. Fluid. Mech.* **33** 289–317

[38] Shaw R, Reade W, Collins L and Verlinde J 1998 Preferential concentration of cloud droplets by turbulence: Effects on the early evolution of cumulus cloud droplet spectra *J. Atmos. Sci.* **55** 1965–76

[39] Smyth W D 2020 Marginal instability and the efficiency of ocean mixing *J. Phys. Ocean.* **50** 2141–50

[40] Sreenivasan K, Tavoularis S and Corrsin S 1982 A test of gradient transport and its generalizations *Turbulent Shear Flows* ed D Bradbury, S Launder and Whitelaw 3rd edn (Springer: Berlin) 96–112

[41] Stapountzis H, Sawford B, Hunt J and Britter R 1986 Structure of the temperature field downwind of a line source in grid turbulence *J. Fluid. Mech.* **165** 401–24

[42] Sutton O 1932 A theory of eddy diffusion in the atmosphere *Proc. R. Soc. Lond.* A **135** 143–65

[43] Talluru K M, Philip J and Chauhan K A 2018 Local transport of passive scalar released from a point source in a turbulent boundary layer *J. Fluid. Mech.* **846** 292–317

[44] Tavoularis S and Corrsin S 1981 Experiments in nearly homogeneous turbulent shear flow with a uniform mean temperature gradient. Part 1 *J. Fluid. Mech.* **104** 311–47

[45] Tavoularis S and Corrsin S 1985 Effects of shear on the turbulent diffusivity tensor *Int. J. Heat. Mass. Trans.* **28** 265–76

[46] Taylor G 1921 Diffusion by continuous movements *Proc. London. Math. Soc.* **s2-20** 196–212

[47] Taylor G I 1915 I. eddy motion in the atmosphere *Philos. Trans. Roy. Soc. Lond. Ser.* A **215** 1–26

[48] Tennekes H and Lumley J 1972 *A First Course in Turbulence* (MIT Press: Cambridge, MA)

[49] Vanderwel C and Tavoularis S 2014 Measurements of turbulent diffusion in uniformly sheared flow *J. Fluid. Mech.* **754** 488–514

[50] Warhaft Z 1984 The interference of thermal fields from line sources in grid turbulence *J. Fluid. Mech.* **144** 363–87

[51] Warhaft Z 2000 Passive scalars in turbulent flows *Annu. Rev. Fluid. Mech.* **32** 203–40

[52] Webster D, Rahman S and Dasi L 2003 Laser-induced fluorescence measurements of a turbulent plume *J. Eng. Mech-ASCE* **129** 1130–7

[53] Webster D R, Roberts P J and Ra'ad L 2001 Simultaneous dptv/plif measurements of a turbulent jet *Exp. in Fluids.* **30** 65–72

[54] Westerweel J, Fukushima C, Pedersen J and Hunt J C R 2009 Momentum and scalar transport at the turbulent/non-turbulent interface *J. Fluid. Mech.* **631** 199–230

[55] Younis B, Speziale C and Clark T 2005 A rational model for the turbulent scalar fluxes *Proc. R. Soc.* A **461** 575–94

Chapter 10

Epilogue

Although turbulence has been studied continually since it was first addressed well over a century ago, there is still no overarching theory which fully accounts for much of its underlying physics. This book has explored perhaps the most important of the basic features of turbulence, but in the specific context of canonical flows, which have attracted the most attention during the history of the subject. For these flows, at least, hypotheses such as those of Kolmogorov (for the small scales), Richardson's (for dispersion), and von Kármán's (leading to the log law), have all provided both far-reaching insights and specific results which, to varying degrees, have been shown to agree reasonably well with experimental data. But the hypotheses are exactly that – hypotheses: they rely on assumptions that are, in some respects, flawed, and it is arguably rather startling that they seem to work at all. Nonetheless, it must be emphasised that turbulent flows of practical interest are usually much more complex than the canonical flows we have discussed in this book and which have, for many decades, formed the basis for testing the hypotheses.

This all means that predictions of particular flows, which are so important in the engineering and environmental contexts (and often quite complex and usually at high Reynolds numbers), remain largely a matter of employing models built on what is known to be at least approximately true. We have not discussed such models, apart from one of the simplest and lowest-order ones – gradient transport (leading to the eddy viscosity and eddy diffusivity). However, it is appropriate here to mention large eddy simulation (LES, see [3] for a modern text) and its more recent derivative, detached eddy simulation (DES, see [4] for a useful review). These are increasingly popular approaches in which the larger scales are resolved directly and only the smallest scales, those less than the mesh size of the computational grid, have to be modelled; they are thus far less demanding of computer power than DNS. Nonetheless, the closure models needed for the subgrid scales inevitably contain assumptions and approximations of one form or another.

Closure models are always developed on the basis of either theoretical ideas or experimental data (or both) and will no doubt become increasingly more sophisticated. However, Davidson's remark made nearly twenty years ago remains true in our view [1]: *we have no right to expect that the complex turbulent flows which face the engineer will ever be encompassed by some grand theory of turbulence.* Eventually, it might just possibly become the case that even if such a grand theory were to be developed – an unlikely contingency, perhaps – the continually increasing power of the computer could make it practically unnecessary. In this book, we have included results from a number of recent direct numerical simulation (DNS) databases to illustrate turbulence features, and there is no doubt that DNS has already led to developments in the understanding of turbulent flows, particularly regarding their structural features. While such simulations so far remain limited to relatively low Reynolds numbers, one might be tempted to **hope** that they may, in the course of time, be able to address directly engineering flows of importance.

Figure 10.1 shows the currently achievable Reynolds numbers at the time of writing for both DNS and laboratory experiments, compared to those which occur in vehicular transport of various kinds and in the atmosphere. Notice how laboratory and computational capabilities have increased over the last couple of decades. The first full DNS of a zero-pressure-gradient turbulent boundary layer was that of Spalart [5] over 30 years ago, which used $\mathrm{Re}_\theta = 1410$. Given the rate of DNS development since then and the gap between current capability and what would be needed to compute typical, complex engineering flows, it seems unlikely that full numerical simulation of practically important flows (over most air or water passenger or freight transport systems at operational Reynolds numbers, for example) will be achieved within the next half-century at the very least (if ever). The **hope** expressed above could well turn out to be far too optimistic. Weather forecasting will certainly rely on approximate models to close the governing equations well beyond the foreseeable future. In the meantime, recourse to physical experiments and further developments in theoretical understanding will inevitably remain important. Nonetheless, DNS will continue to be useful in the validation of experimental data and in revealing yet more of the obscure but perhaps important details of relatively low-Reynolds-number flows.

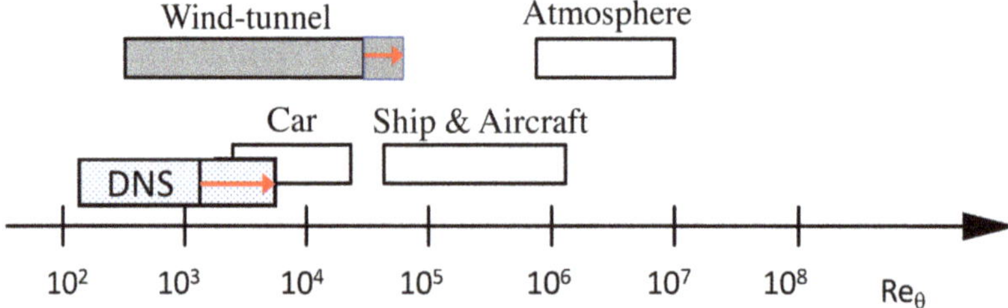

Figure 10.1. Typical Reynolds number ranges. Inspired by [2], but with the DNS and wind tunnel boxes extended (red arrows) because of developments in computational and experimental capability over the last two decades or so.

DNS has also often been touted as a means of providing insights which could be used directly in the development of improved modelling approaches, not least in providing the necessary detailed information that laboratory techniques will perhaps never be able to supply. But, arguably, there is not much evidence that this has happened to any great extent. Although one can confidently expect a continual increase in the upper Reynolds-number limit of DNS, it is perhaps less likely that experimental facilities will develop similarly, and the community may well have to rely, as now, on either sophisticated modelling approaches or results from laboratory experiments at high Reynolds numbers (some of which are already similar to at least some practical engineering situations, see the figure). These could be validated, when possible, by DNS at the highest possible Reynolds numbers and intelligently extrapolated to even higher Reynolds numbers.

Most of the improved experimental and numerical capabilities over the last 20 30 years, indicated above and in figure 10.1, have been made in the context of the simplest wall-bounded flows – channels, pipes, and boundary layers. For such canonical (but actually rather special, even singular) flows, laboratory experiments have confirmed crucial aspects of turbulence theory for high Reynolds numbers, and one could argue that progress over the last 20 years has been greater than over the previous century. Nonetheless, such experiments have also revealed shortcomings in classical understanding, some of which we have identified in this book. There is, in any case, a clear need to move beyond these canonical flows to consider non-equilibrium situations more akin to flows that commonly occur in practice. This would allow the testing of some of the fundamental understandings in more complex situations, for which there are more reasons to doubt the efficacy of assumptions which appear to work reasonably well for the simpler flows.

There is no doubt that focused research on turbulence will continue for decades to come. Opportunities to make good progress are certainly numerous and wide-ranging and will, one has to hope, lead to improvements in closure models that have to be used for real flows for the foreseeable future. However, we must be cautious in expecting any major breakthroughs in the fundamental understanding of turbulent flows of engineering and environmental importance any time soon. Turbulence remains a classically very demanding subject within the overarching field of fluid dynamics.

References

[1] Davidson P 2004 *Turbulence: an introduction for scientists and engineers* (Oxford: Oxford University Press)
[2] Österlund J M 1999 Experimental studies of zero pressure gradient turbulent boundary layer flow *PhD thesis* Kungl Tekniska Högskolan, Stockholm
[3] Sagaut P 2006 *Large Eddy Simulation for Incompressible Flows* 3rd edn (Berlin: Springer)
[4] Spalart P 2009 Detached eddy simulation *Annu. Rev. Fluid Mech.* **41** 181–202
[5] Spalart P R 1988 Direct simulation of a turbulent boundary layer up to $R_\theta=1410$ *J. Fluid Mech.* **187** 61–98

IOP Publishing

Turbulent Flows: an Introduction

Ian P Castro and Christina Vanderwel

Appendix A

Cartesian tensors

Tensors were originally introduced merely as a way of reducing the length of relationships and equations in many fields of mechanics, allowing much more concise presentation. They became increasingly popular from around the beginning of the last century. We have assumed throughout the book that readers are already familiar with vector calculus and thus scalars and vectors. Some may not be so familiar with tensor notation. In this appendix, we therefore present the basics of (Cartesian) tensor notation – no more than is necessary to enable the readers to understand and manipulate the equations given throughout the book. We begin by describing tensor notation, then we show how tensors can be used to effect coordinate transformations, and finally we give some of the basic rules governing tensor operations. Much fuller descriptions of tensors are given in numerous texts; in the context of fluid mechanics in general and turbulence in particular, the reader may wish to consult, e.g., Pope [2].

A.1 Notation

The Cartesian coordinate system has three independent, mutually orthogonal directions with the same unit length in each. We may denote this by (x_1, x_2, x_3) which, for brevity, can be written as $x_i, i = 1, 2, 3$; i is called the *index* and 1, 2, 3 is its range. A scalar (such as pressure or turbulence kinetic energy), being determined by a single component and independent of coordinate transformation, is a *zeroth-order* tensor, whereas a vector (such as velocity), being determined by three components, is a *first-order* tensor. An Nth-order tensor has 3^N components. The notation is considerably simplified by the use of what is called the *Einstein summation convention*, according to which, any repeated index implies summation over all three values, i.e.

$$d_{ijkk} = \sum_{k=1}^{k=3} d_{ijkk}. \tag{A.1}$$

It is therefore immaterial what particular character is used for that repeated (or dummy) index. An index which appears only once is a *free* index. This can all be illustrated using the *Kronecker delta* δ_{ij}, which is defined by

$$\delta_{ij} = \begin{cases} 1 \text{ if } & i = j \\ 0 \text{ if } & i \neq j \end{cases} \tag{A.2}$$

This tensor (since $N = 2$) has nine components, δ_{11}, δ_{12}, etc. and could obviously be written in a 3×3 matrix form. A related tensor is, for example, δ_{ii} which, according to the summation convention, is $\delta_{11} + \delta_{22} + \delta_{33} = 3$. Note that here, the index i in these two examples could be replaced by any character – e.g. $\delta_{ii} \equiv \delta_{pp}$ – the *substitution rule*. (In some texts, Greek characters are used for indices – e.g. α instead of p in the latter expression – when they are *not* to be summed; in this text, we have avoided Greek indices to avoid any possibility of confusion.) Note too that repeating an index in any term more than twice is inadmissible. A term like $x_i y_i z_i$ is thus meaningless. The summation rule can easily be used to demonstrate the substitution rule, as expressed in the useful property of the Kronecker delta

$$\delta_{ij} u_j = u_i, \tag{A.3}$$

using **u** as an example of a first-order tensor (a vector).

A common operation is *contraction* – the operation of equalising two different indices, thus turning them into a repeated index. Consider, for example, the second-order tensor formed by the product of two first-order tensors, $u_i u_j$; contraction turns this into $u_i u_i$, which is a scalar. Generally, the contraction of an Nth-order tensor yields an $(N - 2)$-order tensor. One should be careful about using squares (or other powers) of any expression with an index, as this may be ambiguous. u_i^2, for example, is not the same as $u_i u_i$. The former has just one value, depending on the value of its index, whereas the latter is the sum of all three components of velocity. Consider the Laplacian of some variable f, as a further example. This is usually written as $\nabla^2 f$ which, in tensor notation, can be expressed as

$$\nabla^2 f = \frac{\partial^2 f}{\partial x_i \partial x_i} = \frac{\partial^2 f}{\partial x_1 \partial x_1} + \frac{\partial^2 f}{\partial x_2 \partial x_2} + \frac{\partial^2 f}{\partial x_3 \partial x_3} \tag{A.4}$$

and should not be written (and is not the same) as $\partial^2 f / \partial x_i^2$. In tensor notation, any vector, such as F_i, for example, represents all three components (F_1, F_2, F_3). Likewise, a second-order tensor, as we have seen for the Kronecker delta, can be written as F_{ij} and also in 3×3 matrix form for its nine components.

A.2 Coordinate transformation

Let there be two Cartesian coordinate systems E and E' that have the same origin; the two sets of coordinates are specified by (x_1, x_2, x_3) and (x'_1, x'_2, x'_3), respectively. The coordinates x_i, $i = 1, 2, 3$ of any point P in space can naturally also be expressed in terms of the coordinates x'_j, $j = 1, 2, 3$ (and vice versa) by means of functions of the angles (x_i, x_j') between the two coordinate axes x_i and x_j' of the two systems. We may define *direction cosines* e_{ij} (a second-order tensor) as

$$e_{ij} = \cos(x_i, x_j'), \tag{A.5}$$

representing the cosine of the angle between the i axis in system E and the j axis in system E'. The general coordinate transformation equation is thus

$$x_i' = x_j e_{ji}, \quad i, j = 1, 2, 3. \tag{A.6}$$

As an example, consider a simple rotation of the E Cartesian system about the x_3 axis by an angle θ, as illustrated in figure A.1. In this case, we have

$$e_{ij} = \cos(x_i, x_j') = \begin{pmatrix} e_{11} & e_{12} & e_{13} \\ e_{21} & e_{22} & e_{23} \\ e_{31} & e_{32} & e_{33} \end{pmatrix} = \begin{pmatrix} \cos\theta & \cos\left(\dfrac{\pi}{2} + \theta\right) & \cos\dfrac{\pi}{2} \\ \cos\left(\dfrac{\pi}{2} - \theta\right) & \cos\theta & \cos\dfrac{\pi}{2} \\ \cos\dfrac{\pi}{2} & \cos\dfrac{\pi}{2} & \cos 0 \end{pmatrix}. \tag{A.7}$$

This transformation does not apply if the origins of the E and E' coordinate systems are different, but it does apply to reflections of the system (as well as rotations, like the one considered in the example). There are alternative ways of discussing such transformations, but they all involve the direction cosines e_{ij}; see e.g. [1, 2].

Since a zeroth-order tensor (a scalar), which we will call F, is not dependent on the coordinate system, it clearly transforms from the E to the E' system via

$$F' = F. \tag{A.8}$$

On the other hand, a first-order tensor (a vector), F_i, say, transforms according to

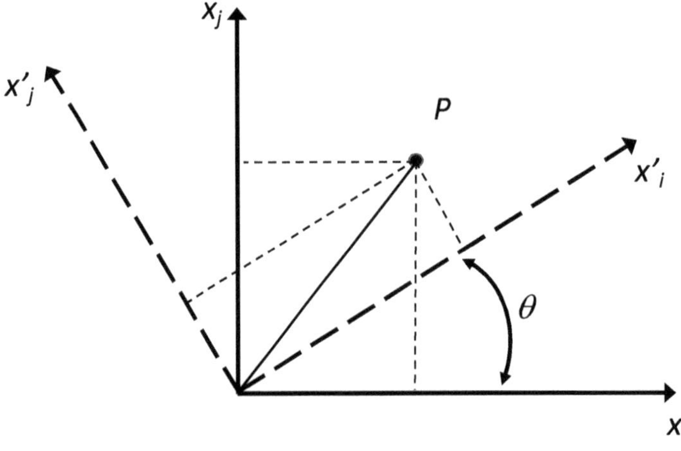

Figure A.1. Rotation of a Cartesian coordinate system E about the x_k ($k = 3$) axis into the E' system. The x_k and x_k' directions are coincident and normal to the page.

$$F_i' = F_j \, e_{ij}, \tag{A.9}$$

whereas a second-order tensor such as F_{ij} transforms according to

$$F_{ij}' = F_{kl} e_{ki} e_{lj}. \tag{A.10}$$

These transformation rules can be extended to the more general case of a tensor of order N (e.g. F with N indices). We emphasise that no Nth-order tensor is a Cartesian tensor if it does *not* obey these coordinate transformation rules.

A.3 Tensor operations

First, we note that two tensors of the same order can be added or subtracted. Thus, for example, the components of the sum s, say, of a_{ij} and b_{ij} are simply $s_{ij} = a_{ij} + b_{ij}$.

Second, it should be obvious that the product of two first-order tensors is a second-order tensor. Choosing, for example, the velocities u and v as two first-order tensors, the components of the product p are given by $p_{ij} = u_i v_j$ (with, naturally, nine components). The product of first-order and second-order tensors is usually a tensor of the third order, and, in general, the product of an Nth-order tensor and an Mth-order tensor is a tensor of order $N + M$. One must be careful, however, in counting the indices. Consider two second-order tensors a_{ij} and b_{jk}, for example. Their product, $a_{ij} b_{jk}$ is *not* a tensor of order four, because there is a repeated suffix j whose role is merely to remind one to perform summations.

Third, note that there is no tensor operation corresponding to division (although a second-order tensor may have an inverse).

Fourth, derivatives are also straightforward to express. For example, the gradient of pressure (a scalar), $\partial P / \partial x_i$, or ∇P in vector notation, is clearly a first-order tensor that has three components. In general, the derivative of an Nth-order tensor is a tensor of order $N + 1$. Thus, for example, a velocity gradient, $\partial u_i / \partial x_j$, is a second-order tensor and has nine components. If the indices are contracted (see above), so that the gradient becomes $\partial u_i / \partial x_i$, the result is a zeroth-order tensor (i.e. the divergence of u_i, a scalar).

In considering the dot and cross products of vectors, it is helpful to define a *unit vector*, e, so that e_i refers to the unit vector in one of the three coordinate directions. Note that since the three vectors are orthogonal, $e_i \cdot e_j = \delta_{ij}$, which is referred to as the orthonormality property. The gradient operator can thus be written as

$$\nabla \equiv e_i \frac{\partial}{\partial x_i}. \tag{A.11}$$

In general, the dot (inner) product of an Nth-order tensor and an Mth-order tensor becomes a tensor of order $N + M - 2$. So, for example, the dot product of two first-order tensors, a and b, becomes a second-order tensor which can be written as

$$c = a \cdot b = (e_i a_i) \cdot (e_j b_j) = \delta_{ij} a_i b_j = a_i b_i, \tag{A.12}$$

which is a scalar. Thus, for example, the divergence of the velocity tensor u can be written in tensor notation as

$$\nabla \cdot \boldsymbol{u} = \left(e_j \frac{\partial}{\partial x_j} \right) \cdot (e_i u_i) = \delta_{ij} \frac{\partial u_i}{\partial u_j} = \frac{\partial u_i}{\partial x_i}. \qquad (A.13)$$

Of course, in this case, since we chose velocity as the first-order tensor, the divergence is zero in incompressible flow (the solenoidal or divergence-free condition, $\nabla \cdot \boldsymbol{u} = 0$). Likewise, the Laplacian (a scalar operator) can be written

$$\nabla^2 = \nabla \cdot \nabla = \left(e_i \frac{\partial}{\partial x_i} \right) \cdot \left(e_j \frac{\partial}{\partial x_j} \right) = \delta_{ij} \frac{\partial^2}{\partial x_i \partial x_j} = \frac{\partial^2}{\partial x_i \partial x_i}, \qquad (A.14)$$

which is written out in full for an arbitrary scalar in equation (A.4).

Finally, we note that strictly there is no genuine tensor equivalent for the cross product of two tensors. It can, however, be written using suffix notation using the *Levi-Civita* symbol defined by

$$\epsilon_{ijk} = \begin{cases} 1 & \text{if} \quad (i, j, k) \quad \text{are cyclic} \\ -1 & \text{if} \quad (i, j, k) \quad \text{are anticyclic} \\ 0 & \text{otherwise.} \end{cases} \qquad (A.15)$$

The cross product of two vectors can then be written as

$$c = a \times b = \epsilon_{ijk} e_i a_j b_k, \qquad (A.16)$$

whose components are $c_i = \epsilon_{ijk} a_j b_k$. Likewise, the curl (cross product) of the velocity – i.e. the vorticity – is

$$\omega = \nabla \times \boldsymbol{u} = e_i \epsilon_{ijk} \frac{\partial U_k}{\partial x_j}, \qquad (A.17)$$

with the components

$$\omega_i = \epsilon_{ijk} \frac{\partial u_k}{\partial x_j}. \qquad (A.18)$$

It can be shown that ϵ_{ijk} is *not* a third-order tensor and thus neither c nor ω above can be tensors; they are more properly called pseudovectors, as they behave differently from a vector under a reflection of the coordinate system. This point is explained fully in the more rigorous and comprehensive discussion of tensors given, for example, by Pope [2].

References

[1] Davidson P 2004 *Turbulence: An Introduction for Scientists and Engineers* (Oxford: Oxford University Press)
[2] Pope S B 2000 *Turbulent Flows* (Cambridge: Cambridge University Press)